Chemical Methods for Processing Nanomaterials

T0199787

Editor

Vidya Nand Singh

National Physical Laboratory (CSIR)
New Delhi
India

CRC Press
Taylor & Francis Group
Boca Raton   London   New York

CRC Press is an imprint of the
Taylor & Francis Group, an **Informa** business

A SCIENCE PUBLISHERS BOOK

Cover images provided by the editor of the book.

First edition published 2021
by CRC Press
6000 Broken Sound Parkway NW, Suite 300, Boca Raton, FL 33487-2742

and by CRC Press
2 Park Square, Milton Park, Abingdon, Oxon, OX14 4RN

© 2021 Taylor & Francis Group, LLC

CRC Press is an imprint of Taylor & Francis Group, LLC

Library of Congress Cataloging-in-Publication Data
Names: Singh, Vidya Nand, 1976- editor.
Title: Chemical methods for processing nanomaterials / editor, Vidya Nand
    Singh.
Description: First edition. | Boca Raton : CRC Press, Taylor & Francis
    Group, 2021. | Includes bibliographical references and index.
Identifiers: LCCN 2020029168 | ISBN 9780367085889 (hardcover)
Subjects: LCSH: Nanostructured materials.
Classification: LCC TA418.9.N35 C479 2021 | DDC 660/.282--dc23
LC record available at https://lccn.loc.gov/2020029168

ISBN: 978-0-367-08588-9 (hbk)

Typeset in Times New Roman
by Radiant Productions

Preface

The purpose of this book is to summarize the recent developments in the important research field of "Chemical Processing of Nanomaterials". Book chapters were invited on different methods for fabricating nanomaterials. Finally, the book is being published with fourteen chapters. The topics are focused on chemical methods for processing nanomaterials. The target audience of change the proposed with this book is academia and researchers in the universities and the research laboratories. Nearly 15% of researchers in different area deal with nanomaterials, and among them 60–70% of them employ chemical methods for processing nanomaterials. The present book gives various aspects of chemical processing of nanomaterials. This book describes latest synthesis methods for all kinds of nanostructures using various chemical methods. It also describes the latest techniques used for synthesizing and characterizing nanomaterials of several kinds, such as active groups, core-shell, quantum dots, metal and metal oxide, perovskite nanocrystals, etc. The chapters deal with chemical methods for chalcogenides, nanostructured materials using microemulsions, wet chemical methods for nanomaterial synthesis, chemical vapor deposition method, sol-gel processing of nanocrystalline metal oxide thin films, electrodeposition— a versatile and robust technique for synthesizing nanostructured materials, synthesis of nanomaterials and nanostructures, low dimensional carbon nanomaterials, synthesis and applications of two dimensional materials, methods of manufacturing composite materials, surface modification of nanomaterials, nanomaterials for gas sensing applications and the synthesis process of the quantum dots.

I would like to thank all those who have kindly contributed chapters for this book. Thanks are also due to Dr. Prashant Ambekar and Dr. Jasmirkaur Randhawa for their help in finalizing and improving the contents of the book.

Vidya Nand Singh

National Physical Laboratory (CSIR), New Delhi, India

Contents

CHAPTER 1

Chemical Methods for Processing Carbon Nanomaterials

Sehmus Ozden[1,2,3]

1. Introduction

1.1 Carbon Nanomaterials

One of the basic elements which has various allotropes is carbon, and it has the ability to create covalent bonds with other carbon atoms in a range of hybridization states, such as sp, sp^2, and sp^3. The most well-known natural allotropes of carbon are graphite and diamond. Even though these allotropes mainly consist of carbon atoms, they have different physicochemical properties because of the hybridization of carbon atoms. For example, even though diamond, which consists of sp^3 carbon hybridization, is transparent, an electrical insulator, and the hardest known material, graphite, which consists of sp^2 carbon hybridization, is opaque, and a soft material with high electrical conductivity.

Fullerene is the first unusual allotrope of carbon that was discovered in 1985 by Robert Curl, Harold Kroto, and Richard Smalley [1]. The discovery of fullerenes opened a new area for carbon allotropes at the nanoscale. The discovery of fullerenes was followed by carbon nanotubes, the straight tubular carbon structure, in 1991 [2]. In 2004, graphene, the one atom thick hexagonal lattice of carbon, was discovered by Andre Geim and Kostya Novoselov at Manchester University (Figure 1.1) [3]. Although various forms of carbon elements have been discovered, there are still many carbon allotropes that need to be experimentally discovered, such as graphyne sheets and graphyne nanotubes [4].

[1] Princeton Institute for the Science and Technology of Materials, Princeton University, Princeton, NJ 08540 USA.

[2] Chemical and Biological Engineering, Princeton University, Princeton, NJ 08540 USA.

[3] Department of Mechanical and Aerospace Engineering, Princeton University, Princeton, NJ 08540 USA.

Email: sozden@princeton.edu

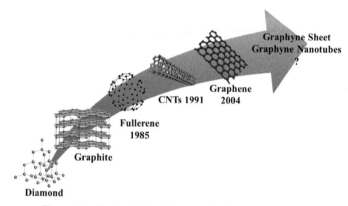

Figure 1.1. Timeline for the discovery of carbon nanostructures.

All of these carbon allotropes at nanoscale can be considered as members of the same group because they consist of sp^2 hybridization of hexagonal carbon lattice network. As a result of this hybridization, their properties, such as electrical conductivity and mechanical properties are similar, although they have significant differences because of their sizes and shapes.

Different classification methods can be used for carbon nanomaterials. The first classification method is based on the hybridization of the carbon element, such as sp^2 and sp^3. Fullerene, carbon nanotubes (CNTs), and graphene can be categorized as sp^2 hybridization, and nanodiamonds can be counted into sp^3 hybridization. The other classification method is based on the structural dimension of carbon element, and in this book we are taking this classification into account. There are zero-dimensional (0D) nanostructures, such as fullerene, one-dimensional (1D) nanostructures, such as nanotubes and nanofibers, graphene-like two-dimensional (2D), and three-dimensional (3D) carbon nanostructures (Figure 1.2).

Carbon nanomaterials can be considered attractive candidates in a broad range of nanotechnology applications, ranging from medicine to aerospace, because of

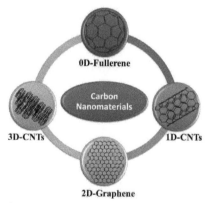

Figure 1.2. Classification of nanomaterials—zero-dimensional (0D) fullerene, one-dimensional (1D) carbon nanotubes, two-dimensional (2D) graphene, and three-dimensional (3D) carbon nanotubes structures.

their attractive properties, but these nanomaterials need to be processed and modified during or after their fabrication for the specific application. In this chapter, we will discuss chemical processing and modification of carbon nanomaterials to control their properties for emerging applications.

2. Carbon Nanotubes

Carbon nanotubes were discovered in 1991 by Iijima, although there are some reports that show tubular structure of CNTs in 1952 and 1976. Ijima et al. and Bethune et al. have reported two separate works about the growth of single-wall carbon nanotubes (SWCNTs). SWCNTs can be described as a rolled-up single-layer graphene. Multi-wall carbon nanotubes (MWCNTs) can be thought as rolling up a multi-layer graphene sheet.

CNTs can be classified based on their wall numbers, such as SWCNTs and MWCNTs, or they can be classified based on their chirality, such as zigzag and armchair. They are semiconductors or metallic based on their chirality. CNTs have extraordinary physical, chemical, and mechanical properties that make them exciting future materials for a broad range of advanced technological applications. For example, their mechanical tensile strength (> 100 GPa) and elastic modulus (~ 1 TPa) are much higher than known materials, such as diamond, with the advantages of low density and flexibility. The thermal conductivity is theoretically reported as 6,600 $Wm^{-1}K^{-1}$. The electrical conductivity of CNTs is 1000 times higher than copper.

CNTs can be produced using various techniques, as shown in Figure 1.3. CNTs were first produced using high temperature fabrication methods, such as arc-discharge and laser ablation methods, but recently these methods have been replaced by low temperature chemical vapor deposition (CVD) techniques. The CVD method is the most commonly used method for producing CNTs. In this process, nanotubes

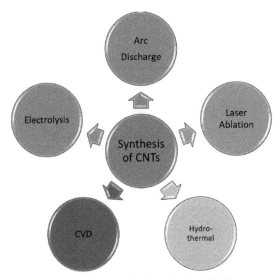

Figure 1.3. Methods for synthesis of CNTs.

are obtained by thermal decomposition of a hydrocarbon vapor in the presence of a metal catalyst. The properties of CNTs, such as density, length, and quality can be controlled by CVD technique, etc. Here, we will look into the CVD method in a more detailed manner. These methods mainly require a carbon source, catalyst, temperature, and inert gases.

The CVD process involves passing a hydrocarbon vapor using methane, ethylene, acetylene, benzene, xylene, and carbon monoxide as hydrocarbon sources through a reactor with a metal catalyst, such as iron, nickel, and cobalt at a proper temperature (Figure 1.4). The growth of CNTs with CVD method depends on different parameters, such as hydrocarbon source, catalyst, temperature, pressure, gas-flow rate, deposition time, reactor geometry, and the location of substrate inside the reactor. All of these factors not only affect the growth of CNTs, but also their yield, type, and quality.

Although the fundamental mechanism of the growth of CNTs is still not understood well, the widely accepted mechanism is that hydrocarbon sources decompose at high temperature and precipitate on the catalyst. When the decomposed carbon precipitates on the surface of the catalyst, the nucleation of CNT growth starts. If the interactions between metal catalyst and the substrate are weak, the growth continues as tip growth (Figure 1.5a). If the interactions between the metal catalyst and the substrate are strong, decomposed carbon deposits on the catalyst from the lower peripheral surface of metal. In both cases, the growth of CNTs stops when the surface of the catalyst is covered by amorphous carbon. To avoid the catalyst poisoning, which stops the growth, a small amount of water vapor can be purged into the furnace during the growth. The water vapor introduced into the CVD reactor oxidizes the amorphous carbon that covers the surface of the catalyst and reactivates the metal catalyst, and thus the growth of CNTs continues for producing longer nanotubes. The carbon source affects the morphology of nanotubes. If small and aliphatic carbons, such as CO, CH_4, and CH_3CH_2OH are used, usually straight tubular structures are obtained. If aromatic carbon sources, such as benzene and xylene are used, usually buckled tubular structures form.

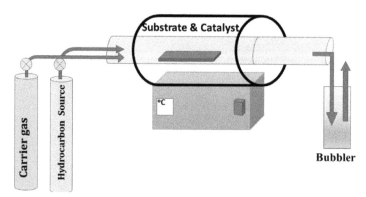

Figure 1.4. A schematic figure of chemical vapor deposition (CVD) setup with fundamental needs for the growth of CNTs.

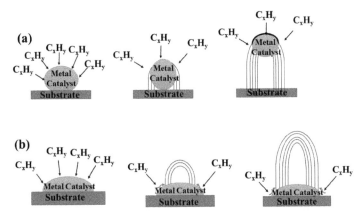

Figure 1.5. The schematic representation of CNTs growth mechanism—(a) Tip-growth, (b) Base-growth mechanism.

Catalysts also play an important role in the growth of CNTs. Fe, Co, and Ni are the most commonly used catalysts for CNT growth because of the high solubility of carbon and high carbon diffusion rate in these metals. In addition, these metals have high melting point and low-equilibrium vapor pressure with advantages of a wide range of carbon precursors. The diameter of CNTs can be controlled by the size of the metal catalyst. In addition to Fe, Co, and Ni, which are commonly used catalysts for CNT growth, highly active crystallographic phases of Co-Mo, Ni-Mo alloys, Cu, Au, Pt, Ag, and Pd were used as catalysts for CNTs growth.

2.1 Processing of Carbon Nanotubes by Chemical Functionalization

Although CNTs have extraordinary physical and chemical properties, some limitations prevent them from real-world technological applications. For example, all CNTs fabrication methods produce a mixture of various diameters and chirality of nanotubes, and they are contaminated with metallic and amorphous impurities. In addition, their solubility in common solvents and their seamless surface morphology block out their dispersion in common solvent media due to the surface energy and weak physical interaction between CNTs and the matrix. To overcome these limitations, the surface structure of nanotubes need to be modified for desirable processibility and applications.

Chemical surface functionalization is a unique method for controlling their properties and their integration into the practical applications. To date, different types of surface modification methods have been developed for nanotube functionalization, which have been focused on the interfacial molecular engineering, improving interaction with macromolecules, separation based on diameter and chirality, defect engineering, etc. However, developing facile and controllable processes of CNTs functionalization is a key step toward their potential applications. Nanotubes functionalization approaches can be basically divided into covalent and noncovalent functionalization.

The covalent functionalization is based on linkage of functional groups onto the surface of CNTs (Figure 1.7). There are a few reactive chemicals, which can be directly connected to the inert hexagonal lattice of CNTs. For example, fluorine is able to link up to a conjugated π-π system because of its high reactivity. Another highly reactive functional group that can attach covalently on the surface of nanotubes is diazonium salt, which is associated with the change of hybridization from sp^2 to sp^3 and a simultaneous loss of conjugation. The other covalent functionalization conducts by creation of defects on the surface of CNTs via acid treatment, which is one of the most common methods to process CNTs for the desired properties and applications. In this method, acid generates carboxyl groups onto surface of CNTs following conversion to acid chloride, and so can easily be functionalized with other functional groups, such as amine and alkyl groups. All of these functionalization methods have advantages and disadvantages. For example, while the acid treatment method reduces the conductivity and mechanical properties of CNTs, these properties are not as affected by the diazonium functionalization.

The second functionalization approach is non-covalent functionalization, which is the most commonly used method to modify CNTs without disturbing their straight tubular structures. This process relies on weak physical interactions, such as van der Waals, hydrophobic, and π-π interactions (Figure 1.6). Non-covalent functionalization is a method for CNTs dispersion in aqueous and non-aqueous solvents without changing their unique physical and mechanical properties, and it is the most popular method for separation of nanotubes depending on their diameter and chirality. For example, polymer molecules selectively wrap on different single-wall CNTs (SWCNTs), leading them to be dissolved in organic media, while others are insoluble. Thus the different structure of SWCNTs can be separated via centrifugation.

In addition to chemical surface functionalization, properties of CNTs can be tailored by doping with heteroatoms, such as boron, nitrogen, and sulfur atom. In this

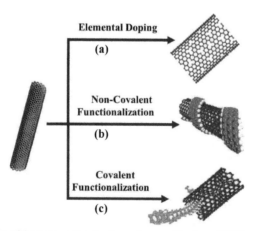

Elemental Doping

(a)

Non-Covalent
Functionalization

(b)

Covalent
Functionalization

(c)

Figure 1.6. Schematics of the main approaches for surface modification of CNTs—(a) Elemental doping of CNTs Ref. [5], (b) Non-covalent functionalization of CNTs via polymer wrapping, and (c) Covalent functionalization CNTs. Ref. [7].

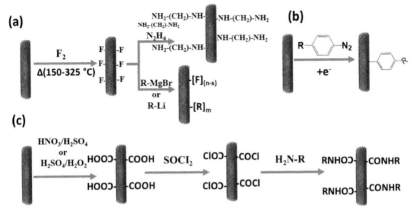

Figure 1.7. Schematic representation of (a) Depiction of fluorination of CNTs and reductive defluorination with hydrazine, and displacement of fluorines, (b) Oxidative etching of CNTs followed by treatment with thionyl chloride, and subsequent amination, (c) Oxidation of CNTs following chlorination and amination.

approach, thermal decomposition of heteroatom source (N-, B-, S-containing) along with the carbon source and metal catalyst results in the formation of heteroatom-doped CNTs. Here, the heteroatom inserts into the hexagonal lattice of the tubular structure by creating B-C, N-C, S-C bonding. The heteroatom in CNTs structure can be seen as a regular defect, which alters the chemical behavior of the nanotube structure. The reactivity of these heteroatom-doped nanotubes is usually more reactive than undoped nanotubes with the same diameters. For example, theoretical calculations predict a localization of the unpaired electrons around the N-doped sites of semiconducting nanotubes.

There are different forms of N substitutions in CNTs, such as pyridinic N (N atoms in six-member ring bonded with two carbon atoms), pyrolic N atoms in five member ring, graphitic N atoms within the graphene lattice bonded to three carbon atoms, and oxidized N species where N atoms bonded with oxygen atoms (Figure 1.8a). Each N substitution has different functionalities, so it is desirable to control the N-substitutions, which still remain a challenge. Toward controlling

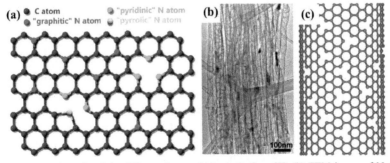

Figure 1.8. (a) Schematic of the different forms of N substitutions [8], (b) TEM image of N-doped bamboo-like nanotube structures, (c) Molecular structure of N-doped nanotube with pyridinic N substitutions. Reproduced with permission from ref. [7]. Copyright© 2006 Nature Publishing Group.

the doping, metallic N-doped CNTs were produced as bamboo-like structures by pyrolysis of ferrocene–melamine mixtures at 1050°C, which contains 3–4% N atoms [7, 5]. Theoretical predictions show that the metallic behavior of N-doped CNTs originates from the pyridine-like N structures (Figure 1.8b) [5].

2.2 Properties of Carbon Nanotubes by Chemical Processing

Although surface modifications improve some properties of CNTs, functionalization reactions have undesirable effects on the structures and intrinsic properties. In this section, we will discuss how functionalization approaches affect some properties of nanotube structures.

2.2.1 Mechanical Properties

The exceptional mechanical properties of CNTs attracted a great interest for a broad range of applications, especially because of their extremely high strength to weight ratio, but their relatively smooth graphene-like surface prevents them from dispersion in polymer matrixes due to the lack of interactions. The design of functionalization determines the influence of atomic structure, and hence the mechanical properties of nanotubes and their composites. Theoretical calculations predict that Young's Modulus varies between 0.5 to 5.5 TPa and experimental results are reported between 0.32 and 1.47 TPa. The key difficulties to improve mechanical properties of nanotubes are dispersion, alignment, and interfacial load transfer. Understanding the interface interactions between functional group and surface of nanotubes is one of the key parameters for developing an optimized nanotube material for mechanical application. Covalent and non-covalent functionalization, which improve the interfacial bonding, are attractive processes to enhance their dispensability and interface interactions with polymer matrixes, and hence improve the mechanical properties of composite structures. For example, Velasco-Santos et al. incorporated non-functionalized and functionalized CNTs into the thermoplastic polymer matrix for investigation of the role of nanotube functionalization on the mechanical behavior of composites [9]. Zhang and coworkers showed that the storage modulus (E') of composite samples contains 1 wt% of the non-functionalized (MWCNT) and 1 and 1.5 wt% functionalized (f-MWCNT). In terms of E', 1 wt% f-MWCNT composite samples show better behavior than 1 wt% of non-functionalized MWCNT composite samples [10]. They display that the tensile strength and modulus are increased in the sample with the addition of 1 wt% MWCNT compared to pure PMMA. The behavior of 1 wt% MWCNT reinforced composite is strong, but not as tough as that of sample with 1 wt% f-MWCNT compared to addition of 1 wt% MWCNT, which enhances the tensile strength and modulus because of the improved dispersibility and interface interactions between nanotubes and polymer structure. In addition, the elastic behavior of composites increases with the addition of f-MWCNT. While surface modifications improve mechanical properties of CNTs reinforced composite materials, functionalization processes decrease the mechanical properties of nanotubes because functionalization creates defects on the perfect hexagonal honeycomb structures of a CNT, and hence decreases their own mechanical properties. For example, Zhang and coworkers performed atomistic simulations to show the effect of hydrogenization on the mechanical properties of CNTs [10]. They showed that the elastic modulus of CNTs decreases with the increasing rate of functionalization, and both functionalized and non-functionalized structures show a rapid drop after the strength is reached. The strength and fracture strain of the functionalized nanotubes is lower than that of the non-functionalized ones. The reduction of stress peak is due to the functionalization, which creates localized deformation and early fracture of the CNT [10].

2.2.2 Thermal and Electrical Conductivity

The electrical and thermal conductivity of CNTs as fillers in composites is affected by their structure quality, loading and alignment, and the resistance of the interface between CNTs and polymer matrix. For example, Pan et al. investigated the thermal conductivity of (10, 0) single-SWNTs at 300 K by covalent functionalization with hydrogen atoms using a non-equilibrium molecular dynamics (MD) method [11]. They showed that the thermal conductivity of nanotubes drop by attaching hydrogen atoms by chemical functionalization. When 5% of carbon atoms are hydrogenated, the thermal conductivity decreases by about a factor of 1.5.

The preservation of the exceptional electrical properties of individual nanotubes in macroscopic carbon nanotubes assemblies still remains a challenging issue because of the non-uniformity of as-produced nanotubes in terms of chirality, diameter, length, and number of walls. In addition, the electrical conductivity of the macroscopic nanotubes is sensitive to defects on the hexagonal tubular structure. Their resistance increases with impurities, such as amorphous carbon and topological defects on as-synthesized CNTs, which creates further scattering points for electrical transport. The theoretical electrical conductivity of CNTs can be as high as 10^6 to 10^7 S/m for pure CNTs. These values are comparable to the two best known metal conductors, silver and copper, which are 6.3×10^7 and 5.96×10^7, respectively. In experiments, the intrinsic electrical conductivities of nanotubes are usually reported between 10^3 to 10^5 S/m. Covalent functionalization reduces the electrical conductivity of CNTs because of the disorder on the perfect hexagonal honeycomb grid sheet, but using covalently functionalized nanotubes as a filler in polymer composites improves the electrical conductivity. Certain concentrations of nanotubes can improve the electrical conductivity dramatically for several orders of magnitude, and thus the entire composite goes from almost insulating to highly conductive.

2.2.3 Electronic and Optical Properties

The electronic band structure of CNTs is basically obtained from the graphene by the zone-folding method [12]. CNTs structures can be semiconductors and have a direct bandgap, which depends on their diameter. O'Connell et al. shows the single-wall CNTs structure with the index by two integers (n, m), which describe the length (π times the tube diameter, dt) and chiral angle (α) of the nanotubes with the roll-up vector on a graphene sheet. They indicate the qualitative pattern of sharp van Hove peaks that arise from quasi–one dimensionality, which are predicted for electronic state densities of semiconducting single-wall CNTs. It shows that the light absorption at photon energy E_{22} is followed by fluorescence emission near E_{11}, which varies with nanotube structure [13].

The electronic structure of single-wall CNTs can be controlled by chemical functionalization processes [14, 13, 15]. For example, Strano et al. used diazonium functionalization, which involves electron transfer from CNTs to the adsorbed diazo groups, and creates covalent bonding between aryl and CNTs, for controlling the electronic structure of single-wall CNTs [15]. This bond forms with extremely high affinity for electrons with energies, ΔE_r, near the Fermi level, E_f of the single-wall CNTs [15].

Photoluminescence (PL) of single-wall CNTs originates in the lowest-energy band edge exciton state, which is E_{11}. PL of single-wall CNTs can be tuned to longer wavelengths by increasing their diameter [16, 17, 18, 19, 20]. For example, Kwon and coworkers modify the surface of single-wall CNTs with various functional groups to tune their optical properties [18]. They showed that the unique chemical surface processing provides exceptional chemical tunability of the near-infrared PL energy (Figure 1.9a). As a result of chemical functionalization, the E_{11}^- emission red-shifted with increasing the number of fluorine atoms along a six-carbon alkyl backbone (Figure 1.9b). The energy shift goes from 133 meV to 190 meV. They observed a consistent trend with the series of partially fluorinated groups by changing the chain length $-(CH_2)_nCF_3$ (n = 0, 1, 2, 3, 4, 5). To obtain larger optical tunability, single-wall CNTs can be functionalized with diiodo-containing precursors (Figure 1.9c). For example, the emission of (6,5)-CNT > CH_2 is at 1125 nm, which is red-shifted

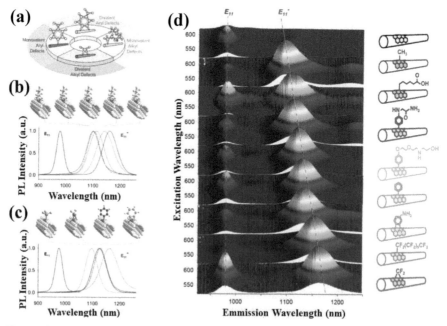

Figure 1.9. (a) Schematic of the four classes of molecularly specific functionalization, which create quantum defects by reaction with halides functional group, (b) The PL spectra of functionalized (6,5)-CNTs, (c) Comparison of monovalent and divalent fluorescent quantum defects by chemical functionalization, (d) Excitation–emission maps of (6,5)-CNTs with chemically tailored fluorescent quantum defects. Reproduced with permission from ref. [18].

by 31 meV more than its monovalent counterpart, (6,5)-CNT–CH$_3$. In (6,5)-CNT > CF$_2$, the defect emission further shifted to 1164 nm. Divalent aryl defects created by reaction with, for instance, o-diiodoaniline and o-diiodobenzene also yield novel emission peaks, which are more red-shifted from the parent single-wall CNTs [18]. Figure 1.9d shows nine chemical functional groups attached on the surface of single-wall CNTs which continuously tune near-infrared emissions [18].

3. Graphene

Graphene is a single layer of hexagonal lattice of sp^2-bonded carbon structure [3]. It naturally exists as a building block of graphite, and π-π stacking of graphene sheets holds the lamellar graphite structure strongly in place with an interlayer spacing of 3.41 Å between the sheets [21]. Graphene was first exfoliated mechanically using a scotch tape by Geim and Novoselov, and they investigated the electronic properties of graphene, which lead to the Nobel Prize in Physics in 2010 [3]. Scientists and researchers were actively involved in growth and processing techniques using mechanical exfoliation, epitaxial growth, and chemical exfoliation, to understand and tune the properties of graphene [22, 23, 24, 25, 26, 27]. The high electron mobility of graphene in the range of 15,000 cm^2 V^{-1}s^{-1} held it as a highly promising material for electronic applications [22]. The mechanical and chemical properties of graphene further expanded its potential areas of application, and it gained popularity in the fields of electronic devices, multifunctional composites, sensors, and energy storage systems [28, 29]. However, a reproducible and large-scale production of graphene is needed for such a broad range of emerging applications. Various methods have been used for producing graphene, and these methods generally can be categorized as top-down and bottom-up approaches, such as mechanical exfoliation [30, 31], solvent assistant exfoliation [32, 33], oxidation of graphite [34], epitaxial growth on SiC surfaces [35], and chemical vapor deposition (CVD) [36, 37] (Figure 1.10).

3.1 Top-Down Synthesis of Graphene

One of the most facile methods for production of graphene is exfoliation. The stacked parallel layers of graphite with the 3.41 Å distance have weak van der Waals attraction between the layers, and this interaction lets them slide on each other perpendicularly, but the interaction between layers is strong enough to exfoliate graphite into single layer graphene. The van der Waals interactions between the graphene layers needs to be overcome to exfoliate graphite successfully. Expanding the distance between the layers by oxidation or chemical intercalating reactions is a method which reduces the interaction forces [38, 39]. To reduce interaction forces between layers and separate graphene layers, various exfoliation methods have been developed, such as using mechanical exfoliation [3], sonication assistant liquid phase exfoliation [40, 41, 42], surfactant assistant [43, 44], and oxidation and reduction process [45].

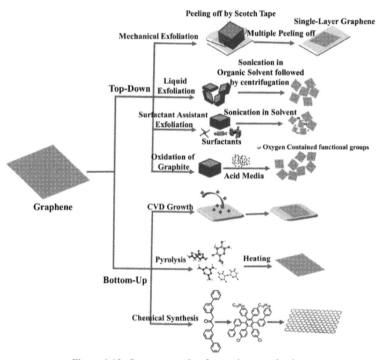

Figure 1.10. Some approaches for graphene production.

3.1.1 Mechanical Exfoliation

The scotch tape method is the most used mechanical exfoliation method, in which graphene is detached from a graphite crystal using adhesive tape. After peeling off the graphite, multiple-layer graphene can be obtained on the adhesive tape. By repeating the peeling off process, the multiple-layer graphene is cleaved into various flakes of few-layer and single-layer graphene sheets. The exfoliated graphene flakes are in the form of different sizes and thickness—from nanometers to micrometers in size for single-layer graphene with very high quality, which is dependent on the preparation of the used wafer.

3.1.2 Sonication Assistant Liquid Phase Exfoliation

Sonication assistant liquid phase exfoliation of graphite is one of the most popular large scale production of graphene [41, 46]. N-methylpyrrolidone (NMP) and N,N'-dimethylformamide (DMF) are the most widely used organic solvents for graphene and other 2D-nanosheets exfoliation mediums because of their well-matched surface tension and ability to stabilize graphene suspension [47, 48, 49]. The first successful attempt of sonication assistant graphene exfoliation was achieved by Coleman group, in which they achieved 0.01 mg ml^{-1} graphene by 30 minutes of graphite sonication in NMP, followed by centrifugation to remove unexfoliated graphite particles [41]. Very high-quality defects-free graphene can be produced by liquid phase exfoliation of graphite. The graphene-solvent interactions need to be strong enough to achieve high-yield exfoliation. It is necessary to understand the interactions between solvent and graphene to improve the quality and yield of the graphene. Toward this aim, Shih and coworkers reported the mechanism of stabilization of liquid-phase-exfoliated graphene sheets in polar solvents using molecular-dynamic (MD) simulations [49]. They investigated the interactions between graphene and various solvents, NMP, DMF, DMSO, GBL, and water. The potential of mean force between graphene layers in each of these solvents was simulated to investigate the thermodynamic stability of the graphene dispersion in each solvent. Solvent-solvent interactions decrease when the solvent molecules are trapped between graphene layers because of a reduction in the number of adjacent solvent molecules (Figure 1.11a). The calculated interaction energies between solvent molecules and graphene layers are negative. Under these conditions, the solvent molecules are in favor of being trapped, since the negative solvent-graphene interaction energy compensates to increase the solvent-solvent interaction energy. However, compression of the Van der Waals interactions between graphene and solvent molecules causes an increase in the potential energy of the trapped solvent molecules.

Authors studied the interactions between graphene and five polar solvents [water, DMF, NMP, dimethyl sulfoxide (DMSO), and γ-butyrolactone (GBL)] using MD simulations. The potential of mean force (PMF) between graphene layers in each of these solvents was simulated to investigate the thermodynamic stability of the graphene dispersion in these solvents (Figure 1.11b). Among these solvents, water

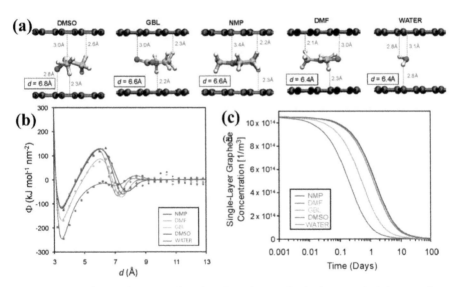

Figure 1.11. (a) Simulated representation of the five solvent molecules between single-layer graphene layers at the corresponding most confined intersheet separation d. Each sphere represents an atom using the following color scheme: white, H; green, solvent-molecule C; red, O; blue, N; yellow, S; black, graphene C. (b) Potentials of mean force per unit area, Φ, between two parallel graphene sheets in the five solvents considered here (NMP, DMF, GBL, DMSO, and water) as functions of the intersheet separation d. (c) Numerically predicted concentrations of single-layer graphene sheets in the five solvents as functions of time. Reproduced with permission from ref. [49].

shows the lowest energy barrier, the deepest interlayer Van der Waals interactions, and the least efficient solvent to stabilize graphene colloids, which clearly explains the reduction process of graphene oxide in water, when the oxygen functionality is removed. The graphene layers lose their dispersibility, aggregation, and precipitate in a very short time. The other four solvents, GBL, NMP, DMF, and DMSO, are significantly more efficient at stabilizing graphene colloids since they are providing higher energy barriers (Figure 1.11b). Figure 1.11c shows the stability of calculated concentrations of single-layer graphene sheets for all five solvents as functions of time. As seen in Figure 1.12c, the dispensability and stability of single-layer is as follows: NMP $\approx$ DMSO > DMF > GBL > water [49].

3.1.3 Surfactant Assistant Exfoliation

Exfoliation of graphene in water is mainly challenging because of the hydrophobic nature of graphene and the lack of colloidal stability. To overcome these challenges, surfactants can be used since they are amphiphilic molecules active at the surface/ interface and able to improve the interface interactions and stabilize suspension in water [50, 51, 43, 52, 53]. The first surfactant assistant graphene exfoliation was reported by Coleman group using the sodium dodecyl benzene sulfonate (SDBS) [43]. The obtained graphene mainly consists of multilayer graphene with < 5 layers and smaller quantities of monolayer graphene. The exfoliated graphene suspension is stable because of the relatively large potential barrier as a result of Coulomb repulsion between surfactant and graphene sheets.

To date, a variety of surfactants have been used for graphene exfoliation [54, 55, 51, 56, 52, 57]. Exfoliation yield, flake size, and dispersibility of graphene are different from surfactant to surfactant. As shown in Figure 1.12a, the concentration of graphene in aqueous dispersions is a result of exfoliation using various ionic and non-ionic surfactants. Non-ionic surfactants are more efficient for exfoliation and stabilization of graphene compared to ionic surfactants [51]. The mechanism of surfactant assistant exfoliation is shown in the Figure 1.12b. The hydrophobic faces of surfactants associated with the superhydrophobic graphene layers expand graphene sheets by intercalating between layers. The graphene dispersion is stabilized since

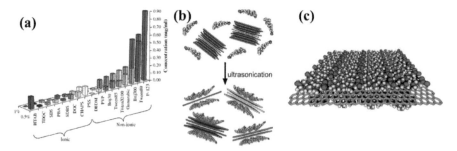

Figure 1.12. (a) Concentration of exfoliated graphene in aqueous dispersions in different surfactants media. Reproduced with permission from ref. [51], (b) The schematic diagram of the graphene exfoliation mechanism using surfactants, (c) Schematic illustration of an ordered single layer graphene and surfactants. Reproduced with permission from ref. [55].

the hydrophobic region of surfactants interact with the graphene surface and the hydrophilic faces interact with the surrounding aqueous environment (Figure 1.12c) [55].

3.1.4 Graphene Synthesis by Oxidation and Reduction Process

Graphene Oxide (GO) is a derivative of graphene and is a widely used approach for the chemical synthesis of graphene [58, 59, 60, 61, 62, 63]. GO is a derivative of graphene that contains oxygen containing functional groups, which allows tuning the properties and integrating GO with other materials to create novel composites (Figure 1.13). The degree of functional groups depends on the oxidation reaction method and synthesis conditions [64]. The structure and interlayer spacing of graphene oxide changes depending on the synthesis method and has intense effects on the properties of GO.

The first synthesis of GO dates back to the 19th century, when B. C. Brodie used potassium chlorate and fuming nitric acid on graphite. After about 60 years, Hummers and Offeman used a novel process of oxidizing graphite using potassium permanganate (KMnO$_4$), sodium nitrate (NaNO$_3$), and concentrated sulfuric acid (H$_2$SO$_4$) (Figure 1.13). As a result of the reaction between H$_2$SO$_4$ and KMnO$_4$, the active species Mn$_2$O$_7$ forms, and it gets more reactive above 55°C.

Hummer's method is the fundamental approach for recent GO synthesis methods that have improved the quality and yield of GO. In the late 1990s, modified Hummer's method was developed by Kovtyukhova (Figure 1.13). Hummer's method was preceded by pretreating graphite with sulfuric acid, potassium sulfate, and phosphorous pentoxide around 80°C for several days [61]. The pretreatment was conducted for the proper expansion of graphite before the real oxidation step, which can enhance the degree of oxidation [61, 65]. After the discovery of graphene and its potential broad range of applications, researchers focused on large quantity fabrication, which naturally brought chemical derivation methods into prominence. GO has been used as a starting material for producing graphene

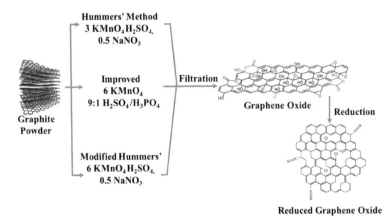

Figure 1.13. Schematic representation of graphene synthesis using graphene oxide by Hummer's, modified Hummer's, and improved methods followed by reduction to remove oxygen functionalities.

by thermal or chemical reduction. Tour group at Rice University improved the Hummer's method by replacing sodium nitrate or phosphorus pentoxide with the phosphoric acid, which improved the degree of graphite oxidation, produced larger graphene flakes, and evolved less toxic gases [45]. The oxygen functionalities on the surface of GO increase the interspace between 6–12 Å depending on moisture absorbed. GO is dispersible in water because of the hydrophilicity generated by oxygen functionalities on the surface of GO. The dispersibility of GO in various solvents and improved interaction with various compounds are key advantages for developing monolith and hybrid multifunctional materials. Another advantage of GO structures is the presence of both hydrophilic and hydrophobic part that makes it an amphiphile, which increases the interfacial interactions and minimizes the interfacial energy [66, 67, 68]. The defects from oxygen functionalities and disruption in the sp^2 lattice makes GO electrically insulating. However, reduction of GO can restore the sp^2 carbon network and hence increase the conduction of electrons [69, 59, 70]. The material after reduction is called reduced graphene oxide (rGO) instead of graphene, since it contains the structural defects and small amounts of oxygen functionalities.

The reduction of GO is an attractive approach for the mass production and applications of graphene. Although the complete reduction of GO to graphene still remains a big challenge, the partial reduction of GO has been improved a lot toward the aim of high quality graphene production. Various methods have been developed to get rid of the oxygen functionalities on the surface of GO and obtain high quality reduced rGO, such as thermal method and chemical reduction methods [71, 72, 73]. GO can be reduced exclusively by deoxygenation using thermal annealing [74]. Thermal reduction of GO is one of the most promising strategies for the mass production of rGO, and it has a lower cost and higher quality material. During the thermal reduction of GO, sp^3 carbon atoms are fully oxidized and turn to carbonaceous gases, such as CO_2, and the rest is reduced to the sp^2 graphene. Most of the functional groups can be removed above 300°C, and the thermal deoxygenation is more controllable at lower temperatures [75, 76, 74].

The chemical reduction of GO is based on the chemical reactions between GO and the reduction agent. One of the most used reduction agents is hydrazine, which was used before the discovery of graphene [77]. The reduction of GO using hydrazine and its derivatives can be achieved by the GO aqueous dispersion, and it results in agglomeration graphene sheets because of the increased hydrophobicity. The strong reactivity with water of metal hydrides, such as sodium borohydride ($NaBH_4$) and lithium aluminium hydride, which are string reducing reagents in organic chemistry, is the main obstacle for the use of GO reduction [78, 79].

3.2 Bottom-Up Synthesis of Graphene

3.2.1 Chemical Vapor Deposition (CVD) Growth of Graphene

Chemical vapor deposition (CVD) method is one of the most attractive approaches for the preparation and production of graphene for various applications [80, 81, 82]. Even though various strategies have been developed to produce graphene, the CVD method is the most promising approach, with the advantages of being inexpensive and producing large-area graphene [37, 83, 84, 85]. CVD grown graphene usually consists of two steps—the pyrolysis of precursor to carbon, and disassociation of carbon on the surface of the substrate in the form of the hexagonal carbon structure. In the first stage, hydrocarbon gas species are purged into the CVD reactor and hydrocarbon precursors decompose to carbon radicals at hot zone and at the second stage, decomposed carbon atoms disassociated on the surface of metal substrate form single-layer and few-layers graphene (Figure 1.14). The metal substrate (e.g., Cu, Ni) works as a catalyst to lower the energy barrier of the reaction and determine the graphene deposition mechanism. In general, polycrystalline metal substrate is annealed under the Ar/H$_2$ atmosphere around 900–1000°C, and then H$_2$/CH$_4$ gas mixture purges into the CVD reactor. In this step, hydrocarbon decomposes to carbon atoms and then dissolves and disassociates on the surface of metal substrate. Compared to the other metal substrate Cu, Ni has a higher solubility of carbon at elevated temperatures, and the solubility decreases when the temperature decreases. During the cooling down process of the furnace, carbon atoms diffuse out from the Ni-C solid solution and precipitate on the Ni/Cu surface to form graphene films.

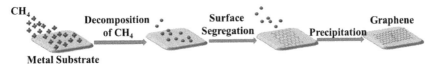

Figure 1.14. The illustration of the two-stage growth mechanism of graphene.

3.3 Chemical and Physical Properties of Graphene

Carbon atoms are densely packed in a hexagonal honeycomb crystal lattice of graphene with a bond length of 0.141 nm. The thickness of graphene has been measured by various research groups, ranging from 0.35 nm to 1.00 nm [86]. For example, Novoselov and coworkers have reported the thicknesses of graphene to be between 1.00 nm and 1.60 nm, and Gupta and coworkers have measured the thickness of single layer graphene as 0.33 nm using an atomic force microscope (AFM) [87, 3]. The strength of graphene is higher than steel by about 200 times, and this makes graphene the strongest material that has been tested until now. The Young's modulus and Poisson's ratio of graphene reported by Li and coworkers are 1.02 TPa and 0.149, respectively [88].

The electrical conductivity of graphene at room temperature is 7200 S m^{-1} with mobilities about 200,000 cm^2 V s^{-1}. The thermal conductivity of single layer graphene at room temperature has been determined between $(4.84 \pm 0.44) \times 10^3$ and $(5.30 \pm 0.48) \times 10^3$ WmK^{-1} by Balandin and coworkers [89].

3.4 Processing of Graphene by Chemical Functionalization

The nature of graphene is chemically inert and detailed reactivity of graphene in terms of the size, shape, and functionalities needs to be understood well. The chemical functionalization of graphene is one of the promising approaches for controlling its electronic properties. The chemical functionalization of graphene can be categorized as (i) covalent functionalization, (ii) non-covalent functionalization, iii) doping with heteroatoms.

Figure 1.15 shows the schematic representation of possible covalent functionalization methods of graphene. GO is more reactive and can be tailored more easily because of the oxygen functionalities, such as carboxyl, epoxy, and hydroxyl groups on the surface of GO. In terms of the covalent functionalization of graphene, acylation reactions are the most common approaches used for linking molecular moieties on the surface after oxidizing graphene. Amine functionalization can be performed after the acylation reaction between the carboxyl acid at the edge of GO and amine functional groups to modify GO by long alkyl chains [91]. In another example, Vinod et al. used the advantage of epoxy functional groups on the surface of GO and opened up the possibility of using GO as the major epoxy matrix constituent rather than just as a filler material [92]. They mixed up GO with polymercaptan-based hardener, which a resin composite when thermal energy is applied. They conducted control experiments using graphite powder and reduced GO, which confirms that the functional groups in GO instigates the reaction because the epoxy does not react with the graphite powder and rGO [92]. Authors used the first principles of density functional theory (DFT) to investigate the interaction between graphene oxide functional groups (–OH, -O-, and -COOH) and epoxy functional groups, which is mainly –SH group. The calculations show that –SH has a stronger interaction with oxygen of the functional groups -O- and –OH, compared to the graphene sheet. The interaction between functional groups and -SH breaks the C-O, C-O-C, and C-OH bonds, and forms X-SH (where X = -O- and –OH) molecules, as shown in Figure 1.16. Beside these common covalent functionalization, some well-known

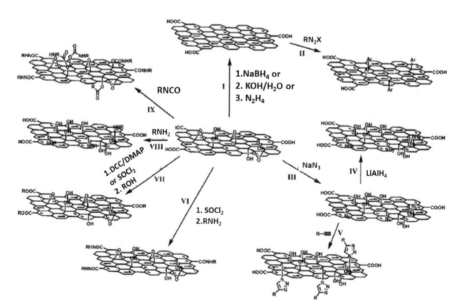

Figure 1.15. Schematic of various covalent functionalization chemistry of graphene or GO. Reproduced with permission from ref. [90].

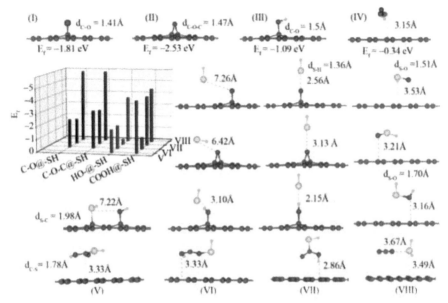

Figure 1.16. DFT calculations showing the enthalpy of formation of various functional groups in GO with mercaptan group. Among the different functional groups, the interaction of C-O-C (epoxide) group with –SH is found to be high in all configurations, indicating a favorable reaction between the two, provided the thermal barrier is overcome. Reproduced with permission from ref. [92].

organic chemistry reactions, such as cycloadditions, click reactions, and carbene insertion reactions can be applied to pure graphene [93, 94, 95].

Noncovalent functionalization of graphene and GO primarily involve physical interactions, such as hydrophobic, van der Waals, hydrogen bonding, and electrostatic interactions with the additive molecules [96, 97, 24]. Similar to other carbon nanomaterials, various materials, such as polymer, surfactants, small aromatic molecules, and biomolecules, which lead to enhanced dispersibility, biocompatibility, reactivity, binding capacity, sensing properties of graphene, have been used for non-covalent functionalization of graphene [96, 97, 98, 99]. The most important advantage of non-covalent functionalization is protecting sp^2 hybrid structure of carbon, and thus outstanding physical and chemical properties of graphene can be saved.

In addition to the surface chemical functionalization, doping with heteroatoms is another approach to control the properties of graphene. Heteroatom-doped (e.g., B, N, and S) graphene nanosheets have been fabricated using the CVD system. Heteroatom-doped graphene can be synthesized by introducing solid, liquid, or gaseous precursors containing the desired heteroatom into the CVD furnace during the growth together with the carbon sources [100, 101, 8, 102, 103]. Boron (B) and nitrogen (N) are the most commonly used heteroatoms for doping graphene, and these atoms prefer to be replaced with a carbon atom within the hexagonal lattice of graphene because they have similar sizes and valence electron numbers as carbon. In addition, graphene can be doped with multiple atoms for more effectively tuning its properties. For example, boron and nitrogen atoms can be co-doped into graphene

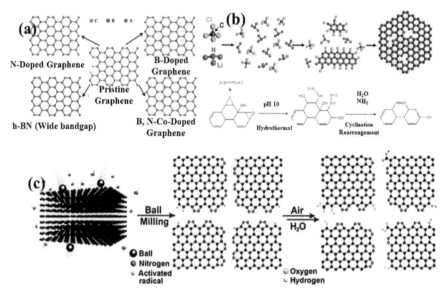

Figure 1.17. Schematic of graphene and heteroatom-doped (B and N) graphene with possible band-gap. Reproduced with permission from ref. [159], (b) Doping graphene with heteroatoms using hydrothermal method. Reproduced with permission from ref. [109], (c) Doping graphene by ball-milling technique. Reproduced with permission from ref. [106].

lattices to form a variety of semiconducting layered structures with variable stoichiometry and a tunable bandgap (Figure 1.17) [104, 105]. Besides the CVD method, other approaches can be used for doping graphene, such as hydrothermal method, ball milling, and thermal annealing methods (Figures 1.17b,c) [106, 107, 108, 109].

4. 3D Engineered Carbon Nanomaterials Processing

Although carbon nanomaterials exhibit attractive physical and chemical properties, which could be exploited for a broad range of emerging applications, the real challenge lies in integrating them to macroscale platforms. To overcome this challenge, it is highly desirable to create various macrostructures, and thus utilize their excellent mechanical and physical properties. Moreover, fabrication of 3D macrostructures using nanoscale elements allows tailoring of their properties. Although various nanostructures in different dimensionalities can be used, the fascinating physical phenomena in carbon nanomaterials, such as nanotubes, graphene, and other nanostructures are extremely attractive to create building blocks of macroscopic assemblies. Various approaches have been developed for creating carbon-based macrostructures [110, 111, 112, 113, 114, 115]. Here, we will review 3D engineered carbon nanostructures based on materials they are processing as self-assembly 3D carbon based macrostructures and covalently interconnected 3D carbon based macrostructures [116, 117, 118, 119].

4.1 Self-Assembly 3D Carbon-Based Architectures

One of the most promising methods to create 3D hierarchical architectures, such as aerogels, foam, and sponges using individual honeycomb carbon nanostructures is self-assembly strategy, which relies on physical interactions between the chemical functionalities on the surface of carbon nanostructures and their additives as a result of hydrogen bonding, Van der Waals interactions, π-π interactions, electrostatic interactions, hydrophobic interactions, and so on [120, 110, 111, 121, 122]. The unique structure of the 3D macroscopic self-assemblies of carbon nanostructures could provide excellent opportunities to integrate their properties into the critically needed applications. Therefore, it is highly required to develop non-expensive and highly efficient preparation techniques.

The main precursors for 3D nanocarbon-based structural aerogels, foam, and sponges are CNTs, graphene, and graphene derivatives, such as graphene oxide (GO), reduced graphene oxide (rGO), etc. In an usual process, carbon nanostructures and their composite additives are dispersed in a solvent, followed by the lyophilization process for removing solvents trapped in the pore structures. Thereby, the interconnected porous structures can be retained. Generally, there are two key factors for the formation of 3D nanocarbon structures and their 3D macrostructure stabilization. The first one is their proper dispersion in suspensions and the other one is strong physicochemical interactions between individual nanostructures and additives that cause self-assembling into 3D macrostructures during and after removal of the solvent. Alternatively, the self-assembly behavior of carbon nanomaterials can be created by their chemical functionalization using techniques shown above, thus improving their interface interactions. Thus, the assembly process becomes easier and more effective since functionalization makes interactions stronger.

The specific chemical surface functionalization of carbon nanomaterials, such as oxidation or amidation are able to produce a uniform suspension in water since chemical functionalities convert the hydrophobic nature of CNTs and graphene to hydrophilicity. Here, the major improved interface interactions between CNTs/graphene as a result of chemical surface modification comes from hydrogen bonding, Van der Waals interactions, hydrophobic interactions, etc. For example, Hu et al. reported ultra-lightweight and highly compressible 3D graphene aerogel amine functionalized self-assembled 3D GO after the lyophilization process [123]. After the free-drying process, they conducted microwave treatment to remove chemical functional groups that shaped ultra-lightweight 3D graphene aerogel. The resultant lightweight (3 mg/cm^3) 3D macrostructure of graphene sponge exhibited superior resilience and could recover 90% compression (Figure 1.18). They demonstrate that the lightweight 3D graphene aerogel with high elasticity has a great potential for energy-absorbing applications [123].

4.2 Covalently Interconnected 3D Carbon-Based Architectures

If atomic-scale junctions between individual carbon nanostructures can be created, they can also be ordered as covalently bonded 3D solid networks with control over properties, such as density, porosity, and mechanical properties. The covalent

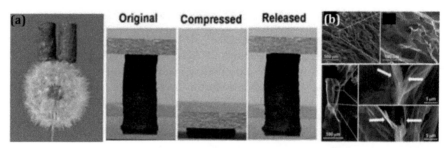

Figure 1.18. (a) Images of ultra-lightweight and highly compressible 3D GO structure, (b) SEM images of porous 3D graphene aerogels. Reproduced with permission from ref. [123].

bonding between individual nanostructures can be formed by several methods, such as chemical vapor deposition (CVD) method [124, 117, 114] surface chemistry [125, 111, 113, 119, 92, 126], and welding techniques [127, 128, 112].

Chemical vapor deposition (CVD) method is one of the exceptional approaches to create covalent junctions between individual nanotubes and/or graphene for shaping covalently interconnected 3D carbon-based macrostructures [129, 130, 124, 117, 131, 132, 114]. To create covalently interconnected 3D nanotube and graphene-based architectures via CVD method, several approaches have been used. Using additive elements, such as boron (B), nitrogen (N), and sulfur (S) during growth is one of the most commonly used methods for fabricating covalently interconnected carbon nanostructures, which is especially used for 3D CNTs. The role of additive elements is to create pentagonal, heptagonal topological defects in the hexagonal honeycomb structure of carbon nanomaterials and promote the growth of covalent junctions [133, 130, 134, 132]. For example, Shan et al. reported a covalently interconnected 3D nitrogen and sulfur induced CNTs sponge using CVD method (Figure 1.19). As shown in the Figure 1.19, the nitrogen and sulfur doped CNTs sponge consist of

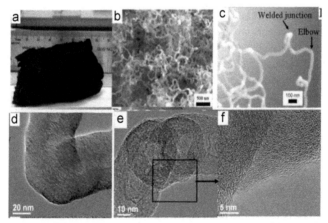

Figure 1.19. (a) Image of nitrogen-doped CNTs sponge, (b–c) SEM images of nitrogen-doped 3D CNTs structures, (d) TEM image of an "elbow" junction in 3D CNTs sponge, (e–f) TEM images of nitrogen-doped 3D CNT junctions. Reproduced with permission from ref. [132].

CNTs with "elbow" structures as well as "welded" junctions that forms because of the synergistic effect of additive elements, nitrogen and sulfur [132].

Beside using additive elements to create covalent junction between individual nanostructures, we developed a novel approach by using multiple catalysts during the growth [117]. In this novel approach, 10 nm Al, and 1.5 nm Fe were deposited on silicon substrate and placed in the CVD furnace. In addition, we used 1% (mg/ml) ferrocene/xylene solution as a carbon and secondary catalyst source. The ferrocene/ xylene solution was transferred into the reactor chamber, which used Ar/H$_2$ (15% H$_2$) carrier gas at a flow rate of 0.2 ml/min. The furnace temperature was set to 790°C in the chamber where the precursor solution was vaporized. As a result, 3D CNTs solid material with nanoscale intermolecular junctions was obtained (Figure 1.20a). The density of 3D solid CNTs can be produced between 0.13 mg/mm^3 and 0.32 mg/mm^3. The 3D architectures consist entirely of entangled CNTs with different orientations producing spatially varying morphologies, such as Y-type, X-type, multi-branched, and ring-like configurations (Figures 1.20b–k).

Chen et al. reported a nickel template-directed CVD process for synthesis of covalently interconnected 3D graphene structures (Figure 1.21) [129]. The fabricated 3D graphene structure is seamlessly interconnected into a 3D porous, and flexible network because of the Ni template (Figure 1.21). The seamless interconnection between graphene layers give the material porosity, outstanding electrical conductivity, mechanical properties, and high specific surface area.

The other approach for synthesis of 3D nanocarbon structures is the surface chemistry method. This approach essentially depends on the chemical surface

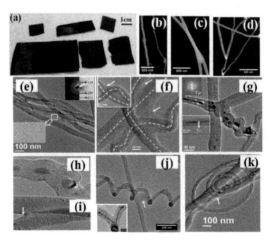

Figure 1.20. (a) Images of 3D CNT block of different dimensions, (b–d) High magnification SEM image of the sample showing different kinds of CNTs junctions, (e) Bright field (BF) TEM image of the CNT bundle with SAD showing polycrystalline diffraction of BCC iron (shown as top inset), bottom inset showing HRTEM image of the interconnected region. (f) BF TEM image showing two CNT interconnected with a thin region of carbon. Inset showing lower magnification image again with similar morphology. (g) BF TEM image of a CNT showing the Fe filled core. The top inset showing SAD of iron and bottom inset showing HRTEM of the same. (h) BF TEM image of "elbow" junction, (i) "Y" junction, (j) Helical structure, and (k) Coiled structure. Reproduced with permission from ref. [117].

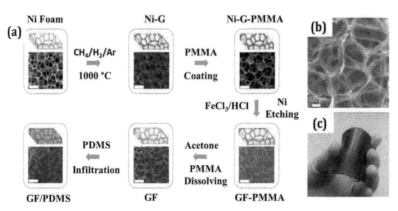

Figure 1.21. (a) CVD growth of 3D graphene architecture using nickel as templates, (b) SEM image of 3D graphene structure shows the uniform porosity, (c) Image of flexible 3D graphene/PDMS composite. Reproduced with permission from ref. [129].

functionalization of carbon nanostructures. Surface modification by functionalization is one of the fundamental requirements for fabrication of 3D Carbon-based building blocks. The essence of a chemical functionalization approach to create 3D carbon frameworks relies on covalent junctions between individual carbon nanostructures. In general, chemical cross-linking process is following by post process, such as freeze-drying. The properties of carbon nanostructures can be manipulated easily using appropriate functionalities for desirable applications. For example, we used Suzuki cross-coupling reaction, which is a well-known organic chemistry reaction to create carbon-carbon bonding, for the synthesis of covalently connected 3D macroscopic solids of nanotubes (Figure 1.22) [116].

CNTs were found to form a solid macrostructure when they have enough functional groups to cross-link with covalent interconnection. The image of a

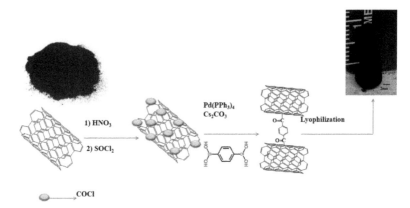

Figure 1.22. Schematic of the 3D CNT solid macrostructure synthesis from CNT powder. Initial CNT powder was oxidized in HNO_3 and then converted to acid chloride. CNTs were covalently interconnected via Suzuki coupling reaction, which is a palladium catalyst-based coupling reaction. After lyophilization (Freeze Drying), 3D CNTs solid structure formed. Reproduced with permission from ref. [116].

3D-CNT solid network is shown in Figure 1.23, which contains various network morphologies in different regions. The 3D CNT solid structure is highly porous, as can be seen in Figure 1.23, and CNT bundles are interconnected as a network, with the microporosity.

Chemical cross-linking nanomaterials via solution chemistry has been applied to graphene structures [135, 113, 118, 136, 119]. Similar strategies, which are used for cross-linking of nanotubes can be used for graphene cross-linking as well, and this approach is easier for GO compared to CNTs, since GO already contains functional groups. For example, Sudeep and coworkers treated dispersed GO and fluorinated GO (FGO) with borax and glutaraldehyde, followed by sonication and freeze-drying process (Figure 1.24). The obtained structure was made of a completely cross-linked 3D network of GO as a result of polymerization between GO nanosheets using glutaraldehyde and resorcinol [118]. The representative process that shows the structural evaluation of GO is shown in Figure 1.24. In general, glutaraldehyde, which contains two aldehyde functionalities, reacts with the alcohol (–OH) functional groups on the surface of GO with the help of borax catalyst. The resultant 3D macro-GO has high porosity, which shows promising results for CO_2 gas absorption. Such 3D porous macroscopic structures are found to keep their structural integrity up to 100°C [118].

Another approach for building interconnected 3D carbon-based macrostructures is the welding method [127, 128, 137, 112, 138, 139, 140]. Welding techniques are important for joining metals or remediating structural defects. These methods have started to merge with some nanomaterials, such as nanotubes, nanowires, and nanoparticles under particular conditions, such as heating, beam irradiation, sonication, and spark plasma, etc. For example, Terrones et al. performed electron beam to form intermolecular junctions between individual single-wall carbon nanotubes (SWCNTs). They created intermolecular junctions of SWCNTs

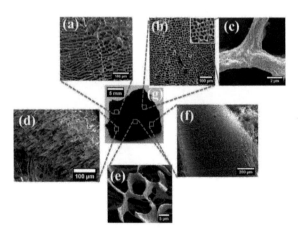

Figure 1.23. Scanning electron microscopy (SEM) images of different regions of the porous 3D CNT solid structure. (a, d) Layer by layer CNT thin films, (b, c, f) CNT bundles as a network, (e) 3D CNT after lyophilization of the covalently connected CNTs via Suzuki coupling reaction formed as tubular structure, (g) 3D CNT solid macrostructure. Reproduced with permission from ref. [116].

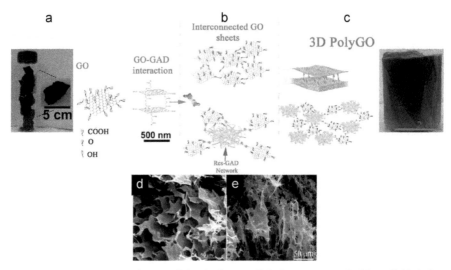

Figure 1.24. (a–c) Schematic images of chemically cross-linked 3D macroscopic GO scaffolds before and after interconnection. (d–e) High and low SEM images of 3D GO and 3D FGO scaffolds. Reproduced with permission from ref. [118].

with various geometries using *in situ* transmission electron microscope (TEM) (Figure 1.25). Cross-linking dangling bonds between SWCNTs form as a result of electron beam exposure at high temperatures [141].

In the other example, we used spark plasma sintering for creation of 3D-welded carbon nanostructures. The spark plasma sintering (SPS) process gives rise to the properties of carbon structure in the form of 3D macrostructure because of the interconnection created by welding (Figure 1.26) [127, 128, 112]). We performed SPS at three different temperatures (1000, 1200, and 1400°C) and three different holding times (5, 15, and 30 minutes) (Figures 1.26a,b). The low magnification of SEM shows a smooth surface morphology (Figure 1.26c). The high magnification

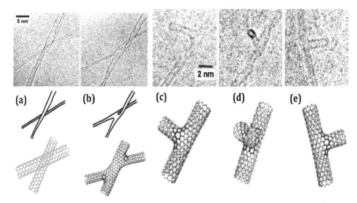

Figure 1.25. TEM images of high resolution TEM and schematic representation of various geometries of intermolecular junction between individual SWCNTs (a) Non-interconnected SWCNTs, (b) "X" Junction, and (c–e) "T"-Junction of SWCNTs structures. Reproduced with permission from ref. [141].

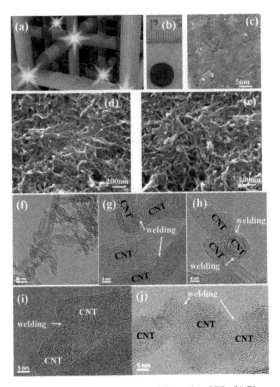

Figure 1.26. (a) Schematic of the process of CNT welding using SPS. (b) Picture of a representative sample processed using the current method. (c) Low magnification SEM image of the sample. (d, e) High magnification SEM image of different regions depicting interconnected nanotubes. (f) Low magnification, bright-field TEM of the interconnected CNT 3D structure. (g–j) High-resolution TEM images of CNTs welded in different orientations. Reproduced with permission from ref. [112].

SEM images of the sample display the interconnected 3D CNT solid structure (Figures 1.26d,e), which revealed that the structure is highly porous. The low magnification bright field TEM image in Figure 1.26f shows that nanotube structures are welded together to form the interconnected structure. The HRTEM images of the welded structure reveal various possible orientations of the welding joints (Figures 1.26g–j). The HRTEM images clearly resolve multi-walled CNTs that are welded at their tips in different possible orientations. We observed nanotubes connecting to each other through the surface of the CNTs and end-to-end welding. Additionally, the morphology of individual nanotubes can also change to nano-onion-like structures, as seen in Figure 1.26h.

5. Application of Carbon Nanomaterials

Carbon nanomaterials have found a wide range of applications because of their exceptional properties (Figure 1.27). Carbon-based nanomaterials have integrated some commercial applications [142]. Carbon nanomaterials have been integrated into structural and composite materials for developing lightweight, stiff, and strong

Figure 1.27. Some application areas of carbon nanomaterials.

composites due to their excellent mechanical properties, specifically the high tensile strength, electrical conductivity, and chemical properties [121, 143]. Taking the advantage of their dispersion, functionalization, and large-area deposition methods, they can be used as a multifunctional coating material on the surfaces [144, 145]. They have been widely used in energy storage and conversion applications as an electrode, catalyst, and catalyst support, such as Li-ion batteries, supercapacitors, photovoltaics, and fuel cells [146, 147, 148, 149, 150]. These materials have been used in a wide range of biotechnology applications, such as controlled drug delivery, biosensors, and neural and orthopedic implants, etc. [151, 128, 152, 153]. The use of carbon nanomaterials, CNTs and graphene, functionalization and their hybrid structures allow novel properties to be exploited in the areas of electronics, optoelectronics, and photonics [154, 22, 155]. To date, various materials have been used for aerospace applications. Carbon nanomaterials, nanotubes, and graphene, especially their porous 3D macrostructures, are ideal materials for the next generation aerospace vehicles because they are ultra lightweight, flexible, radiation resistant, and have good mechanical properties [142, 156, 157]. The high surface area, hydrophobicity, high thermal and electrical conductivity, and tunability of their properties by surface functionalization make them excellent materials for environmental applications, such as oil absorption, water purification, CO_2 storage, etc. [158, 110, 113]. Beside these aforementioned applications of carbon nanomaterials, they have wider potential applications, such as near-infrared (NIR), plasmonics, photonics, rechargeable batteries, automotive parts, and sporting goods to boat hulls, etc.

6. Challenges on Carbon Nanomaterials

Although it has been more than three decades since the discovery of novel carbon nanomaterials, they have still not been integrated into the daily-life technologies. A significant progress has been achieved regarding the synthesis, processing, and functionalization methods, but there are still a lot of challenges that need to be overcome to see these amazing materials in current technologies. Although various methods have been developed for their synthesis, their mass production is the biggest challenge for their commercialization. The other challenge is their control over synthesis, such as chirality, diameter, and geometry, which are key parameters affecting their properties (e.g., electronic band gap, electrical conductivity). For example, the electronic structure of CNTs depends on their chirality and diameter, but there is no report that shows the synthesis of single diameter CNTs. In addition, the properties of these structures need to be modified for specific applications during synthesis or post-processing methods. Although there is a significant progress regarding the chemical surface functionalization or elemental doping to tailor their properties, novel approaches need to be developed for their control over modification so that their properties can be tuned depending on the desired application.

7. Summary

The discovery of fullerene in 1985 opened a new era that is called nanocarbon, followed by the discovery of CNTs and graphene. Carbon nanomaterials can be considered as attractive candidates in a broad range of nanotechnology applications, ranging from medicine to aerospace because of their outstanding properties, but these nanomaterials need to be processed and modified during or after their fabrication for the specific applications. This chapter discusses chemical processing of carbon nanomaterials, especially CNTs and graphene, which covers various strategies of their synthesis, functionalization for tuning their properties, and the effect of functionalization on their properties. In addition, applications and current challenges on the integration of these materials into the real life technologies are discussed.

References

[1] Kroto, H.W., J.R. Heath, S.C. Obrien, R.F. Curl and R.E. Smalley. 1985. C-60-buckminsterfullerene. Nature, 318(6042): 162–163. doi:DOI 10.1038/318162a0.
[2] Iijima, S. 1991. Helical microtubules of graphitic carbon. Nature, 354(6348): 56–58. doi:DOI 10.1038/354056a0.
[3] Novoselov, K.S., A.K. Geim, S.V. Morozov, D. Jiang, Y. Zhang, S.V. Dubonos, I.V. Grigorieva and A.A. Firsov. 2004. Electric field effect in atomically thin carbon films. Science, 306(5696): 666–669. doi:10.1126/science.1102896.
[4] Coluci, V.R., S.F. Braga, S.B. Legoas, D.S. Galvao and R.H. Baughman. 2003. Families of carbon nanotubes: Graphyne-based nanotubes. Phys. Rev. B, 68(3). doi:ARTN 03543010.1103/PhysRevB.68.035430.
[5] Terrones, M., N. Grobert and H. Terrones. 2002. Synthetic routes to nanoscale BxCyNz architectures. Carbon, 40(10): 1665–1684. doi:Pii S0008-6223(02)00008-8. Doi 10.1016/S0008-6223(02)00008-8.
[6] Ajayan, P.M. and J.M. Tour. 2007. Nanotube composites. Nature, 447(7148): 1066–1068.
[7] Terrones, M., P.M. Ajayan, F. Banhart, X. Blase, D.L. Carroll, J.C. Charlier, R. Czerw, B. Foley, N. Grobert, R. Kamalakaran, P. Kohler-Redlich, M. Rühle, T. Seeger and H. Terrones. 2002. N-doping and coalescence of carbon nanotubes: synthesis and electronic properties. Appl. Phys. A Mater. Sci. Process, 74(3): 355–361. doi:10.1007/s003390201278.
[8] Wei, D.C., Y.Q. Liu, Y. Wang, H.L. Zhang, L.P. Huang and G. Yu. 2009. Synthesis of N-doped graphene by chemical vapor deposition and its electrical properties. Nano Lett., 9(5): 1752–1758. doi:10.1021/nl803279t.

[9] Velasco-Santos, C., A.L. Martinez-Hernandez, F.T. Fisher, R. Ruoff and V.M. Castano. 2003. Improvement of thermal and mechanical properties of carbon nanotube composites through chemical functionalization. Chem. Mater., 15(23): 4470–4475. doi:10.1021/cm034243c.

[10] Zhang, Z.Q., B. Liu, Y.L. Chen, H. Jiang, K.C. Hwang and Y. Huang. 2008. Mechanical properties of functionalized carbon nanotubes. Nanotechnology, 19(39). doi:Artn 39570210.1088/0957-4484/19/39/395702.

[11] Pan, R.Q., Z.J. Xu, Z.Y. Zhu and Z.X. Wang. 2007. Thermal conductivity of functionalized single-wall carbon nanotubes. Nanotechnology, 18(28). doi:Artn 28570410.1088/0957-4484/18/28/285704.

[12] Saito, T. and T. Ikoma. 1989. Effect of stacking-sequence on valence bands in Ga/as/Ge (001) monolayer superlattices. Appl. Phys. Lett., 55(13): 1300–1302. doi:Doi 10.1063/1.101638.

[13] O'Connell, M.J., S.M. Bachilo, C.B. Huffman, V.C. Moore, M.S. Strano, E.H. Haroz, K.L. Rialon, P.J. Boul, W.H. Noon, C. Kittrell, J. Ma, R.H. Hauge, R.B. Weisman and R.E. Smalley. 2002. Band gap fluorescence from individual single-walled carbon nanotubes. Science, 297(5581): 593–596. doi:DOI 10.1126/science.107263.

[14] Kim, K.S., D.J. Bae, J.R. Kim, K.A. Park, S.C. Lim, J.J. Kim, W.B. Choi, C.Y. Park and Y.H. Lee. 2002. Modification of electronic structures of a carbon nanotube by hydrogen functionalization. Adv. Mater., 14(24): 1818–1821. doi:DOI 10.1002/adma.200290008.

[15] Strano, M.S., C.A. Dyke, M.L. Usrey, P.W. Barone, M.J. Allen, H.W. Shan, C. Kittrell, R.H. Hauge, J.M. Tour and R.E. Smalley. 2003. Electronic structure control of single-walled carbon nanotube functionalization. Science, 301(5639): 1519–1522. doi:DOI 10.1126/science.1087691.

[16] Bachilo, S.M., M.S. Strano, C. Kittrell, R.H. Hauge, R.E. Smalley and R.B. Weisman. 2002. Structure-assigned optical spectra of single-walled carbon nanotubes. Science, 298(5602): 2361–2366. doi:10.1126/science.1078727.

[17] He, X., H. Htoon, S.K. Doorn, W.H.P. Pernice, F. Pyatkov, R. Krupke, A. Jeantet, Y. Chassagneux and C. Voisin. 2018. Carbon nanotubes as emerging quantum-light sources. Nat. Mater., 17(8): 663–670. doi:10.1038/s41563-018-0109-2.

[18] Kwon, H., M. Furmanchuk, M. Kim, B. Meany, Y. Guo, G.C. Schatz and Y.H. Wang. 2016. Molecularly tunable fluorescent quantum defects. J. Am. Chem. Soc., 138(21): 6878–6885. doi:10.1021/jacs.6b03618.

[19] Ma, X., N.F. Hartmann, K.A. Velizhanin, J.K.S. Baldwin, L. Adamska, S. Tretiak, S.K. Doorn and H. Htoon. 2017. Multi-exciton emission from solitary dopant states of carbon nanotubes. Nanoscale, 9(42): 16143–16148. doi:10.1039/c7nr06661a.

[20] Saha, A., B.J. Gifford, X.W. He, G.Y. Ao, M. Zheng, H. Kataura, H. Htoon, S. Kilina, S. Tretiak and S.K. Doorn. 2018. Narrow-band single-photon emission through selective aryl functionalization of zigzag carbon nanotubes. Nat. Chem., 10(11): 1089–1095. doi:10.1038/s41557-018-0126-4.

[21] Chen, X.B., F.Y. Tian, C. Persson, W.H. Duan and N.X. Chen. 2013. Interlayer interactions in graphites. Sci. Rep., 3. doi:ARTN 304610.1038/srep03046.

[22] Geim, A.K. 2009. Graphene: Status and prospects. Science, 324(5934): 1530–1534. doi:10.1126/science.1158877.

[23] Georgakilas, V., M. Otyepka, A.B. Bourlinos, V. Chandra, N. Kim, K.C. Kemp, P. Hobza, R. Zboril and K.S. Kim. 2012. Functionalization of graphene: Covalent and non-covalent approaches, derivatives and applications. Chem. Rev., 112(11): 6156–6214. doi:10.1021/cr3000412.

[24] Huang, X., Z.Y. Yin, S.X. Wu, X.Y. Qi, Q.Y. He, Q.C. Zhang, Q. Yan, F. Boey and H. Zhang. 2011. Graphene-based materials: synthesis, characterization, properties, and applications. Small, 7(14): 1876–1902. doi:10.1002/smll.201002009.

[25] Kim, K.S., Y. Zhao, H. Jang, S.Y. Lee, J.M. Kim, K.S. Kim, J.-H. Ahn, P. Kim, J.-Y. Choi and B.H. Hong. 2009. Large-scale pattern growth of graphene films for stretchable transparent electrodes. Nature, 457(7230): 706–710. doi:10.1038/nature07719.

[26] Rao, C.N.R., A.K. Sood, K.S. Subrahmanyam and A. Govindaraj. 2009. Graphene: The new two-dimensional nanomaterial. Angewandte Chemie-International Edition, 48(42): 7752–7777. doi:10.1002/anie.200901678.

[27] Stankovich, S., D.A. Dikin, R.D. Piner, K.A. Kohlhaas, A. Kleinhammes, Y. Jia, Y. Wu, S.B.T. Nguyen and R.S. Ruoff. 2007. Synthesis of graphene-based nanosheets via chemical reduction of exfoliated graphite oxide. Carbon, 45(7): 1558–1565. doi:10.1016/j.carbon.2007.02.034.

[28] Avouris, P. and M. Freitag. 2014. Graphene photonics, plasmonics, and optoelectronics. IEEE Journal of Selected Topics in Quantum Electronics, 20(1). doi:Artn 600011210.1109/Jstqe.2013.2272315.

[29] Bonaccorso, F., Z. Sun, T. Hasan and A.C. Ferrari. 2010. Graphene photonics and optoelectronics. Nat. Photonics , 4(9): 611–622. doi:10.1038/Nphoton.2010.186.

[30] Cai, M.Z., D. Thorpe, D.H. Adamson and H.C. Schniepp. 2012. Methods of graphite exfoliation. J. Mat. Chem., 22(48): 24992–25002. doi:10.1039/c2jm34517j.

[31] Yuan, L., J. Ge, X.L. Peng, Q. Zhang, Z.F. Wu, Y. Jian, X. Xiong, H. Yin and J.F. Han. 2016. A reliable way of mechanical exfoliation of large scale two dimensional materials with high quality. AIP Adv., 6(12): 125–201. doi:Artn 12520110.1063/1.4967967.

[32] Coleman, J.N. 2013. Liquid exfoliation of defect-free graphene. Acc. Chem. Res., 46(1): 14–22. doi:10.1021/ar300009f.

[33] Paton, K.R., E. Varrla, C. Backes, R.J. Smith, U. Khan, A. O'Neill, C. Boland, M. Lotya, O.M. Istrate, P. King, T. Higgins, S. Barwich, P. May, P. Puczkarski, I. Ahmed, M. Moebius, H. Pettersson, E. Long, J. Coelho, S.E. O'Brien, E.K. McGuire, B.M. Sanchez, G.S. Duesberg, N. McEvoy, T.J. Pennycook, C. Downing, A. Crossley, V. Nicolosi and J.N. Coleman. 2014. Scalable production of large quantities of defect-free few-layer graphene by shear exfoliation in liquids. Nat. Mater., 13(6): 624–630. doi:10.1038/Nmat3944.

[34] Pei, S.F. and H.M. Cheng. 2012. The reduction of graphene oxide. Carbon, 50(9): 3210–3228. doi:10.1016/j.carbon.2011.11.010.

[35] Strupinski, W., K. Grodecki, A. Wysmolek, R. Stepniewski, T. Szkopek, P.E. Gaskell, A. Grüneis, D. Haberer, R. Bozek, J. Krupka and J.M. Baranowski. 2011. Graphene epitaxy by chemical vapor deposition on SiC. Nano Lett., 11(4): 1786–1791. doi:10.1021/nl200390e.

[36] Alnuaimi, A., I. Almansouri, I. Saadat and A. Nayfeh. 2017. Toward fast growth of large area high quality graphene using a cold-wall CVD reactor. RSC Adv., 7(82): 51951–51957. doi:10.1039/c7ra10336k.

[37] Chen, X.P., L.L. Zhang and S.S. Chen. 2015. Large area CVD growth of graphene. Synth. Met., 210: 95–108. doi:10.1016/j.synthmet.2015.07.005.

[38] Ramanathan, T., A.A. Abdala, S. Stankovich, D.A. Dikin, M. Herrera-Alonso, R.D. Piner, D.H. Adamson, H.C. Schniepp, X. Chen, R.S. Ruoff, S.T. Nguyen, I.A. Aksay, R.K. Prud'Homme, and L.C. Brinson. 2008. Functionalized graphene sheets for polymer nanocomposites. Nat. Nanotechnol., 3(6): 327–331. doi:10.1038/nnano.2008.96.

[39] Stankovich, S., D.A. Dikin, G.H.B. Dommett, K.M. Kohlhaas, E.J. Zimney, E.A. Stach, R.D. Piner, S.B.T. Nguyen and R.S. Ruoff. 2006. Graphene-based composite materials. Nature, 442(7100): 282–286. doi:10.1038/nature04969.

[40] Chaban, V.V., E.E. Fileti and O.V. Prezhdo. 2017. Exfoliation of graphene in ionic liquids: pyridinium versus pyrrolidinium. J. Phys. Chem. C., 121(1): 911–917. doi:10.1021/acs.jpcc.6b11003.

[41] Hernandez, Y., V. Nicolosi, M. Lotya, F.M. Blighe, Z.Y. Sun, S. De, I.T. McGovern, B. Holland, M. Byrne, Y.K. Gun'Ko, J.J. Boland, P. Niraj, G. Duesberg, S. Krishnamurthy, R. Goodhue, J. Hutchison, V. Scardaci, A.C. Ferrari and J.N. Coleman. 2008. High-yield production of graphene by liquid-phase exfoliation of graphite. Nat. Nanotechnol., 3(9): 563–568. doi:10.1038/nnano.2008.215.

[42] Liu, W.W. and J.N. Wang. 2011. Direct exfoliation of graphene in organic solvents with addition of NaOH. Chem. Commun. (Camb.), 47(24): 6888–6890. doi:10.1039/c1cc11933h.

[43] Lotya, M., Y. Hernandez, P.J. King, R.J. Smith, V. Nicolosi, L.S. Karlsson, F.M. Blighe, S. De, Z. Wang, I.T. McGovern, G.S. Duesberg and J.N. Coleman. 2009. Liquid phase production of graphene by exfoliation of graphite in surfactant/water solutions. J. Am. Chem. Soc., 131(10): 3611–3620. doi:10.1021/ja807449u.

[44] Lotya, M., P.J. King, U. Khan, S. De and J.N. Coleman. 2010. High-concentration, surfactant-stabilized graphene dispersions. ACS Nano, 4(6): 3155–3162. doi:10.1021/nn1005304.

[45] Marcano, D.C., D.V. Kosynkin, J.M. Berlin, A. Sinitskii, Z.Z. Sun, A. Slesarev, L.B. Alemany, W. Lu and J.M. Tour. 2010. Improved synthesis of graphene oxide. ACS Nano, 4(8): 4806–4814. doi:10.1021/nn1006368.

[46] Wang, J., Y. Hernandez, M. Lotya, J.N. Coleman and W.J. Blau. 2009. Broadband nonlinear optical response of graphene dispersions. Adv. Mater., 21(23): 2430–2435. doi:10.1002/adma.200803616.

[47] Blake, P., P.D. Brimicombe, R.R. Nair, T.J. Booth, D. Jiang, F. Schedin, L.A. Ponomarenko, S.V. Morozov, H.F. Gleeson, E.W. Hill, A.K. Geim and K.S. Novoselov. 2008. Graphene-based liquid crystal device. Nano Lett., 8(6): 1704–1708. doi:10.1021/nl080649i.

[48] Hernandez, Y., M. Lotya, D. Rickard, S.D. Bergin and J.N. Coleman. 2010. Measurement of multicomponent solubility parameters for graphene facilitates solvent discovery. Langmuir, 26(5): 3208–3213. doi:10.1021/la903188a.

[49] Shih, C.J., S.C. Lin, M.S. Strano and D. Blankschtein. 2010. Understanding the stabilization of liquid-phase-exfoliated graphene in polar solvents: molecular dynamics simulations and kinetic theory of colloid aggregation. J. Am. Chem. Soc., 132(41): 14638–14648. doi:10.1021/ja1064284.

[50] Di Crescenzo, A., P. Di Profio, G. Siani, R. Zappacosta and A. Fontana. 2016. Optimizing the interactions of surfactants with graphitic surfaces and clathrate hydrates. Langmuir, 32(26): 6559–6570. doi:10.1021/acs.langmuir.6b01435.

[51] Guardia, L., M.J. Fernandez-Merino, J.I. Paredes, P. Solis-Fernandez, S. Villar-Rodil, A. Martinez-Alonso and J.M.D. Tascon. 2011. High-throughput production of pristine graphene in an aqueous dispersion assisted by non-ionic surfactants. Carbon, 49(5): 1653–1662. doi:10.1016/j.carbon.2010.12.049.

[52] Smith, R.J., M. Lotya and J.N. Coleman. 2010. The importance of repulsive potential barriers for the dispersion of graphene using surfactants. New J. Phys., 12. doi:Artn 12500810.1088/1367-2630/12/12/125008.

[53] Wang, S., M. Yi and Z.G. Shen. 2016. The effect of surfactants and their concentration on the liquid exfoliation of graphene. RSC Adv., 6(61): 56705–56710. doi:10.1039/c6ra10933k.

[54] De, S., P.J. King, M. Lotya, A. O'Neill, E.M. Doherty, Y. Hernandez, G.S. Duesberg and J.N. Coleman. 2010. Flexible, transparent, conducting films of randomly stacked graphene from surfactant-stabilized, oxide-free graphene dispersions. Small, 6(3): 458–464. doi:10.1002/smll.200901162.

[55] Green, A.A. and M.C. Hersam. 2009. Solution phase production of graphene with controlled thickness via density differentiation. Nano Lett., 9(12): 4031–4036. doi:10.1021/nl902200b.

[56] Hasan, T., F. Torrisi, Z. Sun, D. Popa, V. Nicolosi, G. Privitera, F. Bonaccorso and A.C. Ferrari. 2010. Solution-phase exfoliation of graphite for ultrafast photonics. Phys. Status Solidi B Basic Res., 247(11-12): 2953–2957. doi:10.1002/pssb.201000339.

[57] Vadukumpully, S., J. Paul and S. Valiyaveettil. 2009. Cationic surfactant mediated exfoliation of graphite into graphene flakes. Carbon, 47(14): 3288–3294. doi:10.1016/j.carbon.2009.07.049.

[58] Dikin, D.A., S. Stankovich, E.J. Zimney, R.D. Piner, G.H.B. Dommett, G. Evmenenko, S.B.T. Nguyen and R.S. Ruoff. 2007. Preparation and characterization of graphene oxide paper. Nature, 448(7152): 457–460. doi:10.1038/nature06016.

[59] Dimiev, A., D.V. Kosynkin, L.B. Alemany, P. Chaguine and J.M. Tour. 2012. Pristine graphite oxide. J. Am. Chem. Soc., 134(5): 2815–2822. doi:10.1021/ja211531y.

[60] Dimiev, A.M. and J.M. Tour. 2014. Mechanism of graphene oxide formation. ACS Nano, 8(3): 3060–3068. doi:10.1021/nn500606a.

[61] Dreyer, D.R., S. Park, C.W. Bielawski and R.S. Ruoff. 2010. The chemistry of graphene oxide. Chem. Soc. Rev., 39(1): 228–240. doi:10.1039/b917103g.

[62] Li, C., Y.X. Shi, X. Chen, D.F. He, L.M. Shen and N.Z. Bao. 2018. Controlled synthesis of graphite oxide: Formation process, oxidation kinetics, and optimized conditions. Chem. Engg. Sci., 176: 319–328. doi:10.1016/j.ces.2017.10.028.

[63] Perrozzi, F., S. Prezioso and L. Ottaviano. 2015. Graphene oxide: from fundamentals to applications. J. Phys. Condens Matter, 27(1). doi:Artn 01300210.1088/0953-8984/27/1/013002.

[64] Krishnamoorthy, K., M. Veerapandian, K. Yun and S.J. Kim. 2013. The chemical and structural analysis of graphene oxide with different degrees of oxidation. Carbon, 53: 38–49. doi:10.1016/j.carbon.2012.10.013.

[65] Dreyer, D.R., A.D. Todd and C.W. Bielawski. 2014. Harnessing the chemistry of graphene oxide. Chem. Soc. Rev., 43(15): 5288–5301. doi:10.1039/c4cs00060a.

[66] Kozbial, A., Z.T. Li, C. Conaway, R. McGinley, S. Dhingra, V. Vahdat, F. Zhou, B. D'Urso, H. Liu and L. Li. 2014. Study on the surface energy of graphene by contact angle measurements. Langmuir, 30(28): 8598–8606. doi:10.1021/la5018328.

[67] Mazzocco, R., B.J. Robinson, C. Rabot, A. Delamoreanu, A. Zenasni, J.W. Dickinson, C. Boxall and O.V. Kolosov. 2015. Surface and interfacial interactions of multilayer graphitic structures with local environment. Thin Solid Films, 585: 31–39. doi:10.1016/j.tsf.2015.04.016.

[68] Wang, S.R., Y. Zhang, N. Abidi and L. Cabrales. 2009. Wettability and surface free energy of graphene films. Langmuir, 25(18): 11078–11081. doi:10.1021/la901402f.

[69] Compton, O.C. and S.T. Nguyen. 2010. Graphene oxide, highly reduced graphene oxide, and graphene: versatile building blocks for carbon-based materials. Small, 6(6): 711–723. doi:10.1002/smll.200901934.

[70] Guex, L.G., B. Sacchi, K.F. Peuvot, R.L. Andersson, A.M. Pourrahimi, V. Strom, S. Farris and R.T. Olsson. 2017. Experimental review: chemical reduction of graphene oxide (GO) to reduced graphene oxide (rGO) by aqueous chemistry. Nanoscale, 9(27): 9562–9571. doi:10.1039/c7nr02943h.

[71] Becerril, H.A., J. Mao, Z. Liu, R.M. Stoltenberg, Z. Bao and Y. Chen. 2008. Evaluation of solution-processed reduced graphene oxide films as transparent conductors. ACS Nano, 2(3): 463–470. doi:10.1021/nn700375n.

[72] Gao, X.F., J. Jang and S. Nagase. 2010. Hydrazine and thermal reduction of graphene oxide: reaction mechanisms, product structures, and reaction design. J. Phy. Chem. C, 114(2): 832–842. doi:10.1021/jp909284g.

[73] McAllister, M.J., J.L. Li, D.H. Adamson, H.C. Schniepp, A.A. Abdala, J. Liu, M.H.-Alonso, D.L. Milius, R. Car, R.K. Prud'homme and I.A. Aksay. 2007. Single sheet functionalized graphene by oxidation and thermal expansion of graphite. Chem. Mater., 19(18): 4396–4404. doi:10.1021/cm0630800.

[74] Kampars, V. and M. Legzdina. 2015. Thermal deoxygenation of graphite oxide at low temperature. 12th Russia/Cis/Baltic/Japan Symposium on Ferroelectricity and 9th International Conference on Functional Materials and Nanotechnologies (Rcbjsf-2014-Fm&Nt), 77. doi:Artn 01203310.1088/1757-899x/77/1/012033.

[75] Ding, J.N., Y.B. Liu, N.Y. Yuan, G.Q. Ding, Y. Fan and C.T. Yu. 2012. The influence of temperature, time and concentration on the dispersion of reduced graphene oxide prepared by hydrothermal reduction. Diam. Relat. Mater., 21: 11–15. doi:10.1016/j.diamond.2011.08.004.

[76] Jeong, H.K., Y.P. Lee, M.H. Jin, E.S. Kim, J.J. Bae and Y.H. Lee. 2009. Thermal stability of graphite oxide. Chem. Phy. Lett., 470(4-6): 255–258. doi:10.1016/j.cplett.2009.01.050.

[77] Kotov, N.A., I. Dekany and J.H. Fendler. 1996. Ultrathin graphite oxide-polyelectrolyte composites prepared by self-assembly: Transition between conductive and non-conductive states. Adv. Mater., 8(8): 637–641. doi:DOI 10.1002/adma.19960080806.

[78] Muda, M.R., M.M. Ramli, S.S.M. Isa, M.F. Jamlos, S.A.Z. Murad, Z. Norhanisah, M.M. Isa, S.R. Kasjoo, N. Ahmad, N.I.M. Nor and N. Khalid. 2017. Fundamental study of reduction graphene oxide by sodium borohydride for gas sensor application. 11th Asian Conference on Chemical Sensors (Accs2015), 1808: 020034. doi:Unsp 02003410.1063/1.4975267.

[79] Shin, H.J., K.K. Kim, A. Benayad, S.M. Yoon, H.K. Park, I.S. Jung, M.H. Jin, H.-K. Jeong, J.M. Kim, J.-Y. Choi and Y.H. Lee. 2009. Efficient reduction of graphite oxide by sodium borohydride and its effect on electrical conductance. Adv. Funct. Mater., 19(12): 1987–1992. doi:10.1002/adfm.200900167.

[80] De Arco, L.G., Y. Zhang, A. Kumar and C.W. Zhou. 2009. Synthesis, transfer, and devices of single- and few-layer graphene by chemical vapor deposition. IEEE Trans. Nanotechnol., 8(2): 135–138. doi:10.1109/Tnano.2009.2013620.

[81] Reina, A., X.T. Jia, J. Ho, D. Nezich, H.B. Son, V. Bulovic, M.S. Dresselhaus and J. Kong. 2009. Large area, few-layer graphene films on arbitrary substrates by chemical vapor deposition. Nano Lett., 9(1): 30–35. doi:10.1021/nl801827v.

[82] Yu, Q.K., J. Lian, S. Siriponglert, H. Li, Y.P. Chen and S.S. Pei. 2008. Graphene segregated on Ni surfaces and transferred to insulators. Appl. Phys. Lett., 93(11). doi:Artn 11310310.1063/1.2982585.

[83] Li, H., Z.B. Fu, H.B. Wang, Y. Yi, W. Huang and J.C. Zhang. 2017. Preperetions of bi-layer and multi-layer graphene on copper substrates by atmospheric pressure chemical vapor deposition and their mechanisms. Acta. Phys. Sin., 66(5). doi:ARTN 05810110.7498/aps.66.058101.

[84] Ning, J., D. Wang, D. Han, Y.G. Shi, W.W. Cai, J.C. Zhang and Y. Hao. 2015. Comprehensive nucleation mechanisms of quasi-monolayer graphene grown on Cu by chemical vapor deposition. J. Cryst. Growth, 424: 55–61. doi:10.1016/j.jcrysgro.2015.05.002.

[85] Weatherup, R.S., B.C. Bayer, R. Blume, C. Baehtz, P.R. Kidambi, M. Fouquet, C.T. Wirth, R. Schlögl and S. Hofmann. 2012. On the mechanisms of Ni-catalysed graphene chemical vapour deposition. Chemphyschem, 13(10): 2544–2549. doi:10.1002/cphc.201101020.

[86] Nemes-Incze, P., Z. Osvath, K. Kamaras and L.P. Biro. 2008. Anomalies in thickness measurements of graphene and few layer graphite crystals by tapping mode atomic force microscopy. Carbon, 46(11): 1435–1442. doi:10.1016/j.carbon.2008.06.022.

[87] Gupta, A., G. Chen, P. Joshi, S. Tadigadapa and P.C. Eklund. 2006. Raman scattering from high-frequency phonons in supported n-graphene layer films. Nano Lett., 6(12): 2667–2673. doi:10.1021/nl061420a.

[88] Li, J.L., K.N. Kudin, M.J. McAllister, R.K. Prud'homme, I.A. Aksay and R. Car. 2006. Oxygen-driven unzipping of graphitic materials. Phys. Rev. Lett., 96(17). doi:ARTN 17610110.1103/PhysRevLett.96.176101.

[89] Balandin, A.A., S. Ghosh, W.Z. Bao, I. Calizo, D. Teweldebrhan, F. Miao and C.N. Lau. 2008. Superior thermal conductivity of single-layer graphene. Nano Lett., 8(3): 902–907. doi:10.1021/nl0731872.

[90] Loh, K.P., Q. Bao, P.K. Ang and J. Yang. 2009. The chemistry of graphene. J. Mater. Chem., 2010, 20: 2277–2289.

[91] Niyogi, S., E. Bekyarova, M.E. Itkis, J.L. McWilliams, M.A. Hamon and R.C. Haddon. 2006. Solution properties of graphite and graphene. J. Am Chem. Soc., 128(24): 7720–7721. doi:10.1021/ja060680r.

[92] Vinod, S., C.S. Tiwary, A. Samanta, S. Ozden, T.N. Narayanan, R. Vajtai, V. Agarwal, A.K. Singh, G. John and P.M. Ajayan. 2018. Graphene oxide epoxy (GO-xy): GO as epoxy adhesive by interfacial reaction of functionalities. Adv. Mater. Interfaces, 5(2). doi:ARTN 170065710.1002/admi.201700657.

[93] Choi, J., K.J. Kim, B. Kim, H. Lee and S. Kim. 2009. Covalent functionalization of epitaxial graphene by azidotrimethylsilane. J. Phys. Chem. C, 113(22): 9433–9435. doi:10.1021/jp9010444.

[94] Hamilton, C.E., J.R. Lomeda, Z.Z. Sun, J.M. Tour and A.R. Barron. 2009. High-yield organic dispersions of unfunctionalized graphene. Nano Lett., 9(10): 3460–3462. doi:10.1021/nl9016623.

[95] Salvio, R., S. Krabbenborg, W.J.M. Naber, A.H. Velders, D.N. Reinhoudt and W.G. van der Wiel. 2009. The formation of large-area conducting graphene-like platelets. Chem. Eur. J., 15(33): 8235–8240. doi:10.1002/chem.200900661.

[96] Choi, E.Y., T.H. Han, J.H. Hong, J.E. Kim, S.H. Lee, H.W. Kim and S.O. Kim. 2010. Noncovalent functionalization of graphene with end-functional polymers. J. Mater. Chem., 20(10): 1907–1912. doi:10.1039/b919074k.

[97] Georgakilas, V., J.N. Tiwari, K.C. Kemp, J.A. Perman, A.B. Bourlinos, K.S. Kim and R. Zboril. 2016. Noncovalent functionalization of graphene and graphene oxide for energy materials, biosensing, catalytic, and biomedical applications. Chem. Rev., 116(9): 5464–5519. doi:10.1021/acs.chemrev.5b00620.

[98] Rochefort, A. and J.D. Wuest. 2009. Interaction of substituted aromatic compounds with graphene. Langmuir, 25(1): 210–215. doi:10.1021/la802284j.

[99] Yang, Y.K., C.E. He, R.G. Peng, A. Baji, X.S. Du, Y.L. Huang, X.-L. Xie and Y.W. Mai. 2012. Non-covalently modified graphene sheets by imidazolium ionic liquids for multifunctional polymer nanocomposites. J. Mater. Chem., 22(12): 5666–5675. doi:10.1039/c2jm16006d.

[100] Ito, Y., C. Christodoulou, M.V. Nardi, N. Koch, H. Sachdev and K. Mullen. 2014. Chemical vapor deposition of n-doped graphene and carbon films: the role of precursors and gas phase. ACS Nano, 8(4): 3337–3346. doi:10.1021/nn405661b.

[101] Reddy, A.L.M., A. Srivastava, S.R. Gowda, H. Gullapalli, M. Dubey and P.M. Ajayan. 2010. Synthesis of nitrogen-doped graphene films for lithium battery application. ACS Nano, 4(11): 6337–6342. doi:10.1021/nn101926g.

[102] Wu, T.R., H.L. Shen, L. Sun, B. Cheng, B. Liu and J.C. Shen. 2012. Nitrogen and boron doped monolayer graphene by chemical vapor deposition using polystyrene, urea and boric acid. New J. Chem., 36(6): 1385–1391. doi:10.1039/c2nj40068e.

[103] Zhang, J.T., L.T. Qu, G.Q. Shi, J.Y. Liu, J.F. Chen and L.M. Dai. 2016. N,P-codoped carbon networks as efficient metal-free bifunctional catalysts for oxygen reduction and hydrogen evolution reactions. Angewandte Chemie-International Ed., 55(6): 2230–2234. doi:10.1002/anie.201510495.

[104] Nascimento, R., J.D. Martins, R.J.C. Batista and H. Chacham. 2015. Band gaps of BN-doped graphene: fluctuations, trends, and bounds. J. Phys. Chem. C, 119(9): 5055–5061. doi:10.1021/jp5101347.

[105] Rani, P. and V.K. Jindal. 2013. Designing band gap of graphene by B and N dopant atoms. RSC Adv., 3(3): 802–812. doi:10.1039/c2ra22664b.

[106] Jeon, I.Y., H.J. Choi, M.J. Ju, I.T. Choi, K. Lim, J. Ko, H.K. Kim, J.C. Kim, J.-J. Lee, D. Shin, S.-M. Jung, J.-M. Seo, M.-J. Kim, N. Park, L. Dai and J.B. Baek. 2013. Direct nitrogen fixation at the edges of graphene nanoplatelets as efficient electrocatalysts for energy conversion. Sci. Rep., 3. doi:ARTN 226010.1038/srep02260.

[107] Jung, S.M., E.K. Lee, M. Choi, D. Shin, I.Y. Jeon, J.M. Seo, H.Y. Jeong, N. Park, J.H. Oh and J.B. Baek. 2014. Direct solvothermal synthesis of B/N-doped graphene. Angewandte Chemie-International Edition, 53(9): 2398–2401. doi:10.1002/anie.201310260.

[108] Lai, L.F., J.R. Potts, D. Zhan, L. Wang, C.K. Poh, C.H. Tang, H. Gong, Z. Shen, J. Lin and R.S. Ruoff. 2012. Exploration of the active center structure of nitrogen-doped graphene-based catalysts for oxygen reduction reaction. Energy Environ Sci., 5(7): 7936–7942. doi:10.1039/c2ee21802j.

[109] Liu, M.M., R.Z. Zhang and W. Chen. 2014. Graphene-supported nanoelectrocatalysts for fuel cells: synthesis, properties, and applications. Chem. Rev., 114(10): 5117–5160. doi:10.1021/cr400523y.

[110] Owuor, P.S., O.K. Park, C.F. Woellner, A.S. Jalilov, S. Susarla, J. Joyner, S. Ozden, L.X. Duy, R.V. Salvatierra, R. Vajtai, J.M. Tour, J. Lou, D.S. Galvão, C.S. Tiwary and P.M. Ajayan. 2017. Lightweight hexagonal boron nitride foam for CO_2 absorption. ACS Nano, 11(9): 8944–8952. doi:10.1021/acsnano.7b03291.

[111] Owuor, P.S., C.F. Woellner, T. Li, S. Vinod, S. Ozden, S. Kosolwattana, S. Bhowmick L.X. Duy, R.V. Salvatierra, B. Wei, S.A.S. Asif, J.M. Tour, R. Vajtai, J.L. Douglas, S. Galvão, C.S. Tiwary and P.M. Ajayan. 2017. High toughness in ultralow density graphene oxide foam. Adv. Mater. Interfaces, 4(10): 1700030. doi:ARTN 170003010.1002/admi.201700030.

[112] Ozden, S., G. Brunetto, N.S. Karthiselva, D.S. Galvao, A. Roy, S.R. Bakshi, C.S. Tiwary and P.M. Ajayan. 2016. Controlled 3D carbon nanotube structures by plasma welding. Adv. Mater. Interfaces, 3(13): 1500755. doi:ARTN 150075510.1002/admi.201500755.

[113] Ozden, S., T. Tsafack, P.S. Owuor, Y.L. Li, A.S. Jalilov, R. Vajtai, C.S. Tiwary, J. Lou, J.M. Tour, A.D. Mohite and P.M. Ajayan. 2017. Chemically interconnected light-weight 3D-carbon nanotube solid network. Carbon, 119: 142–149. doi:10.1016/j.carbon.2017.03.086.

[114] Vinod, S., C.S. Tiwary, L.D. Machado, S. Ozden, R. Vajtai, D.S. Galvao and P.M. Ajayan. 2016. Synthesis of ultralow density 3D graphene-CNT foams using a two-step method. Nanoscale, 8(35): 15857–15863. doi:10.1039/c6nr04252j.

[115] Yao, J.Y., B.R. Liu, S. Ozden, J.J. Wu, S.B. Yang, M.T.F. Rodrigues, K. Kalaga, P. Dong, P. Xiao, Y. Zhang, R. Vajtai and P.M. Ajayan. 2015. 3D nanostructured molybdenum diselenide/graphene foam as anodes for long-cycle life lithium-ion batteries. Electrochim. Acta, 176: 103–111. doi:10.1016/j.electacta.2015.06.138.

[116] Ozden, S., T.N. Narayanan, C.S. Tiwary, P. Dong, A.H.C. Hart, R. Vajtai and P.M. Ajayan. 2015. 3D macroporous solids from chemically cross-linked carbon nanotubes. Small, 11(6): 688–693. doi:10.1002/smll.201402127.

[117] Ozden, S., C.S. Tiwary, A.H.C. Hart, A.C. Chipara, R. Romero-Aburto, M.T.F. Rodrigues, J. Taha-Tijerina, R. Vajtai and P.M. Ajayan. 2015. Density variant carbon nanotube interconnected solids. Adv. Mater., 27(11): 1842–1850. doi:10.1002/adma.201404995.

[118] Sudeep, P.M., T.N. Narayanan, A. Ganesan, M.M. Shaijumon, H. Yang, S. Ozden, P.K. Patra, M. Pasquali, R. Vajtai, S. Ganguli, A.K. Roy, M.R. Anantharaman and P.M. Ajayan. 2013. Covalently interconnected three-dimensional graphene oxide solids. ACS Nano, 7(8): 7034–7040. doi:10.1021/nn402272u.

[119] Vinod, S., C.S. Tiwary, P.A.D. Autreto, J. Taha-Tijerina, S. Ozden, A.C. Chipara, R. Vajtai, D.S. Galvao, T.N. Narayanan and P.M. Ajayan. 2014. Low-density three-dimensional foam using self-reinforced hybrid two-dimensional atomic layers. Nat. Commun., 5: 4541. doi:ARTN 454110.1038/ncomms5541.

[120] Koizumi, R., A.H.C. Hart, G. Brunetto, S. Bhowmick, P.S. Owuor, J.T. Hamel, A.X. Gentles, S. Ozden, J. Lou, R. Vajtai, S.A.S. Asif, D.S. Galvão, C.S. Tiwary and P.M. Ajayan. 2016. Mechano-chemical stabilization of three-dimensional carbon nanotube aggregates. Carbon, 110: 27–33. doi:10.1016/j.carbon.2016.08.085.

[121] Owuor, P.S., Y. Yang, T. Kaji, R. Koizumi, S. Ozden, R. Vajtai, R. Vajtai, J. Lou, E.S. Penev, B.I. Yakobson, C.S. Tiwary and P.M. Ajayan. 2017. Enhancing mechanical properties of nanocomposites using interconnected carbon nanotubes (iCNT) as reinforcement. Adv. Eng. Mater., 19(2): 1600499. doi:ARTN 160049910.1002/adem.201600499.

[122] Ozden, S., I.G. Macwan, P.S. Owuor, S. Kosolwattana, P.A.S. Autreto, S. Silwal, R. Vajtai, C.S. Tiwary, A.D. Mohite, P.K. Patra and P.M. Ajayan. 2017. Bacteria as bio-template for 3D carbon nanotube architectures. Sci. Rep., 7: 9855. doi:ARTN 985510.1038/s41598-017-09692-2.

[123] Hu, H., Z.B. Zhao, W.B. Wan, Y. Gogotsi and J.S. Qiu. 2013. Ultralight and highly compressible graphene aerogels. Adv. Mater., 25(15): 2219–2223. doi:10.1002/adma.201204530.

[124] Ozden, S., L.H. Ge, T.N. Narayanan, A.H.C. Hart, H. Yang, S. Sridhar, R. Vajtai and P.M. Ajayan. 2014. Anisotropically functionalized carbon nanotube array based hygroscopic scaffolds. ACS Appl. Mater. Interfaces, 6(13): 10608–10613. doi:10.1021/am5022717.

[125] Lee, C., S. Ozden, C.S. Tewari, O.K. Park, R. Vajtai, K. Chatterjee and P.M. Ajayan. 2018. MoS$_2$-carbon nanotube porous 3D network for enhanced oxygen reduction reaction. Chemsuschem, 11(17): 2960–2966. doi:10.1002/cssc.201800982.

[126] Yao, J.Y., Y.J. Gong, S.B. Yang, P. Xiao, Y.H. Zhang, K. Keyshar, G. Ye, S. Ozden, R. Vajtai and P.M. Ajayan. 2014. CoMoO$_4$ nanoparticles anchored on reduced graphene oxide nanocomposites as anodes for long-life lithium-ion batteries. ACS Appl. Mater. Interfaces, 6(22): 20414–20422. doi:10.1021/am505983m.

[127] Chakravarty, D., C.S. Tiwary, C.F. Woellner, S. Radhakrishnan, S. Vinod, S. Ozden, P. A da, S. Autreto, S. Bhowmick, S. Asif, S.A. Mani, D.S. Galvao and P.M. Ajayan. 2016. 3D porous graphene by low-temperature plasma welding for bone implants. Adv. Mater., 28(40): 8959–8967. doi:10.1002/adma.201603146.

[128] Gautam, C., D. Chakravarty, A. Gautam, C.S. Tiwary, C.F. Woellner, V.K. Mishra, N. Ahmad, S. Ozden, S. Jose, S. Biradar, R. Vajtai, R. Trivedi, D.S. Galvao and P.M. Ajayan. 2018. Synthesis and 3D interconnected nanostructured h-BN-based biocomposites by low-temperature plasma sintering: bone regeneration applications. ACS Omega, 3(6): 6013–6021. doi:10.1021/acsomega.8b00707.

[129] Chen, Z.P., W.C. Ren, L.B. Gao, B.L. Liu, S.F. Pei and H.M. Cheng. 2011. Three-dimensional flexible and conductive interconnected graphene networks grown by chemical vapour deposition. Nat. Mater., 10(6): 424–428. doi:10.1038/Nmat3001.

[130] Hashim, D.P., N.T. Narayanan, J.M. Romo-Herrera, D.A. Cullen, M.G. Hahm, P. Lezzi, J.R. Suttle, D. Kelkhoff, E. M.-Sandoval, S. Ganguli, A.K. Roy, D.J. Smith, R. Vajtai, B.G. Sumpter, V. Meunier, H. Terrones, M. Terrones and P.M. Ajayan. 2012. Covalently bonded three-dimensional carbon nanotube solids via boron induced nanojunctions. Sci. Rep., 2: 363. doi:ARTN 36310.1038/srep00363.

[131] Peng, Q.Y., Y.B. Li, X.D. He, X.C. Gui, Y.Y. Shang, C.H. Wang, C. Wang, W. Zhao, S. Du, E. Shi, P. Li, D. Wu and A.Y. Cao. 2014. Graphene nanoribbon aerogels unzipped from carbon nanotube sponges. Adv. Mater., 26(20): 3241–3247. doi:10.1002/adma.201305274.

[132] Shan, C.S., W.J. Zhao, X.L. Lu, D.J. O'Brien, Y.P. Li, Z.Y. Cao, A.L. Elias, R. C.-Silva, M. Terrones, B. Wei and J. Suhr. 2013. Three-dimensional nitrogen-doped multiwall carbon nanotube sponges with tunable properties. Nano Lett., 13(11): 5514–5520. doi:10.1021/nl403109g.

[133] Gui, X.C., J.Q. Wei, K.L. Wang, A.Y. Cao, H.W. Zhu, Y. Jia, Q. Shu and D.H. Wu. 2010. Carbon nanotube sponges. Adv. Mater., 22(5): 617–621. doi:10.1002/adma.200902986.

[134] Romo-Herrera, J.M., B.G. Sumpter, D.A. Cullen, H. Terrones, E. Cruz-Silva, D.J. Smith, V. Meunier and M. Terrones. 2008. An atomistic branching mechanism for carbon nanotubes:

Sulfur as the triggering agent. Angewandte Chemie-International Edition, 47(16): 2948–2953. doi:10.1002/anie.200705053.

[135] Jia, Z.Q. and Y. Wang. 2015. Covalently crosslinked graphene oxide membranes by esterification reactions for ions separation. J. Mater. Chem. A, 3(8): 4405–4412. doi:10.1039/c4ta06193d.

[136] Sudeep, P.M., S. Vinod, S. Ozden, R. Sruthi, A. Kukovecz, Z. Konya, R. Vajtai, M.R. Anantharaman, P.M. Ajayan and T.N. Narayanan. 2015. Functionalized boron nitride porous solids. RSC Adv., 5(114): 93964–93968. doi:10.1039/c5ra19091f.

[137] Krasheninnikov, A.V., K. Nordlund, J. Keinonen and F. Banhart. 2002. Ion-irradiation-induced welding of carbon nanotubes. Phys. Rev. B, 66(24). doi:ARTN 24540310.1103/PhysRevB.66.245403.

[138] Piper, N.M., Y. Fu, J. Tao, X. Yang and A.C. To. 2011. Vibration promotes heat welding of single-walled carbon nanotubes. Chem. Phys. Lett., 502(4-6): 231–234. doi:10.1016/j.cplett.2010.12.068.

[139] Schauerman, C.M., J. Alvarenga, J. Staub, M.W. Forney, R. Foringer and B.J. Landi. 2015. Ultrasonic welding of bulk carbon nanotube conductors. Adv. Eng. Mater., 17(1): 76–83. doi:10.1002/adem.201400117.

[140] Yang, X.M., Z.H. Han, Y.H. Li, D.C. Chen, P. Zhang and A.C. To. 2012. Heat welding of non-orthogonal X-junction of single-walled carbon nanotubes. Physica E Low Dimens. Syst. Nanostruct., 46: 30–32. doi:10.1016/j.physe.2012.08.015.

[141] Terrones, M., F. Banhart, N. Grobert, J.C. Charlier, H. Terrones and P.M. Ajayan. 2002. Molecular junctions by joining single-walled carbon nanotubes. Phys. Rev. Lett., 89(7). doi:ARTN 07550510.1103/PhysRevLett.89.075505.

[142] De Volder, M.F.L., S.H. Tawfick, R.H. Baughman and A.J. Hart. 2013. Carbon nanotubes: present and future commercial applications. Science, 339(6119): 535–539. doi:10.1126/science.1222453.

[143] Yu, C., K. Choi, L. Yin and J.C. Grunlan. 2013. Light-weight flexible carbon nanotube based organic a composites with large thermoelectric power factors (vol. 5, pg. 7885, 2011). ACS Nano, 7(10): 9506–9506. doi:10.1021/nn404924h.

[144] Jurn, Y.N., M.F. Malek, W.W. Liu, H.K. Hoomod and A.A. Kadhim. 2014. Review—coating methods of carbon nanotubes and their potential applications. 2014 IEEE International Conference on Control System Computing and Engineering, 118–123.

[145] Nine, M.J., M.A. Cole, D.N.H. Tran and D. Losic. 2015. Graphene: a multipurpose material for protective coatings. J. Mater. Chem. A, 3(24): 12580–12602. doi:10.1039/c5ta01010a.

[146] Candelaria, S.L., Y.Y. Shao, W. Zhou, X.L. Li, J. Xiao, J.G. Zhang, Y. Wang, J. Liu, J. Li, G. Cao, and G.Z. Cao. 2012. Nanostructured carbon for energy storage and conversion. Nano Energy, 1(2): 195–220. doi:10.1016/j.nanoen.2011.11.006.

[147] Dai, L.M., D.W. Chang, J.B. Baek and W. Lu. 2012. Carbon nanomaterials for advanced energy conversion and storage. Small, 8(8): 1130–1166. doi:10.1002/smll.201101594.

[148] Figueiredo, J.L. 2018. Nanostructured porous carbons for electrochemical energy conversion and storage. Surf. Coat. Technol., 350: 307–312. doi:10.1016/j.surfcoat.2018.07.033.

[149] Hu, L.B. 2014. Rational nanostructured carbon for energy conversion and storage. Abstracts of Papers of the American Chemical Society, 248.

[150] Lawes, S., A. Riese, Q. Sun, N.C. Cheng and X.L. Sun. 2015. Printing nanostructured carbon for energy storage and conversion applications. Carbon, 92: 150–176. doi:10.1016/j.carbon.2015.04.008.

[151] Bekyarova, E., Y.C. Ni, E.B. Malarkey, V. Montana, J.L. McWilliams, R.C. Haddon and V. Parpura. 2005. Applications of carbon nanotubes in biotechnology and biomedicine. J. Biomed., 1(1): 3–17. doi:10.1166/jbn.2005.004.

[152] Shen, H., L.M. Zhang, M. Liu and Z.J. Zhang. 2012. Biomedical applications of graphene. Theranostics, 2(3): 283–294. doi:10.7150/thno.3642.

[153] Wang, Y., Z.H. Li, J. Wang, J.H. Li and Y.H. Lin. 2011. Graphene and graphene oxide: biofunctionalization and applications in biotechnology. Trends Biotechnol., 29(5): 205–212. doi:10.1016/j.tibtech.2011.01.008.

[154] Avouris, P. and J. Chen. 2006. Nanotube electronics and optoelectronics. Mater. Today, 9(10): 46–54. doi:Doi 10.1016/S1369-7021(06)71653-4.

[155] Jariwala, D., V.K. Sangwan, L.J. Lauhon, T.J. Marks and M.C. Hersam. 2013. Carbon nanomaterials for electronics, optoelectronics, photovoltaics, and sensing. Chem. Soc. Rev., 42(7): 2824–2860. doi:10.1039/c2cs35335k.

[156] Ozden, S., P.A.S. Autreto, C.S. Tiwary, S. Khatiwada, L. Machado, D.S. Galvao, R. Vajtai, E.V. Barrera and P.M. Ajayan. 2014. Unzipping carbon nanotubes at high impact. Nano Lett., 14(7): 4131–4137. doi:10.1021/nl501753n.

[157] Ozden, S., L.D. Machado, C. Tiwary, P.A.S. Autreto, R. Vajtai, E.V. Barrera, D.S. Galvao, and P.M. Ajayan. 2016. Ballistic fracturing of carbon nanotubes. ACS Appl. Mater. Interfaces, 8(37): 24819–24825. doi:10.1021/acsami.6b07547.

[158] Mauter, M.S. and M. Elimelech. 2008. Environmental applications of carbon-based nanomaterials. Environ. Sci. Technol., 42(16): 5843–5859. doi:10.1021/es8006904.

CHAPTER 2

Synthesis of Nanomaterials and Nanostructures

Preeti Kaushik,[1,2] *Amrita Basu*[3] *and Meena Dhankhar*[4,*]

1. Introduction

With the evolution of nanotechnology, new paths for synthesis of nanomaterials have led to the discovery of unconventional nanomaterials. Researchers are trying to develop low cost novel materials with better functionality and properties to be used in various applications, such as electronics, optics, aerospace, defense, etc. Nanomaterials and nanostructures can be 0D, 1D, or 2D. 0D materials, such as nanoparticles or clusters, can be synthesized using sol-gel, colloidal, or hydrothermal methods, and 1D and 2D nanomaterials such as nanotubes or nanowires are synthesized using chemical vapor deposition (CVD), arc discharge, or vapor-liquid-solid methods.

Broadly, there are two main approaches for the fabrication of nanostructures and for the synthesis of nanomaterials:

- Bottom-up approach: This approach uses the force at an atomic scale to sum up into the large complex structures. Typical examples are the formation of nanoparticles from colloidal dispersion, and carbon nanotube (CNT) synthesis using CVD.

- Top-down approach: This approach involves the breaking down of a large structure into a smaller structure, like going from mesoscopic to nanostructures. A typical example is lithography.

[1] RG Plasma Technologies, Central European Institute of Technology, Masaryk University, Brno, Czech Republic.
[2] Department of Physical Electronics, Faculty of Science, Masaryk University, Brno, Czech Republic.
[3] RECETOX, Research Centre for Toxic Compounds in the Environment, Masaryk University Brno, Czech Republic.
[4] CEITEC, Central European Institute of Technology, Brno University of Technology Brno, Czech Republic.
* Corresponding author: meena.dhankhar@ceitec.vutbr.cz

2. Bottom-up Approach

2.1 Carbon Nanotube Synthesis Methods

Discovery of carbon nanotubes by Sumio Iijima in 1991 led to a new class of nanomaterials useful in different commercial applications [1]. CNTs can be described as one atom thick graphene sheets rolled into cylinders. CNTs consisting of one cylinder are termed as single-walled carbon nanotubes (SWNTs), and those with multiple concentric cylinders are called multi-walled carbon nanotubes (MWNTs). Due to their unusual electrical, mechanical, optical, and thermal properties, CNTs are being incorporated into different applications, ranging from optical coatings, adhesives, storage devices, to membranes for water purification. The extraordinary high aspect ratio structure of these carbon nanostructures makes them useful in gas sensors and field emission displays.

Highly reliable synthesis techniques are required for controlled growth of CNTs to integrate these nanostructures into different applications. There is a need to control the purity and structural quality of CNTs. Independent of the growth method, CNTs are always produced with some impurities, such as amorphous carbon, crystalline graphite, or the catalyst used during synthesis. The type and amount of these impurities depends on the synthesis method used for growth. These impurities interfere with most of the desired properties of CNTs, and cause hindrance in characterization and applications. Understanding the actual nucleation and growth process of these nanostructures is still a controversial subject. Figure 2.1a shows the transmission electron microscopy (TEM) image and Figure 2.1b shows the scanning electron microscopy (SEM) images of vertically aligned MWNTs.

CNT synthesis methods can be broadly classified as:

- Solid carbon source-based synthesis techniques
- Gaseous carbon source-based synthesis techniques

Solid carbon source-based methods include arc discharge and laser ablation methods. Here, the source material used for CNT growth is in the solid form. In arc discharge method, graphite electrodes are evaporated in an electric arc at very high temperatures ($\sim 4000°C$). The nanotubes produced by this method have a high amount of impurities in the form of nano-crystalline graphite, catalytic particles, and

Figure 2.1. MWNTs: (a) Transmission electron microscopy (TEM) image and (b) Scanning electron microscopy (SEM) image of cross-section.

amorphous carbon, making them inefficient for use in further applications. In the laser ablation technique, graphite targets of high purity are evaporated by high power lasers at high temperatures.

Both of the above methods are inefficient in terms of purity at the cost of resources used. Gaseous carbon source-based methods include chemical vapor deposition (CVD), which is the most commonly used method nowadays.

2.1.1 Catalytic Chemical Vapor Deposition (CCVD)

Catalytic CVD (CCVD) involves catalytic decomposition of a carbon-containing source on small metallic particles or clusters. CCVD can be either a heterogeneous process, if a solid substrate is involved (supported catalyst), or a homogeneous process, if everything takes place in the gas phase (floating catalyst). Both homogeneous and heterogeneous processes appear very sensitive to the nature and the structure of the catalyst used, as well as to the operating conditions. Metals generally used for these reactions are transition metals, such as Fe, Co, and Ni. This is a low-temperature process compared to arc discharge and laser ablation methods, with the formation of carbon nanotubes typically occurring between 600°C and 1000°C. CNTs prepared by CCVD methods are generally much longer (a few tens to hundreds of micrometers) than those obtained by arc discharge (a few micrometers); depending on the experimental conditions, it is possible to grow dense arrays of nanotubes. MWNTs from CCVD contain more structural defects than MWNTs from arc discharge, due to the lower temperature of the reaction, which does not allow any structural rearrangements. CCVD SWNTs generally gather into bundles of smaller diameter (a few tens of nm) than their arc discharge and laser ablation counterparts. Specifically, when using fluidized bed reactor, CCVD provides reasonably good perspectives on large-scale and low-cost processes. Although CCVD formation mechanisms for SWNTs and MWNTs can be quite different, it is agreed that CNTs form on very small metal particles, typically in nm range size.

CCVD heterogenous process or supported catalyst method uses Fe, Ni, or Co and their alloys in the form of:

- thin film (vacuum evaporation, magnetron sputtering, or CVD)—The key factor is to perform restructuring into active catalyst nanoparticles (NPs): heating in N_2, H_2, NH_3, or plasma treatment. Process depends on time, gas, thickness of the film, its morphology, and material under the catalyst. Particles can coalescence during continuous heating and other material phases can be formed.
- direct deposition of nanoparticles (NPs)—By electrochemical deposition or plasma enhanced chemical vapor deposition (PECVD). Controlling metal particle size is the key issue, and coalescence must be avoided (various supports, such as Al_2O_3, SiO_2, TiN, MgO can be used).

CCVD homogenous process or floating catalyst method uses organo-metallic volatile compounds, such as iron pentacarbonyl $Fe(CO)_5$ or a metallocene, such as ferrocene, nickelocene, or cobaltocene. It differs from the supported catalyst approach because it uses only gaseous species and does not require the presence of any solid phase in the reactor. The basic principle of this technique, like the other

CCVD processes, is to decompose a carbon source (ethylene, xylene, benzene, carbon monoxide, and so on) on nanosized transition metal (generally Fe, Co, or Ni) particles in order to obtain carbon nanotubes. Catalytic particles are formed directly in the reactor and are not introduced before the reaction, as occurs in supported CCVD. The typical reactor used in this technique is a quartz tube placed in an oven into which the gaseous feedstock containing the metal precursor, the carbon source, some hydrogen, and an inert gas (N_2, Ar, or He) are introduced. The main drawback of this type of process is that it is difficult to control the size of the metal NPs, and thus, nanotube formation is often accompanied by the production of undesired carbon forms (amorphous carbon or polyaromatic carbon phases found as various phases or as coatings). Parameters must be controlled in order to finely tune the process and selectively obtain the required structure and morphology of nanotubes. The important parameters are—choice of the carbon source, reaction temperature, residence time, composition of the incoming gaseous feedstock, with attention paid to the role played by the proportion of hydrogen, and the ratio of the metalorganic precursor to the carbon source.

The growth mechanism for nanotubes has been a debated topic, as different authors have different theories based on the growth and pre-treatment conditions. Some authors have described this mechanism based on the catalyst-substrate interaction. If the catalyst-substrate interaction is weak, the growth of CNTs is based on 'tip-growth' model, and if the interaction is strong, then it is the 'base-growth model' [2]. Whereas, others described that carbon nanotube growth mechanisms are correlated with the catalyst nanometric dimension [3]. For the same substrate/catalyst couples, single or few-wall carbon nanotubes follow the "base-growth" mechanism, while the "tip-growth" occurs only for the large multi-walled nanotubes.

2.1.2 Plasma-enhanced Chemical Vapor Deposition (PECVD)

CVD utilizing plasma discharge is called plasma-enhanced CVD (PECVD). In PECVD, various types of low-pressure discharges, as well as atmospheric pressure discharges, such as microwave (MW), MW electron cyclotron resonance, dc glow discharges, and capacitive or inductive radio-frequency glow discharges have been applied. Figure 2.2a shows the schematic view and Figure 2.2b shows the experimental setup of MW plasma torch. Zajíčková et al. have shown that MW plasma torch is capable of high-speed synthesis of vertically aligned CNTs at atmospheric pressure [4]. Due to the spatial non-uniformity of the discharge and a relatively small diameter of the torch with respect to the substrate dimensions, the prepared samples exhibited the density gradient structure of nanotubes. It was also identified that substrate temperature gradients and a size distribution of catalytic particles were also the reasons for non-uniform deposition of CNTs [5]. Later, MW plasma torch was used for the synthesis of catalytic nanoparticles and CNTs, both using the floating catalyst approach [6].

2.2 Nanoparticle Liquid Phase Synthesis

The liquid phase fabrication comprises of a wet chemistry route, where the nanoparticles are usually precipitated from a solution and divided into four major

(a)

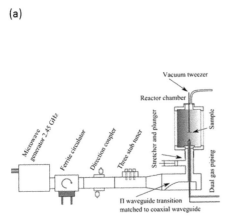

Figure 2.2. MW plasma torch: (a) schematic view and (b) experimental set-up. Zajíčková et al. have shown that MW plasma torch is capable of high-speed synthesis of vertically aligned CNTs at atmospheric pressure [4].

groups, such as colloidal method, sol-gel method, hydrothermal synthesis, and polyol method. The major principle encompasses a metallic salt (chloride, nitride, etc.) which is dissolved in water, wherein the metal cations exist in the form of metal hydrates which are added with basic solutions (NaOH). The hydrolyzed species are condensed and processed via filtration and drying for the formation of the final product. The distribution attributed to the particle size is the actual measure of the liquid phase synthesis technique, and the model constitutes of two modules. The first model is responsible for the nanoparticle growth and Ostwald ripening, and the second model revolves around the growth kinetics in a population balance [7].

There are many advantages associated with the liquid phase preparation, as it is a highly controlled method which is highly developed due to its low cost and continuous operation, resulting in the production of highly sophisticated products. A few challenges associated with this procedure are the requirements of a number of chemical species, which increases the processing steps, such as aging or filtration or washing and drying.

2.2.1 Colloidal Method

The major principle encompassing the colloidal method technique would be the precipitation process, in which different ions under controlled temperature and pressure solutions are mixed to form insoluble precipitates. Usually this method is performed by controlling the nucleation, which is employed due to ultrasonic and sono-chemical effect and the growth kinetics, and furthermore, used in bulk solutions or in reverse micelles [8]. In one of the studies, the importance of light of different energies and how it enhances various chemical events, resulting in the nucleation and growth of metal nanocrystals was discussed, and hence light was regarded as an important component to direct the growth and reproducibility [9]. Other studies have also shown that similar to gold nanoclusters, silver nanoclusters can portray high colloidal stability and fluorescence in the red [10].

2.2.2 Sol-Gel Processing

This process revolves around the basic principle that a solution under the influence of a chemical transformation turns into a gel state, with further treatment transforming it into oxide material. Due to its high purity and uniformity at low temperatures, this is considered to be one of the most established methods for synthesis. The process starts with the precursor solution which undergoes hydrolysis and condensation with continuous stirring, resulting in the formation of a gel, which then enters the drying mode. It results in the formation of xerogel, if it undergoes evaporative drying and aerogel due to the supercritical drying and freeze drying resulting in cryogel, and finally resulting in the final compound [11]. One of the common examples of nanoparticles using the sol-gel method is the investigation revolving around titanium-aluminum [12], and there are various approaches involved for studying the structural changes and the phase analysis with the help of various imaging and spectroscopic techniques, such as XRD, FITR, and SEM [13]. Sol-gel method can be classified into major routes according to the solvent that is being utilized. Aqueous sol-gel method is used when the reaction medium consists of water, and non-aqueous sol-gel method is used when the medium is an organic solvent [14].

2.2.3 Hydrothermal Synthesis

In the hydrothermal synthesis of nanoparticles, the reactants are dissolved or immersed in water in the vessel, most preferably an autoclave. The internal temperature of the autoclave is usually maintained within 200°C because of the internal structure consisting of teflon. The major advantage of this technique is that it does not require any post heat treatment and it also helps in controlling the particle size and shape. It also helps in regulating the morphology and the surface chemistry. One of the major applications of hydrothermal synthesis would be TiO_2 nanoparticles preparation, which is obtained through the hydrothermal treatment of peptized precipitate of titanium precursor with water [15]. The hydrothermal method of synthesis of nanoparticles portrayed different morphology with various co-surfactants, such as floral-like, or wire-like, or sheets, which helped in the application of the same material in various ways [16].

2.2.4 Polyol Method

Luminescent materials, color pigments, and nanoscale functional materials are produced by polyol method [17], and they are usually in the crystalline, spherical form within the range of 20–200 nm in size, which is characterized by SEM, XRD, and other spectroscopic techniques [17]. There are many advantages associated with this polyols as it provides for adaptability and flexibility, and the boiling point increases with increasing molecular weights, and similarly, the polarity and viscosity also increases [18]. There is a wide application of polyols in the form of standard solvents, cosmetics, or in food additives and pharmaceutical industries due to its property of moderate toxicity and being highly biodegradable. In studies done by Byeon and group, it was seen that ultrasonic irradiation played an important role in the polyol method for the synthesis of colloidal silver nanoparticles, resulting in morphological uniformity and better formation kinetics [19].

3. Top-down Approach

3.1 Lithography

In lithography [20, 21], mainly photolithography and electron-beam lithography are the most widely used techniques. They are used to fabricate ICs (Integrated Circuit) and produce structures smaller than 10 nm.

3.1.1 Photolithography

The first and most widely used lithography [22] is photolithography using UV (ultra violet) light ($\lambda \cong 0.2$ μm or $\lambda \cong 0.4$ μm) or deep UV light ($\lambda \leq 0.2$ μm), which is used to transfer patterns from the mask on the thin film. The UV light is transmitted through the transparent part of the mask, and the exposed area will then be soluble or insoluble in the developer based on the type of resist used (positive or negative resist, respectively). This transfers the pattern from the mask to the wafer. The wafer is then processed by the etching process to selectively remove the undesired portion. The step-by-step procedure for patterning is described below, and also shown in Figure 2.3.

- Spin coating of the resist: A circular substrate (wafer) is used for the spin coating of photoresist. The photoresist is a radiation sensitive organic polymer. To spin coat the wafer, a small volume of the resist is dispensed to the wafer placed on

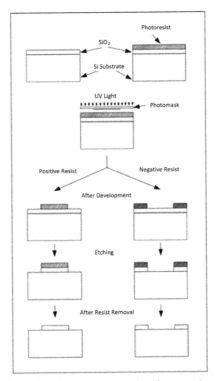

Figure 2.3. Step-by-step procedure for patterning.

the spin coater chuck. The wafer is held by the vacuum and rotated at a high speed to get the desired thickness of resist film. Sometimes adhesion promotor is also used below the resist layer to enhance the adhesion between the resist and very top layer. For the best adhesion possible, make sure that the wafer is properly cleaned.

• Soft bake of the resist: After spin coating of the resist, the resist needs to be soft baked to remove the residual solvent and densify the resist film. It also removes the stress from the resist layer and promotes adhesion, which helps in better processing of later steps.

• Exposure: In the first step of exposure, the wafer does not require to be aligned with the mask, but in case of second and further exposures, the wafer needs to be aligned well with the features of the masks. The lithography in those steps will be valuable only if it is aligned well with respect to the mask. Then the UV light is shined on the mask, and the light will pass through the transparent area or through the gray area of the mask and expose the resist. Due to exposure of photoactive compound of photoresist, the photoresist will undergo selective chemical reaction in the influenced region.

• Post-exposure bake: Sometimes, after exposure, the resist needs to be baked for further chemical reaction, contrast enhancement, or stress relaxation.

• Development: In the development process, selective dissolving of resist takes place. Now this selective dissolving depends on the type of photoresist used. After exposure of positive photoresist, the exposed area will become soluble in developer, and removed during developing process. The pattern after development is the same as on the mask. In the case of negative photoresist exposure, the exposed area undergoes a chemical reaction, which causes cross-linking of the resist molecules, which makes it insoluble in developer. The exposed area remains on the wafer and the unexposed area is dissolved in the developer.

• Hard bake: Hard bake is an optional process. The hard bake is performed to remove the volatile materials and water, and make the resist stiffer and more durable in further etching process (e.g., wet chemical etching).

The main disadvantage is the difficulty to produce the structures below 100 nm directly because of diffraction effect. There are techniques such as dual exposure, or structure splitting, which allow the achievement of higher resolution structures by application of subsequent processes. In R&D applications, this drawback is usually overcome by using electron beam lithography, which is much slower, but can achieve higher resolution.

3.1.2 Electron-beam Lithography

The electron beam lithography [23] (EBL) is quite a common technique in R&D and research, with a limited number of applications in the industry. EBL offers high resolution patterning by high energy electrons (2 to 100 keV). The beam can be focused into a narrow beam in case of so-called gaussian scanning systems, or shaped into various geometric structures in so-called shape-beam systems. This technique offers precise control of energy and the dose delivered to the electron beam is sensitive to the resist-coated wafer. Quite commonly, the masks for photolithography are manufactured by the utilization of this technique. Typically, the writing strategy does not require a hard mask. Instead, the beam position is precisely controlled by computer-controlled pattern generator, according to the design file.

The resolution of e-beam lithography does not depend only on the spot size of the focused beam. It is also affected by the scattering of electrons inside the resist and by the electron back scattering from the substrate. These effects influence a wider area than is the spot size of the focused electron beam. This is called the proximity effect. The proximity effect influence can be suppressed by application of a proximity correction algorithm. The algorithm calculates proximity corrected structures with modified shapes and doses. The resulting pattern may be quite complicated, and processing may be slow, which further prolongs the time of exposure.

References

[1] IJIMA, SUMIO. 1991. "Paper1.Pdf." Nature, 354: 56–58. doi:10.1088/1126-6708/2008/08/073.

[2] Kumar, Mukul and Yoshinori Ando. 2010. Chemical vapor deposition of carbon nanotubes: a review on growth mechanism and mass production. J. Nanosci. Nanotechnol., 10(6): 3739–58. doi:10.1166/jnn.2010.2939.

[3] Gohier, A., C.P. Ewels, T.M. Minea and M.A. Djouadi. 2008. Carbon nanotube growth mechanism switches from tip- to base-growth with decreasing catalyst particle size. Carbon, 46(10): 1331–38. doi:10.1016/j.carbon.2008.05.016.

[4] Zajíčková, Lenka, Marek Eliáš, Ondřej Jašek, Zuzana Kučerová, Petr Synek, Jiřina Matějková, Magdaléna Kadlečíková, Mariana Klementová, Jiří Buršík and Anna Vojačková. 2007. Characterization of carbon nanotubes deposited in microwave torch at atmospheric pressure. Plasma Process Polym., 4: 5245–49. doi:10.1002/ppap.200730710.

[5] Jašek, O., M. Eliáš, L. Zajíčková, Z. Kučerová, J. Matějková, A. Rek and J. Buršík. 2007. Discussion of important factors in deposition of carbon nanotubes by atmospheric pressure microwave plasma torch. J. Phys. Chem. Solids, 68(5–6): 738–43. doi:10.1016/j.jpcs.2007.01.039.

[6] Zajíčková, Lenka, Petr Synek, Ondřej Jašek, Marek Eliáš, Bohumil David, Jiří Buršík, Naděžda Pizúrová, Renáta Hanzlíková and Lukáš Lazar. 2009. Synthesis of carbon nanotubes and iron oxide nanoparticles in MW plasma torch with $Fe(CO)_5$ in gas feed. Appl. Surf. Sci., 255(10): 5421–24. doi:10.1016/j.apsusc.2008.09.003.

[7] Mantzaris, V. Nikos. 2005. Liquid-phase synthesis of nanoparticles: particle size distribution dynamics and control. Chem. Eng. Sci., 60(17). Pergamon: 4749–70. doi:10.1016/J.CES.2005.04.012.

[8] Nagarajan, R. 2008. Nanoparticles: Building blocks for nanotechnology. ACS Symposium Series Vol. 996: 2–14. doi: 10.1021/bk-2008-0996.ch001. Publication Date: September 19, 2008.

[9] Grzelczak, Marek and Luis M. Liz-Marzán. 2014. The relevance of light in the formation of colloidal metal nanoparticles. Chem. Soc. Rev., 43(7): 2089–97. doi:10.1039/c3cs60256g.

[10] Sherry Huang, Christian Pfeiffer, Jana Hollmann, Sebastian Friede, Justin Jin-Ching Chen, Andreas Beyer, Benedikt Haas, Kerstin Volz, Wolfram Heimbrodt, Jose Maria Montenegro Martos, Walter Chang and Wolfgang J. Parak. 2012. Synthesis and characterization of colloidal fluorescent silver nanoclusters. Langmuir, 28(24): 8915–19. doi:10.1021/la300346t.

[11] Rao, G. Bolla, Deboshree Mukherjee and Benjaram M. Reddy. 2017. Novel approaches for preparation of nanoparticles. Nanostructures for Novel Therapy, January. Elsevier, 1–36. doi:10.1016/B978-0-323-46142-9.00001-3.

[12] Ahmed, M.A. and M.F. Abdel-Messih. 2011. Structural and nano-composite features of TiO_2–Al_2O_3 powders prepared by sol–gel method. J. Alloys Compd., 509(5): 2154–59. Elsevier. doi:10.1016/J.JALLCOM.2010.10.172.

[13] Ramesh, S. 2013. Sol-gel synthesis and characterization of nanoparticles. J. Nanosci., 2013(August): 1–8. Hindawi. doi:10.1155/2013/929321.

[14] Ficai, D. and A.M. Grumezescu (eds.). 2017. Nanostructures for Novel Therapy: Synthesis, Characterization and Applications. Elsevier.

[15] Zha, J. and H. Roggendorf. 1991. Sol–gel science, the physics and chemistry of sol–gel processing, Ed. by C.J. Brinker and G.W. Scherer, Academic Press, Boston 1990, xiv, 908, bound—ISBN 0-12-134970-5. Adv. Mater., 3: 522–522. doi:10.1002/adma.19910031025.

[16] Bharti, B. Dattatraya and A.V. Bharati. 2017. Synthesis of ZnO nanoparticles using a hydrothermal method and a study its optical activity. Luminescence, 32(3): 317–20. doi:10.1002/bio.3180.

[17] Feldmann, C. 2003. Polyol-mediated synthesis of nanoscale functional materials. Adv. Funct. Mater., 13(2): 101–7. John Wiley & Sons, Ltd. doi:10.1002/adfm.200390014.

[18] Dong, H., Y.-C. Chen and C. Feldmann. 2015. Polyol synthesis of nanoparticles: status and options regarding metals, oxides, chalcogenides, and non-metal elements. Green Chem., 17(8): 4107–32. doi:10.1039/C5GC00943J.

[19] Byeon, Jeong Hoon and Young-Woo Kim. 2012. A novel polyol method to synthesize colloidal silver nanoparticles by ultrasonic irradiation. Ultrasonics Sonochem., 19(1): 209–15. doi:10.1016/j.ultsonch.2011.06.004.

[20] Harry J. Levinson. 2010. Principles of Lithography. SPIE Press, 3: 51–108. doi:10.1117/3.601520.

[21] Aranzazu del Campo and Eduard Arzt. 2008. Fabrication approaches for generating complex micro- and nanopatterns on polymeric surfaces. Chem. Rev., 108(3): 911–945.

[22] Marc J. Madou. 2002. Fundamentals of Microfabrication. CRC Press, 1: 1–70.

[23] Dhankhar, Meena, Marek Vaňatka and Michal Urbanek. 2018. Fabrication of magnetic nanostructures on silicon nitride membranes for magnetic vortex studies using transmission microscopy techniques. J. Vis. Exp., 137: 1–7. doi:10.3791/57817.

CHAPTER 3

Wet Chemical Methods for Nanoparticle Synthesis

Abhijit Jadhav

1. Introduction: (Nano agglomeration)

Nanomaterials are widely regarded as holding potential answers to challenges in electronics, medicine, biochemistry, environmental, and chemical process areas. A rapid change in properties is observed following the reduction in dimensions of the system. In the nanometer regime, particulates deviate sharply from the properties displayed by their bulk counterparts, as the surface effects become more substantial. The special scaled size-dependent properties hold from many hundred nanometers down to around 1 nm, whence individual differences between atomic clusters or nuclei are many and large [1]. There is now a growth in the understanding of structure and bonding in clusters. Wet chemical synthesis helps to achieve selective surface structures, shapes, phases, and sizes of metal and metal oxide nanoparticles.

Nanomaterials have attracted considerable interest due to their peculiar mechanical, electronic, optical, and magnetic, as well as the thermodynamic properties that differ significantly from those of either of the materials in the bulk of the single molecule. These unique properties can be attributed to their special structure and interactions thereof. Generally, isolable particles between 1 and 50 nm sizes that are prevented from agglomerating by protecting shells are regarded as nanostructured metal colloids. The term nanoparticle is used here in the sense of covering a wide range of nanostructured materials, including colloidal materials, nanoclusters, nanorods, etc.

Nanostructured colloidal metal particles can be obtained by two approaches. The top-down approach uses breaking down bulk materials and subsequent stabilization of the resulting nanosized metal particles by the addition of colloidal protecting agents. The bottom-up approach of wet chemical nanoparticle preparation, on the other hand, relies on electrochemical pathways or the controlled decomposition of

Department of Chemistry, Ewha Womans University, Seoul, Republic of Korea, 03760.

susceptible organometallic compounds, or the more versatile, chemical reduction of metal salts. There is a wealth of information on this type of nanostructured materials. A variety of stabilizers, such as donor ligands, polymers, and surfactants, are used in order to control the growth of the primarily formed nanoparticles and to prevent their agglomeration.

2. Wet Chemical Methods

Wet chemical synthesis for nanoparticle preparation is one of the simplest and one-step synthesis techniques available [2, 3]. The molecular motions and the chemical reactions proceed very smoothly in the liquid phase. Common reactions of the type acid + base → salt often yield precipitate of the desired compound in the form of hydroxides. Wet chemical synthesis is considered one of the promising methods for producing fine, phase pure, chemically homogeneous, and defect-free nanostructures in synthesis conditions. The wet chemical synthesis methods are simple, potentially safe, energy saving, and less time consuming, which can be helpful in producing various metal and metal oxide nanostructures. The most common wet chemical synthesis methods are sol-gel, precipitation, hydrothermal/solvothermal, and polyol synthesis, etc.

2.1 Sol-gel Technique

The sol-gel method is a well-established colloidal chemistry technology, which offers the possibility to produce various materials with novel, predefined properties in a simple and low-cost process. The sol can be a colloidal solution made of solid particles a few hundred nm in diameter suspended in a liquid phase, while the gel is considered a solid macromolecule immersed in a solvent. The sol-gel synthesis process consists of chemical transformation of a liquid (sol) into a gel state, and with subsequent post-treatment, the nanoparticles of metal oxides can be obtained. The high purity and uniform nanostructures achievable at low temperature are one of the main advantages of the sol-gel technique. It involves the formation of a sol-gel containing nanomaterials which later dries, and is heat-treated at a specific temperature and conditions in order to obtain the desired nanomaterials.

Usually, the starting materials can be metal chlorides or metal alkoxides used to form solvated metal precursors (sol). The precursors are hydrolyzed with water or alcohol to produce hydroxide. The reaction can be described by the following equations (1–3), respectively,

$$M\text{-}O\text{-}R + H_2O \rightarrow M\text{-}OH + R\text{-}OH \quad \text{(hydrolysis)} \tag{1}$$

$$M\text{-}OH + HO\text{-}M \rightarrow M\text{-}O\text{-}M + H_2O \quad \text{(water condensation)} \tag{2}$$

$$M\text{-}O\text{-}R + HO\text{-}M \rightarrow M\text{-}O\text{-}M + R\text{-}OH \quad \text{(alcohol condensation)} \tag{3}$$

Sometimes, along with water and alcohol, an acid or base can also help to hydrolyze the precursor. Equation 2 shows the hydrolysis reaction between alkoxides and the acid [4]. The hydroxide molecule forms oxide or alcohol bridged network, which we call gel, through polycondensation or polyesterification reactions.

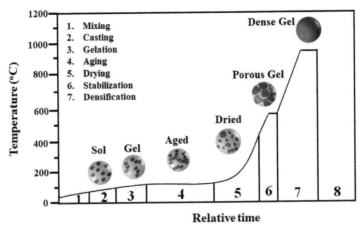

Figure 3.1. Schematic representation of the sol-gel process.

The as-formed gel consists of a three-dimensional skeleton of interconnected pores. The rate of hydrolysis and condensation highly affects the properties of final products, and is dependent on the electronegativity of metal atoms and its coordination number. Higher electronegative metal shows faster hydrolysis rate. Slower and controlled hydrolysis results in smaller particles. According to Srivastava et al., the hydrolysis rate reduces as the coordination number of the metal increases due to the steric hindrance [5].

There are various advantages of the sol-gel technique for the synthesis of nanoparticles. Sol-gel can produce thin bond coating to provide excellent adhesion between the metallic substrate and the top coat. It can produce a thick protective coating to provide corrosion. The sol-gel technique can help to shape materials into complex geometries in a gel form. It can also produce high purity products, as the organometallic precursors of the desired metal/ceramic oxides can be mixed, dissolved in a specific solvent, and later hydrolyzed into a sol, and later to gel form. There are a few limitations to the sol-gel method, such as requirements of expensive raw materials (such as metal alkoxides) compared to mineral-based metal ion sources. The obtained product contains a high carbon content when organic reagents are used in the synthesis process. The excess carbon content could inhibit densification during the calcination process. The high number of processing steps requires close monitoring.

The sol-gel technique has applications for preparing (1) monoliths, (2) powders, grains, and spheres, (3) fibers, (4) composites, (5) porous gels and membranes, (6) thin films and coatings, and (7) nanotechnology, etc.

2.2 Chemical Precipitation

Chemical precipitation routes routinely prepare the nanosized ceramics, metal oxides, and composites. Chemical precipitation is one of the wet chemical synthesis methods for nanoparticles and preparation of complex oxides. It is a highly simple and cost-effective method, and can yield products with high purity and near perfect stoichiometry with or without high-temperature treatment. During the chemical precipitation process, a solution containing a dissolved metal salt and a precipitating agent, usually hydroxide, ammonium acid carbonate, or oxalic acid, are added gradually to the metal salt solution to obtain metal hydroxide precipitates. The chemical precipitation route consists of direct precipitation, homogeneous precipitation, and co-precipitation process. The direct precipitation process usually has only one cation in the solution, such as Y_2O_3 [6, 7]. In the case of co-precipitation, multi-cations are present in the solution. Homogeneous precipitation has advantages over direct precipitation and co-precipitation method, due to excellent homogeneity of nucleation and precipitation. The co-precipitation reaction involves the simultaneous occurrence of a sequential process of nucleation, growth, coarsening, and agglomeration, respectively. The summarized overall process of chemical precipitation has been shown below in Figure 3.2 [8].

The initial mixing of metal precursors or interdispersing of components in the solution has a significant effect on the precipitation process. The homogeneity of the precipitated product highly depends on the proper mixing of precursors in the solution. The rate of stirring/centrifugal force primarily affects the nucleation, while the growth rate is less affected by this. The stirring rate can affect the aggregation of precipitated nanoparticles. Nucleation occurs when the solution is in supersaturation phase with respect to the components that need to be precipitated.

The various parameters that affect supersaturation during co-precipitation have been shown in Figure 3.3. In the supersaturation region, the system is unstable, and little disturbance can lead to precipitation. The supersaturation can be achieved by reducing the solution volume by evaporation or by lowering the temperature and/or increasing pH value. The solubility of a component increases with temperature, as shown in Figure 3.3. The solubility curve is also a function of the solution pH. With

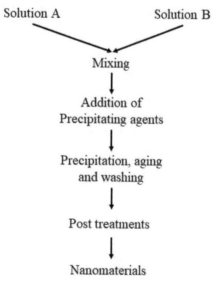

Figure 3.2. A summary of the steps involved in the chemical precipitation of nanoparticles [8].

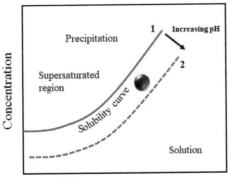

Temperature

Figure 3.3. Parameters affecting supersaturation during co-precipitation reaction.

the increase of pH value, solubility decreases, and the curve shifts from position 1→2. The point shifts from the solution region to supersaturation region with an increase in pH. The increase in solution pH is the most convenient method for precipitation reaction. The phenomenon can be explained by using a reaction, M^{n+} + $nOH^- \rightleftharpoons M(OH)_n$.

Nucleation can proceed spontaneously through the formation of metal hydroxide entities. Addition of seed or particle fragment can enhance the rate of nucleation. The nucleus can be defined as the smallest solid particle/phase, which is formed during precipitation, and is capable of spontaneous growth. Several attempts were made to evaluate the nucleation rate [9–13].

$$\frac{dN}{dt} = \beta exp\left[\frac{-16\pi\sigma^3 v^2}{3(kT)^3 ln^2 s}\right]$$

where β is the pre-exponential term, σ is solid-fluid interfacial energy, v is solid molecular volume, and T is the temperature. The super saturation 's' is defined as the ratio of actual concentration to solubility; $s = \dfrac{actual\ concentration}{solubility}$.

The equation can be written as $\dfrac{dN}{dt} = \beta exp\left[\dfrac{-A}{ln^2 s}\right]$ $A = \dfrac{16\pi\sigma^3 v^2}{3(kT)^3}$.

Thus, the nucleation depends strongly on the concentration and temperature of the solution. There is a critical super saturation concentration below which nucleation is very slow and above which it is very fast.

In co-precipitation reaction, the products are generally insoluble species prepared under the conditions of supersaturation. The primary step of nucleation produces a large number of tiny particles, and later they grow through Ostwald ripening and aggregation. The rate of nucleation and Ostwald ripening affect the size, morphology, and properties of the nanomaterials. The precipitated nanoparticles (nuclei) undergo an aging process by Ostwald ripening process. After giving required aging time, the precipitate later undergoes washing treatment with distilled water and ethanol, or another particular solvent to wash out the impurities. After drying the precipitate, post treatment, including annealing or calcination is carried out to prepare nanoparticles with desired morphology and high purity crystal structure.

The exact mechanism of chemical precipitation is difficult to understand due to the limitations of isolation of each process for independent studies [14]. The most important step in precipitation is nucleation of nanoparticles that is dependent on the degree of supersaturation. During nucleation, a large number of very small crystallites are initially formed and they grow subsequently through secondary process of Ostwald ripening and aging to form highly stable particles. Control over the nucleation and growth process is required to control the characteristics of the synthesized nanoparticles. There are various parameters, such as pH, surface charge or zeta potential, reaction concentration, reaction temperature, degree of mixing, and recrystallization of the precipitates, etc. Control over these parameters helps to achieve desired particle size, stoichiometry, and particle size distribution. This method also has a few limitations, such as the requirement of lots of chemicals, and it can generate potentially hazardous waste products. The nanoparticles synthesized by chemical precipitation mostly require stabilizers and post processing treatments to improve the phase purity of the products. Co-precipitation method is not suitable for the preparation of high, pure, accurate, and stoichiometric phase. Variable solubility of reactants and different precipitation rates can be major limitations for the synthesis.

Nanoparticle synthesis by urea precipitation method involves controlled release of anions, which helps to understand the kinetics of nucleation and particle growth in homogeneous solutions. Urea $((NH_2)_2CO)$ incorporation helps to form homogeneity of solutions, slow precipitation, and easy control of the final pH, which is a basic requirement for homogeneous precipitation. The reaction mechanism for $Y_2O_3:Eu^{3+}$ synthesis by urea precipitation has been shown below [15].

$$H_2N - CO - NH_2 \leftrightarrow NH_4^+ + NCO^- \tag{4}$$

$$NCO^- + OH^- + H_2O \rightarrow CO_3^{2-} + NH_3 \tag{5}$$

$$[(Y_{1.83-x}Eu_x)(H_2O)_n]^{3+} + H_2O \leftrightarrow [(Y_{1.83-x}Eu_x)(OH)(H_2O)_{n-1}]^{2+} + H_3O^+ \tag{6}$$

$$[(Y_{1.83-x}Eu_x)(OH)(H_2O)_{n-1}]^{2+} + CO_3^{2-} \rightarrow (Y_{1.83-x}Eu_x)(OH)CO_3.H_2O + (n-2) H_2O \tag{7}$$

Finally, upon the calcination process, the desired oxide nanostructures can be obtained with uniform shape and size distribution.

2.3 Hydrothermal/Solvothermal Method

The hydrothermal/solvothermal method is a very simple and highly versatile method for the synthesis of inorganic materials under high temperature and high-pressure conditions. The term hydrothermal/solvothermal process is defined as performing chemical reactions of metal salts dissolved in solvents contained in a sealed vessel in which the temperature of the solvent has been brought to around their critical points via heating simultaneously with autogenous pressures [16]. When water is used as a solvent, the process is referred to as a hydrothermal method, and when organics are used as a solvent, it is referred to as the solvothermal method.

The hydrothermal/solvothermal reactions are carried out in a sealed reactor called autoclave or high-pressure bomb. Mostly, hydrothermal/solvothermal reactors are metal (stainless steel) autoclaves with Teflon or alloy linings with a tube or cup made up of Teflon. The Teflon cup helps to protect the autoclave body from highly corrosive solvents at high temperature and pressure. Hydrothermal synthesis is carried out in a closed container using precursor solutions, which undergo stepwise transformation to form the nanoparticles, as shown below,

Hydrolysis: $MX_x(aq) + xH_2O\ (l) \rightarrow M(OH)_x\ (s) + xHX(aq)$ $\qquad$ (8)

Dehydration: $M(OH)_x\ (s) \rightarrow MO_{\frac{x}{2}}\ (s) + \frac{x}{2}\ H_2O\ (l)$ $\qquad$ (9)

Here, M denotes the metal and X the anions. The hydrated metal ions are first hydrolyzed to form the metal hydroxide, which later undergoes dehydration to form metal oxide [17]. Material synthesis carried out by hydrothermal/solvothermal method consists of crystallization process directly from solutions through nucleation and subsequent growth. By controlling various parameters, such as reactant concentration, pH, temperature, and additives, the morphologies and size of the final product can be varied. The phenomena responsible for controlling the size and morphology through tuning the process variables are the overall nucleation and growth rates, which depend on supersaturation [18]. The term supersaturation is described as the ratio of the actual concentration to the saturation concentration of the species in the solution [19]. Nucleation takes place at the point of supersaturation, and this reaction is irreversible. The solute precipitates into clusters of tiny crystals that can grow to microscopic size [20].

Other than aqueous solutions, nonaqueous organic solvents have also been used in the solvothermal reaction, which is similar to the hydrothermal synthesis. The commonly used organic solvents in solvothermal synthesis are methanol [21], toluene [22], 1,4-butanediol [23], and amines [24]. As a substitute to hydrothermal reaction, the solvothermal reaction can provide products at relatively lower temperatures and pressures. The precursors, which are sensitive to water, can also be handled in solvothermal reactions easily [25]. The products obtained from the solvothermal reaction are phase pure, free from foreign anions with controlled morphology [26–29]. There are a few more factors that can significantly affect the crystal nucleation and nanomaterials growth during hydrothermal/solvothermal synthesis. They are-precursors [30, 31], reaction time [32], additives [33], and filling factor, i.e., the ratio of the volume filled with solution to the total reactor volume [34]. The advantage of hydrothermal/solvothermal process is that the product can be formed directly from the solution. Different starting materials (precursor) and reaction conditions can help to control particle size and shapes. The resulting powers are highly reactive, which aid in low temperature sintering. The limitation of the hydrothermal/solvothermal method is that prior knowledge of solubility of precursor materials is required. The slurries in hydrothermal/solvothermal reactions are potentially corrosive. Accidental explosion of the high-pressure vessel (hydrothermal bomb) cannot be ruled out if proper care is not taken.

There are many other wet chemical processes available for the synthesis of nanostructures, multicomponent oxides powders, such as polyol synthesis [35, 36] micro emulsion [37–39], sonochemical [40–42], microwave heating [43], and spray pyrolysis [44, 45].

3. Conclusion

The wet chemical synthesis routes enable size and shape control of metal oxide nanoparticles. It makes it necessary to study the effects of various reaction parameters, such as size, shapes, specific properties, and compositions of nanomaterials. It was found that the growth and nucleation kinetics are highly sensitive to these reaction parameters, which can be correlated with changes in sizes, compositions of nanoparticles, and shapes. Through wet chemical synthesis, one can obtain nanoparticles with selective surface structures, phases, shapes, and sizes. Wet chemical routes allow fine control on reaction parameters, such as temperature, concentrations of the reactants, additives, surfactants, and pH, etc. to obtain desired nanomaterials. Wet chemical methods also allow control over stoichiometry composition, and it helps to scale-up the reaction. On the other hand, phase purity of the synthesized nanoparticles remains a major issue as surfactants and different additives are required to prevent agglomeration.

References

[1] Su, S.S. and I. Chang. 2018. Commercialization of Nanotechnologies—A Case Study Approach. Springer International Publishing AG 2018, pp. 15–29.
[2] Pandey, C., J.K. Bhasin, S.M. Dhopte, P.L. Muthal and S.V. Moharil. 2009. Wet chemical preparation of nano-crystalline complex fluorides. Mater. Chem. Phys., 115: 804–807.
[3] Wang, F., X. Fan, D. Pi and M. Wang. 2005. Synthesis and luminescence behavior of Eu^{3+}-doped CaF_2 nanoparticles. Solid State Commun., 133: 775–779.
[4] Tavakoli, A., M. Sohrabi and A. Kargari. 2007. A review of methods for synthesis of nanostructured metals with emphasis on iron compounds. Chem. Pap., 61(3): 151–170.
[5] Srivastava, A.K. 2013. Oxide Nanostructures: Growth, Microstructure, and Properties. Taylor & Fraincis Group, Florida.
[6] Li, J., W.B. Liu, B.X. Jiang, J. Zhou, W. Zhang, L. Wang, Y. Shen, Y. Pan and J. Guo. 2012. Synthesis of nanocrystalline yttria powder and fabrication of Cr, Nd:YAG transparent ceramics. J. Alloy Compd., 515: 49–56.
[7] Liu, B.L., J. Li, R. Yavetskiy, M. Ivanov, Y. Zeng, T. Xie, H. Kou, S. Zhuo, Y. Pan and J. Guo. 2015. Fabrication of YAG transparent ceramics using carbonate precipitated yttria powder. J. Eur. Ceram. Soc. 35(8): 2379–2390.
[8] Zhong, W.H., B. Li, R.G. Maguire, V.T. Dang, J.A. Shatkin, G.M. Gross and M.C. Richey. 2012. Nanoscience and Nanomaterials: Synthesis, Manufacturing and Industry Impacts. DEStech Publications Inc, USA.
[9] de Jong, K.P. 2009. Synthesis of Solid Catalysts, Wiley-VCH.
[10] Andrew, S.P.S. 1981. Theory and practice of the formulation of heterogeneous catalysts. Chem. Eng. Sci., 36: 1431–1445.
[11] Richardson, J.T. 1989. Principle of Catalysts Development, Plenum Press.
[12] Ertl, G., H. Knozinger and J. Weitkamp. 1997. Handbook of Heterogeneous Catalysis Vol. 1, Wiley-VCH.

[13] Farrauto, R.J. and C.H. Bartholomew. 1997. Fundamentals of Industrial Catalytic Processes, Blackie Academic & Professional.
[14] Cushing, B.L., V.L. Kolesnichenko, J. Charles and C.J. O'Connor. 2004. Recent advances in the liquid-phase syntheses of inorganic nanoparticles. Chem. Rev., 104(9): 3893–3946.
[15] Jadhav, A.P., T.D.T. Dinh, S. Khan, S.Y. Lee, J.K. Park, S.W. Park, J.H. Oh, B.K. Moon, K. Jang, S.S. Yi, J.H. Kim, S.H. Cho and J.H. Jeong. 2016. Enhanced photoluminescence due to $Bi^{3+} \rightarrow Eu^{3+}$ energy transfer and re-precipitation of RE doped homogeneous sized Y_2O_3 nanophosphors. Mater. Res. Bull., 83: 186–192.
[16] Byrappa, K. and M. Yoshimura. 2001. Handbook of Hydrothermal Technology. William Andrew, Norwich.
[17] Adschiri, T., K. Kanazawa and K. Arai. 1992. Rapid and continuous hydrothermal crystallization of metal oxide particles in supercritical water. J. Am. Ceram. Soc., 75(4): 1019–1022.
[18] Sue, K., M. Suzuki, K. Arai, T. Ohashi, H. Ura, K. Matsui, Y. Hakuta, H. Hayashi, M. Watanabe and T. Hiaki. 2006. Size-controlled synthesis of metal oxide nanoparticles with a flow-through supercritical water method. Green Chem., 8: 634.
[19] Andelman, T., M.C. Tan and R.E. Riman. 2010. Thermochemical engineering of hydrothermal crystallisation processes. Mater. Res. Innov., 14: 9.
[20] Kashchiev, D. 1982. On the relation between nucleation work, nucleus size, and nucleation rate. J. Chem. Phys., 76: 5098.
[21] Yin, S., Y. Fujishiro, J. Wu, M. Aki and T. Sato. 2003. Synthesis and photocatalytic properties of fibrous titania by solvothermal reactions. J. Mater. Process. Technol., 137: 45.
[22] Kim, C.-S., B.K. Moon, J.-H. Park, S.T. Chung and S.-M. Son. 2003. Synthesis of nanocrystalline TiO_2 in toluene by a solvothermal route. J. Crystal Growth, 254: 405.
[23] Kang, M. 2003. Synthesis of Fe/TiO_2 photocatalyst with nanometer size by solvothermal method and the effect of H_2O addition on structural stability and photodecomposition of methanol. J. Mol. Catal. A: Chem., 197: 173.
[24] Zhu, Y., T. Mei, Y. Wang and Y. Qian. 2011. Formation and morphology control of nanoparticles via solution routes in an autoclave. J. Mater. Chem., 21: 11457.
[25] Yu, S.-H. 2001. Hydrothermal/solvothermal processing of advanced ceramic materials. J. Ceramic Soc. Japan, 109(5): S65–S75.
[26] Konishi, Y., T. Kawamura and S. Asai. 1993. Preparation and characterization of fine magnetite particles from iron(III) carboxylate dissolved in organic solvent. Ind. Eng. Chem. Res., 32(11): 2888–2891.
[27] Chen, D. and R. Xu. 1998. Solvothermal synthesis and characterization of $PbTiO_3$ powders. J. Mater. Chem., 8: 965–968.
[28] Lu, J., P. Qi, Y. Peng, Z. Meng, Z. Yang, W. Yu and Y. Qian. 2001. Metastable MnS crystallites through solvothermal synthesis. Chem. Mater., 13: 2169–2172.
[29] Deng, Z.-X., C. Wang, X.-M. Sun and Y.-D. Li. 2002. Structure-directing coordination template effect of ethylenediamine in formations of ZnS and ZnSe nanocrystallites via solvothermal route. Inorg. Chem., 41: 869–873.
[30] Yuan, Z.-Y. and B.-L. Su. 2004. Titanium oxide nanotubes, nanofibers and nanowires. Colloid. Surf. A: Phys. Eng. Aspect., 241: 173–183.
[31] Nakahira, A., W. Kato, M. Tamai, T. Isshiki, K. Nishio and H. Aritani. 2004. Synthesis of nanotube from a layered $H_2Ti_4O_9 \cdot H_2O$ in a hydrothermal treatment using various titania sources. J. Mater. Sci., 39: 4239–4245.
[32] Lu, R., J. Yuan, H. Shi, B. Li, W. Wang, D. Wang and M. Cao. 2013. Morphology-controlled synthesis and growth mechanism of lead-free bismuth sodium titanate nanostructures via the hydrothermal route. CrystEngComm, 15: 3984–3991.
[33] Gao, Y., M. Fan, Q. Fang and W. Han. 2013. Controllable synthesis, morphology evolution and luminescence properties of $YbVO_4$ microcrystals. New J. Chem., 37: 670–678.
[34] Safaei, M., R. Sarraf-Mamoory, M. Rashidzadeh and M. Manteghian. 2010. A Plackett–Burman design in hydrothermal synthesis of TiO_2-derived nanotubes. J. Porous Mater., 17: 719–726.
[35] Silvert, P.Y. and K. Tekaia Elhsissen. 1995. Synthesis of monodisperse submicronic gold particles by the polyol process. Solid State Ionics, 82: 53–60.

[36] Wiley, B., Y. Su, B. Mayers and Y. Xia. 2005. Shape-controlled synthesis of metal nanostructures: the case of silver. Chem. Eur. J., 11: 454–463.

[37] Kim, B.H., J.H. Kim, I.H. Kwon and M.Y. Song. 2007. Electrochemical properties of $LiNiO_2$ cathode material synthesized by the emulsion method. Ceram. Int., 33: 837–841.

[38] Balint, I., Z. Youb and K. Aika. 2002. Morphology and oxide phase control in the micro-emulsion mediated synthesis of barium stabilized alumina nanoparticles. Phys. Chem. Chem. Phys., 4: 2501–2503.

[39] Hana, D.Y., H.Y. Yang, C.B. Shena, X. Zhou and F.H. Wang. 2004. Synthesis and size control of NiO nanoparticles by water-in-oil micro-emulsion. Powder Tech., 147: 113–116.

[40] Gedanken, A. 2004. Using sonochemistry for the fabrication of nanomaterials. Ultrason. Sonochem. 11: 47–55.

[41] Sivakumar, M., A. Gedanken, W. Zhong, Y.H. Jiang, Y.W. Du, I. Brukental, D. Bhattacharya, Y. Yeshurun and I. Nowik. 2004. Sonochemical synthesis of nanocrystalline $LaFeO_3$. J. Mater. Chem., 14: 764–769.

[42] Ganesh Kumar, V. and K.B. Kim. 2006. Ultrasonic enhancement of the supercritical extraction from ginger. Ultrason. Sonochem., 13: 549–556.

[43] Fua, Y.P., Y.H. Su, S.H. Wu and C.H. Lin. 2006. $LiMn_2-yMyO_4$ (M = Cr, Co) cathode materials synthesized by the microwave-induced combustion for lithium ion batteries. J. Alloys Compd., 426: 228–234.

[44] Tok, A.I.Y., F.Y.C. Boey and X.L. Zhao. 2006. Novel synthesis of Al_2O_3 nano-particles by flame spray pyrolysis. J. Mater. Pro. Tech., 178: 270–273.

[45] Ju, S.H., D.Y. Kim, H.Y. Koo, S.K. Hong, E.B. Jo and Y.C. Kang. 2006. The characteristics of nano-sized manganese oxide particles prepared by spray pyrolysis. J. Alloys. Compd., 425: 411–415.

CHAPTER 4

Electrodeposition
A Versatile and Robust Technique for Synthesizing Nanostructured Materials

Pravin S. Shinde* and *Shanlin Pan**

1. Introduction

The nanostructured materials have a wide range of applications in the fields of energy conversion and storage, environmental, medical, and automobile technologies. The nanostructured materials serve as important role as electrodes in several devices, such as lithium batteries, solar cells, fuel cells, and supercapacitors [1]. The nanostructured materials (grain size < 100 nm) [2, 3] exhibit unique mechanical, optical, electronic, and magnetic properties [4] due to the ultra-small building block units and high surface-to-volume ratio. All size-related effects in nanostructured materials can be integrated by monitoring the sizes of the constituent components [5]. Moreover, the synthesis method is very critical in controlling such constituent components. Several synthesis techniques have been used to develop nanostructured metal and semiconductor materials, such as sol-gel process [6], chemical deposition [7], hydrothermal method [8], pyrolysis method [9], chemical vapor deposition (CVD) [10], and electrodeposition [11]. In recent years, the development of inexpensive solution-based chemical methods to synthesize the controlled nanostructures for various technological applications has gained enormous attention of the research community. Among all the solution-based methods, electrodeposition is regarded as one of the most versatile techniques to synthesize nanostructures on complex substrates, and has been widely used at lab-scale as well as industry-level studies [12]. Electrodeposition plays an essential role in the development of sustainable energy conversion technologies, both at the portable and on a global scale. Also, the principles governing the scale-up of electrodeposition processes are well understood, facilitating the development of large-scale manufacturing processes [13].

Department of Chemistry and Biochemistry, The University of Alabama, Tuscaloosa, AL, USA 35487.
* Corresponding authors: psshinde@ua.edu; span1@ua.edu

Electrodeposition offers an inexpensive and simple alternative for the growth of thin films and nanostructures at ambient temperature and pressure as compared to energy-intensive processes. Therefore, electrodeposition is highly regarded as a viable and highly efficient method in synthesizing various nanostructured thin films due to its simplicity, environmental friendliness, absence of any sophisticated fabrication equipment, low-cost, and scalability. It has the ability to control the particle shape, size, crystallographic orientation, mass, film thickness, and surface morphology of the nanostructured materials by tuning the electrodeposition operating conditions and bath chemistry [14, 15, 16, 17, 18, 19].

The chapter presents a brief introduction to the fundamental electrochemical principals and mechanisms involved in conventional electrodeposition techniques, such as anodic electrodeposition, metal anodization, cathodic electrodeposition, and potential cycling/pulsed electrodeposition with a few examples. The key electrodeposition parameters controlling the nanostructured morphology of deposited materials, such as electrolyte precursor, electrode substrate, applied potential, and the annealing treatment are discussed. The significant portion of the chapter focuses on various electrodeposition approaches to synthesize various nanostructures, such as nanorods, nanowires, nanotubes, nanosheets, dendritic nanostructures, and composite nanostructures, by tailoring the electrodeposition parameters. The chapter highlights, not exclusively, the representative studies of nanostructures prepared by various electrodeposition techniques, focusing on the versatility of the techniques. The electrochemical synthesis of nanostructured materials with or without additives and templates in the electrolyte, with particular attention to the effects on surface morphologies, structures, and, more importantly, their performance is discussed. Finally, the perspectives on the electrodeposition processes and applications of electrochemically synthesized nanomaterials are outlined.

2. Fundamental Concepts

2.1 Principles of Electrochemistry

The term electrochemistry deals with the chemical and electrical phenomena in chemical reactions that exchange energies among the participating species. The chemical reactions involving ionic species can readily give rise to an electromotive force (emf) or potential, as in batteries or fuel cells. Conversely, the application of emf in a solution of ions or molten salts brings about the chemical reactions that may form or deposit a new material. This process is known as electrodeposition. In a typical electrodeposition process, an electrode (a solid, electronically conductive support or substrate) is immersed into an electrolyte (ionic conductor) containing positive or negative ions of the material to be deposited. Application of a certain emf or potential across the electrode/electrolyte interface brings about a charge transfer reaction known as a half-cell electrochemical reaction that results in the precipitation/deposition of material onto the substrate. Similarly, another half-cell electrochemical reaction occurs at the other electrode, forming an electrochemical cell in which an electrical charge involving the transfer of electrons and ions, is passed. The driving force for such an electrochemical process is the applied potential, which dictates the material formation process, its nano-to-micro-structural evolution, and other physicochemical properties. The efficiency of such electrochemical reactions depends on several factors, such as the thermodynamics and kinetics of electrolyte, charge and electron transfer processes, and the interfacial electrode processes. This chapter is precisely focussed on the electrochemical synthesis overview of metals and semiconductor nanostructures.

2.2 Electrodeposition Technique

Thermodynamic considerations govern the electrodeposition of metals or semiconductor compounds. The overall free energy change (ΔG) of the electrochemical reaction is positive, and hence, thermodynamically unfavorable, and often requires additional electrical energy supply or the potential to drive the reaction. Figure 4.1 shows the schematic representation of an electrochemical cell to electroplate or electrodeposit a metal "M" from an aqueous electrolyte of metal salt, "MA". As shown in Figure 4.1, a cathode (working electrode, WE) and an anode (counter electrode, CE) are immersed in an aqueous electrolyte containing positive (M^+ cations) and negative ions (A^- anions). When two electrodes are connected using a battery (source of electrons), the electric current flows, and an ion M^{+n} is reduced to M:

$$M^{+n} + ne^- \rightarrow M \tag{1}$$

where n represents the number of electrons taking part in the reaction. The electrodeposition of metallic film occurs on the cathode due to the discharge of metal ions in the electrolyte. The metal ion accepts one or n number of electrons from the electrically conducting material at the electrode-electrolyte interface, and then gets deposited onto the surface as a metal atom. The electrons necessary for this process are either supplied from an external source, or are donated by a reducing agent present in the electrolyte. The metal ions can be derived either from the metal salt (MA) added into the electrolyte, or by the anodic dissolution of the sacrificial anodes, made of the same metal as that of the cathode. The thickness of the electrodeposited layer on the substrate is determined by the electrodeposition time [13].

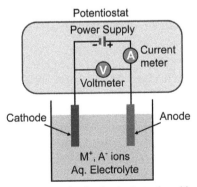

Figure 4.1. A schematic of a simple electrodeposition setup.

3. Electrochemical Growth of Nanostructured Materials

Nanoparticles are the basic building blocks in nanotechnology, and are the starting point in producing nanostructured materials. The morphology, as well as the composition of electrodeposited material, vary significantly and depend on the flowing current density, the nature of the anions or the cations from the electrolyte, electrolyte composition and temperature, electrolyte concentration, power supply of the current waveform, the presence of impurities, and physical-chemical nature of the substrate surface. Some of the common conducting substrates used for electrochemical synthesis include metals (Au, Ag, Cu, Ni, Pt, and stainless steel), metal oxides (ITO, FTO, $SrTiO_3$, Cu_2O, and ZnO), semiconductors (Si, InP, GaAs, GaSb, and InAs), and glassy carbon. The following section discusses various approaches to synthesize nanostructured materials using different electrodeposition processes.

3.1 Electrochemical Methods of Nanostructured Materials

In this section, we focus on the electrochemical synthesis of nanostructures of important semiconductor materials. Iron-based nanostructures have promising applications in solar fuel generation and magnetic devices. Let us consider a case study involving the electrochemical synthesis of iron oxide or hydroxide nanostructures, which can be performed anodically, cathodically, or both (Figure 4.2) [20], whereby the as-synthesized amorphous iron (Fe) film will undergo annealing treatment for crystallization to produce nanostructured hematite [11].

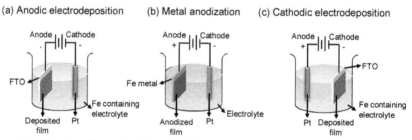

Figure 4.2. Electrodeposition methods for synthesizing nanostructured hematite (a) anodic electrodeposition, (b) metal anodization, and (c) cathodic electrodeposition [20].

3.1a Anodic Electrodeposition

Anodic electrodeposition is performed at a positive applied potential in an electrolyte precursor containing Fe ions using a three-electrode electrochemical setup, consisting of working, counter, and reference electrodes. During this process, oxidation of Fe^{2+} to Fe^{3+} ions, followed by precipitation of Fe^{3+} ions in the form of iron oxyhydroxide (FeOOH) films takes place on the conductive substrate, such as FTO or ITO (Figure 4.2a) due to the limited solubility of Fe^{3+} ions in the precursor solution [21]. The reactions are expressed as follows:

$$Fe^{2+} \rightarrow Fe^{3+} + e^- \qquad (2)$$

$$Fe^{3+} + 2H_2O \rightarrow FeOOH + 3H^+ \tag{3}$$

The overall reaction is represented as,

$$Fe^{2+} + 2H_2O \rightarrow FeOOH + 3H^+ + e^- \tag{4}$$

Since the Fe^{2+} ions barely dissolve in the precursor solution, thus complexing reagents, such as ammonium fluoride (NH_4F) and ammonium sulfate (($NH_4)_2SO_4$) need to be added to stabilize the Fe^{2+} ions in the solutions [22].

3.1b Metal Anodization

Similarly, metal anodization can also be carried out at a positive applied potential, but by using a conventional two-electrode electrochemical setup, where the metal and Pt act as the anode and cathode, respectively (Figure 4.2b). The application of positive voltage forms a thick compact oxide layer on top of the metal substrate, which enhances the corrosion resistance of the iron surface [23]. Typically, the formation of metal oxide through metal anodization is carried out via two competing reactions, namely: (1) the applied potential causes oxidation of Fe metal substrate to form metal oxide, and (2) chemical dissolution of oxide due to the presence of highly complexing F^- ions, which can be substantially improved by the presence of H^+ ions [24]. The reactions are defined as below:

$$2Fe + 2H_2O \rightarrow Fe_2O_3 + 6H^+ \tag{5}$$

$$Fe_2O_3 + 12F^- + 6H^+ \rightarrow 2[FeF_6]^{3-} + 3H_2O \tag{6}$$

The metal oxide formation continuously takes place until an equilibrium is reached between the two competing reactions, which stops the development of nanostructured metal oxide further [25].

3.1c Cathodic Electrodeposition

The cathodic electrodeposition is accomplished at a negative applied potential in a three-electrode electrochemical cell (Figure 4.2c). Generally, a typical cathodic electrodeposition process involves the reduction of H_2O_2 to hydroxide (OH^-) ions in an electrolyte containing cations (Fe^{3+} ions). Upon application of potential, the pH level in the vicinity of the working electrode increases, while the solubility of Fe^{3+} ions in the solution decreases, causing precipitation of Fe^{3+} ions to form amorphous iron oxyhydroxide (FeOOH) films, which then converts to nanostructured α-Fe_2O_3 upon annealing [26]. The reactions involved in cathodic electrodeposition of FeOOH are as follows:

$$H_2O_2 + 2e^- \rightarrow 2OH^- \tag{7}$$

$$FeF^{2+} + 3OH^- \rightarrow FeOOH + 2F^- + H_2O \tag{8}$$

The overall reaction becomes

$$3H_2O_2 + 2FeF^{2+} + 6e^- \rightarrow 2FeOOH + 2F^- + 2H_2O \tag{9}$$

The stability of Fe^{3+} ions in the solution can be improved by adding complexing agents, such as sodium fluoride (NaF) and potassium fluoride (KF) to form FeF^{2+}

complex, which secures the potential for the reduction of Fe^{3+} to Fe^{2+} ions to be more cathodic than the reduction potential of H_2O_2 to OH^- ions [27].

3.1d Cyclic Voltammetry and Pulse Electrodeposition

The potential cycling or cyclic voltammetry (CV) and the pulsed electrodeposition methods are advantages over the anodic and cathodic electrodeposition synthesis methods in terms of constantly maintaining both anodic (E_A) and cathodic (E_C) limit of potentials (Figures 4.3a, b) [28]. In CV or the potential cycling electrodeposition processes, the potential is varied between anodic and cathodic potential limits with a certain scan rate and a different number of cycles. On the other hand, the potential pulsed electrodeposition involves the periodic application of an anodic pulse for a certain time interval (t_A) during the constant anodic potential, and a cathodic pulse for certain time intervals (t_C) during the constant cathodic potential. Thus, each pulse cycle consists of an anodic pulse and a cathodic pulse of anodic and cathodic potentials. The mechanism of potential cycling/pulsed electrodeposition is based on the redox reaction at the semiconductor-liquid junction (SCLJ) interface. The nucleation and growth of the as-synthesized materials are directly governed by the processes involving adsorption and desorption of cations or anions, which can be systematically controlled by tuning the pulsed electrodeposition synthesis parameters. These parameters include the scan rate (Vs^{-1}), cycling window of cathodic and anodic potentials (V), scan direction, etc. for potential cycling electrodeposition, and the pulse frequency (Hz), duty cycle (%), cathodic and anodic pulse potentials (V), and the cathodic and anodic pulse durations (typically in ms) for pulsed electrodeposition [29].

The potential cycling electrodeposition method was previously used to synthesize adherent and uniform hematite thin films from a non-aqueous dimethyl sulfoxide (DMSO) electrolyte [30]. High-quality nanostructured hematite films (nanopetals) were grown using pulsed reverse electrodeposition (PRED) method by applying relatively high pulse potentials between −6 V and +4 V for short pulse durations. The PRED-synthesis from iron sulfate resulted in a compact metallic iron thin film, which was converted to hematite consisting of nanopetals (Figure 4.3c) upon annealing treatment [31]. Such a PRED method can also be used to selectively etch the naturally formed thin oxide-layer surface to get a purely metallic surface for further nanostructure-formations. The stable nanocoral nanostructures can be

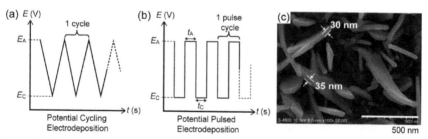

Figure 4.3. The potential waveforms for (a) cyclic voltammetry/potential cycling and (b) pulsed potential electrodeposition. (c) SEM image of Fe_2O_3 nanopetals synthesized by pulsed electrodeposition on FTO.

obtained on iron foil by giving a PRED-treatment in acidic electrolytes, followed by controlled annealing, as shown in Figure 4.4 [29].

Thus, it is viable to synthesize different nanostructures of a single material (iron oxide as discussed above) for various applications of interest just by employing different approaches of electrodeposition technique with ease, thereby demonstrating its versatility.

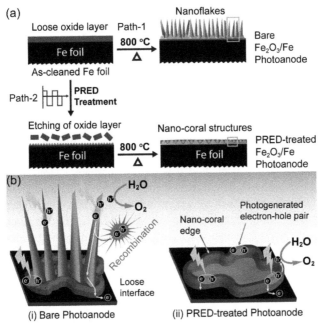

Figure 4.4. Schematic showing the application of (a) PRED-treatment to form (b) nanocoral-structured hematite on Fe foil.

3.2 Synthesis of Metallic Nanostructures using Electrodeposition

3.2a Metal Nanoparticles (NPs)

The few metal nanostructures and their oxides synthesized using different electroanalytical techniques of electrodeposition, such as cyclic voltammetry (CV), chronoamperometry, and chronopotentiometry were reviewed recently [32]. The pulsed electrodeposition technique has been used as a universal technique to prepare Au nanostructures of desirable properties. According to the nucleation theory [33], the size and number of nuclei during the onset of electrodeposition depend on the overpotential (η) by a relation:

$$r = 2\sigma V/ze_0|\eta| \tag{10}$$

where r is the critical nucleation radius, σ is the specific surface energy, V is the atomic volume in the crystal, and z is the number of elemental charges (e_0). The high-nucleation density on the surface of an electrode can be achieved at a higher η value.

This high overpotential has to be maintained for a few milliseconds (ms) because the cation concentration in the vicinity of the cathode depletes. Consequently, the process becomes diffusion-controlled [34]. During the intervals between the ON pulses (i.e., t_{off}), metal ions diffuse from the bulk of the electrolyte to the cathode surface and compensate for the metal ion depletion. Since the exchange current processes are energetically more preferred, the larger crystallites grow at the expense of smaller crystallites. This process can be prevented by surface-active substances that work as grain refiners when added directly to the electrolyte. The crystallite size of Au NPs can be altered down to 16 nm through proper choice of current density, pulse duration, and bath temperature, and further down to 7 nm with the help of sulfur- or arsenic-containing additives [35]. A potential step electrolysis method was adopted to deposit Au nanoparticles (Au NPs) with preferred orientation on a glassy carbon electrode (GCE) from an electrolytic bath that comprised of $NaAuCl_4$ and H_2SO_4 [36]. The Au NP electrodeposited using relatively short potential step width (5–60 s) exhibited the predominant crystal orientation along (111) plane. At longer deposition time (> 60 s), the surface of Au NPs consisted of Au (100) and Au (110) with a relatively larger particle size (> 100 nm) and low particle density. A simple and more convenient electrochemical method was used to prepare a large number of well-dispersed Au NPs with novel dumbbell-shaped structures (Figure 4.5a) from aqueous medium using surfactants, such as cetyl trimethyl ammonium bromide (CTAB), and tetradecyltrimethylammonium bromide (TTABr) [37]. The further shape of such Au NPs can be altered to rod-shaped and dumbbell-shaped by controlling the rate of the appropriate amount of acetone addition during the electrodeposition. Compton group [38] for the first time prepared Au NPs directly on a conducting ITO (indium tin oxide) substrate from a solution of $AuCl_4^-$ in 0.5 M H_2SO_4 by applying a potential

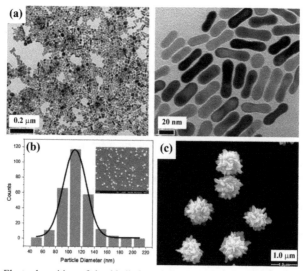

Figure 4.5. (a) Electrodeposition of dumbbell-shaped Au nanostructures (low and high magnification TEM images). (b) Electrodeposition of Au NPs on ITO (particle size distribution and SEM image). (c) SEM images of electrodeposited Au NCs on GCE.

step from +1.055 V (vs SCE) to −0.045 V for a fixed duration (15, 50, 150, or 300 s). In another study, a CV technique was used to electrodeposit 20–60 nm-sized Au NPs on ITO substrate in the potential range of −1.15 and −1.25 V for 20 cycles at 50 mV s⁻¹ scan rate [39]. Recently, our group deposited Au NPs onto planar ITO using the electrochemical multistep deposition method. Nucleation potentials of −0.6, −0.5, −0.4, −0.3, −0.2, −0.1, and 0 V (vs Ag/AgCl) were used, and 0 V (vs Ag/AgCl) was used as growth potential to optimize Au NP size and coverage. The Au NPs with an average particle size of 117 (±18) nm and a density of 1.4 × 10⁹ particles per cm² were observed, as shown in Figure 4.5b [40]. The nanoclusters (NCs) of Au were prepared on GCE [41] using electrodeposition, by running CV curves from −0.8 to 0.6 V at a scan rate of 50 mV s⁻¹ for 30 cycles in a fresh 1 mM HAuClO₄ solution, and the Au-coated GCE was dipped in 0.1 M KCl overnight to obtain Au NCs-modified GCE. These flower-like Au NCs consisted of Au nanorods, with an average diameter of 20 nm and a length of 80 nm (Figure 4.5c).

Recently, Pan group demonstrated a selective electrodeposition of a single Au NP (as small as 86 nm) at the tiny tip of a tungsten (W) wire from Au plating solution (KAu(CN)₂, 0.0687 wt%) by employing a dual step chronoamperometry method that includes a higher initial negative potential (−1.4 V) for a short duration (0.25 s), followed by a smaller negative potential (−0.8 V) for a longer duration (1 s). By such a process, a single Au NP with a particle size as small as 86 nm can be selectively deposited at the tip of a W wire, as shown in Figures 4.6a, b, c [42].

Forster group employed the 10–250 ms nucleation pulse to the conducting fluorine-doped tin oxide (FTO) glass working electrode within the potential range from −0.4 to −2 V, followed by a growth pulse at 0.3–0.7 V until a set charge was passed [43]. Penner's group also used the application of a high overpotential nucleation pulse for a brief time (5–10 ms), followed by a long-lasting low overpotential growth pulse that yielded particle size uniformity [44, 45]. Therefore, high-overpotential nucleation pulse encourages instantaneous nucleation, yielding high particle density and increasing particle size monodispersity. Synthesis of platinum (Pt) NPs for use in electrocatalytic applications has attracted a great deal of attention. Electrodeposition of Pt NPs (ranging from 4.5 to 9.5 nm) has been achieved on dense CNTs grown on carbon cloth (CC)—CNT/CC with the help of ethylene glycol (EG) as a reducing

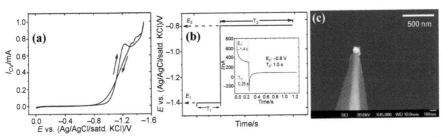

Figure 4.6. (a) CV of W wire from the Au plating solution at 50 mV s⁻¹ scan rate. (b) Electrochemical current response by application of chronoamperometric dual-potential step (E₁ = −1.4 V, T₁ = 0.25 s, E₂ = −0.8 V, T₂ = 1 s). (c) SEM image showing a selective electrochemical deposition of single Au NP (~ 86 nm) at the tip of W wire.

agent, which also acts as a stabilizing surfactant, and prevents the particles from agglomeration during the electrodeposition [46].

Uniform Pt NPs were electrodeposited on GC/PANI substrates by a potentiostatic pulsed electrodeposition process from 5 mM K_2PtCl_6 prepared in 0.5 M H_2SO_4. The deposition conditions involved the application of negative pulse potential (E_{on} = −150 mV), negative pulse potential (E_{off} = +100 mV) for respective pulse durations of t_{on} = 10 ms and t_{off} = 3.33 ms, respectively, with a duty cycle (DC) of 75% [47]. About 300–400 nm-sized Pt NPs were cost-effectively electrodeposited on a GCE using water-immiscible ionic liquid (IL) droplet supported on the electrode surface at −1.5 V vs Ag/AgCl for 500 s from an electrolytic bath composed of KPF_6 and H_2PtCl_6 [48]. A potentiostatic deposition was employed to synthesize Pt NPs from $PtCl_2$ solution on highly oriented pyrolytic graphite (HOPG) from chloride-based electrolytes at open-circuit potential [49]. Highly uniform Pt NPs were electrodeposited on MWCNTs/carbon paper grown *in situ* at −0.6 vs SCE from a bath comprising of chloroplatinic acid (60 g L^{-1}) and hydrochloric acid (10 g L^{-1}) when the electrodeposition voltage was maintained at −0.6 V vs SCE [50]. Recently, an electrochemical synthesis of Pt (acac)$_2$ to Pt NPs having average particle sizes ~ 3.0 nm (Figures 4.7a, b) were obtained on glassy carbon via two-electron reduction process using cyclic voltammetry and RDE methods at −1.8 V with a rotation rate of 1,000 rpm in aprotic TFSA$^-$-based ionic liquids, such as BMPTFSA, HMPTFSA, and DMPTFSA [51]. Highly porous Pt NPs were obtained by two-step potentiostatic electrodeposition with two different nucleation potentials of −0.4 and −0.6 V, and two corresponding growth potentials of −0.1 and −0.2 V, respectively as shown in Figure 4.7c and Figure 4.7d [52]. The chain of nickel (Ni) NPs were also electrodeposited inside the TiO_2 nanotubes by employing a potentiostatic mode of deposition at potentials ranging from −1.3 to −1.7 V from an electrolytic bath composed of titanium fluoride and nickel chloride [53]. In another study, Ni NPs were electrodeposited onto Ti substrate from the electrolyte consisting of 10 mM $NiCl_2$ in 1 mM H_2SO_4, and the Ni^{2+} ions were electrochemically reduced to Ni^0 at −0.26 V [54].

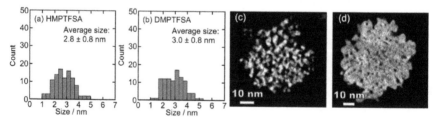

Figure 4.7. The particle size distribution of Pt NPs synthesized by potentiostatic electro-reduction at −1.8 V using a GCRDE in (a) HMPTFSA and (b) DMPTFSA ionic liquids containing a 5 mM Pt(acac)$_2$ at a bath temperature of 50°C. Electron tomography of porous Pt NPs electrodeposited with (c) En = −0.4 V and (d) En = −0.6 V.

3.2b Metal Nanowires (NWs)

Nanowires are one-dimensional nanostructures that offer opportunities to control the electronic and optical properties of materials due to their unique morphology. They are considered the building blocks for next-generation electronics, sensors, photonics, and energy materials. High-density and high-aspect-ratio Au, Cu, and Ag metal nanowire (NW) arrays with diameters ranging from 40 to 250 nm were synthesized by potentiostatic deposition using anodic alumina oxide (AAO) templates. The Au NWs were 20–30 mm in length and 40–50 nm in diameter [55]. Wang et al. electrodeposited Au NWs using AAO templates from an electrolytic bath composed of $HAuCl_4 \cdot 3H_2O$, EDTA, Na_2SO_3, and K_2HPO_4. These Au NWs were ductile, with an average diameter of ~ 180 nm [56]. Similarly, Au NWs which were 250 nm in diameter and 10 mm in length were synthesized using nanopore polycarbonate (PC) membrane [57]. The electrodeposition method was also effectively used to prepare composites of metal NPs and conducting polymers with core-shell structure. For instance, Au/polypyrrole (Au/PPy) core-shell nanocomposites on the ITO surface were synthesized using the electrochemical technique [58]. Au NPs with core-shell structure sizes in the range of 250–300 nm fill coverage of 30 nm polyaniline shell present on the gold surface. A pulse electrodeposition method was used to synthesize PPy-Au NP composites on the graphite surface from a solution comprising of Au-salt and monomer [59]. The electrodeposition technique can still be used to obtain nanostructures on non-conducting dielectric substrates with the help of a versatile lithographically patterned nanowire electrodeposition (LPNE) technique. LPNE can be used to fabricate linear NWs of metals, semiconductors, or chalcogenide directly on the trenched dielectric surfaces, ranging from glass to flexible plastic [60, 61, 62]. It is also possible to obtain wafer-scale fabrication of nanofluidic Au nanochannels arrays/networks with the help of nanoimprint lithography and LPNE techniques [63]. In LPNE, the horizontal trenches were first obtained onto glass, oxidized silicon, or Kapton polymer film substrates using photolithography [62]. Such trenches were then used to obtain the linear NWs of Au, Pt, Pd, or Bi by potentiostatic electrodeposition by selecting particular deposition potential from CVs (Figures 4.8a–h).

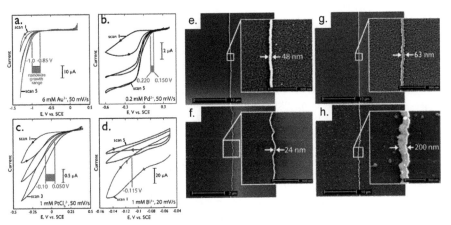

Figure 4.8. Cyclic voltammograms acquired in the metal plating solutions employed for NW growth: (a) Au, (b) Pd, (c) Pt, and (d) Bi, and the corresponding SEM images of linear NWs of (e) Au, (f) Pd, (g) Pt, and (h) Bi. The range of electrodeposition potentials employed for potentiostatic electrodeposition of nanowires is indicated in the CVs.

3.2c Metal Nanorods (NRs)

The gold nanorods (NRs) were first electrodeposited by the Wang group at a current density of 3 mA cm^{-2} from an electrolytic bath composed of hexadecyltrimethylammonium bromide (C16TAB) and tetradodecylammonium bromide (TC12AB), which stabilized the synthesized Au nanostructures by preventing them from further aggregation. The length of the Au NRs was dependent on the concentration of gold ions and their release rates [64]. Martin and co-workers employed a template-based method to synthesize gold nanorods using nanoporous polycarbonate or aluminum template membranes [65, 66, 67]. A few authors electrochemically synthesized ordered Au NRs arrays (Figure 4.9a) using AAO or PC membrane templates of pore size of 100 nm [68]. 1D crooked gold nanorods (CGNRs) and 2D gold network structures have also been synthesized electrochemically in the presence of isopropanol solvent with an ionic surfactant solution [69].

Vertically standing Ag NRs were fabricated on TiO$_2$-modified FTO substrates using a template-free electrodeposition method from dilute aqueous 0.1 mM AgNO$_3$ solution at a DC voltage of -5 V in a two-electrode system (Figure 4.9b). The diameter, length, and surface coverage of Ag NRs are dependent on the electrodeposition time and choice of substrates [17]. Other nanostructures, such as the vertically-aligned Ag nanoplates array on ITO (Figure 4.9c) were also reported using a two-step electrodeposition method. The method involved application of a short nucleation step at a high negative overpotential (-0.4 V vs SCE for 20 ms) and a longer particle growth step at a mild deposition potential (0.250 V vs SCE for 1800 s) from aqueous electrolyte containing 5 mM AgNO$_3$, 200 mM KNO$_3$, and 1 mM Na$_3$Cit electrolyte. The nucleation potential, growth potential, and capping agent/precursor ratio were found to influence the size and morphology of the Ag nanoplates [70].

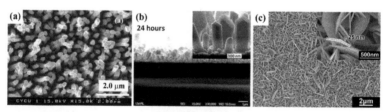

Figure 4.9. (a) SEM image of Au NRs by template electrodeposition. (b) Cross-section SEM image of Ag NRs on TiO$_2$-modified FTO glass at a growth time of 24 h. (c) SEM image of Ag nanoplates array on ITO by two-step electrodeposition.

3.3 Nanostructured Semiconductor Materials by Electrodeposition

In the following section, we will review and discuss some of the other technologically important nanostructured semiconductor materials synthesized by electrochemical methods regarding different synthesis conditions.

3.3a Semiconductor Nanorods

The most interesting NR structures of zinc oxide (ZnO) on FTO were first reported by cathodic electrodeposition from oxygen-bubbled 5 mM ZnCl$_2$ electrolyte at 80°C at a deposition potential of -1.3 V for 30 minutes [71]. Since then, several attempts were made to synthesize ZnO NRs using the ZnO buffer layer and seed-layer free electrodeposition [72, 73] (Figures 4.10a, b). A simple dc electrodeposition can be used to selectively coat ZnO NRs on small electrodes, such as microelectrodes, very effectively [74]. The electrodeposition mechanism of ZnO NRs generally involves both electrochemical and solution chemical processes. During the electrochemical process, the O_2 ions are reduced on the surface of the substrate, causing the formation of hydroxide (OH^-) ions. Then, zinc and hydroxide ions combine to form zinc hydroxide precipitate on the substrate, which can then be dehydrated to convert into ZnO at sufficiently high solution temperatures (60–70°C) [75]. The reactions are represented as below:

$$O_2 + 2H_2O + 4e^- \rightarrow 2OH^- \tag{11}$$

$$Zn^{2+} + 2OH^- \rightarrow Zn(OH_2) \rightarrow ZnO + H_2O \tag{12}$$

Similar approaches have been employed for electrochemical synthesis of oxide-based nanostructures. The electrodeposited ZnO NRs can be used as a template to synthesize Cu_2O NRs (Figure 4.10c) on FTO by potentiostatic electrodeposition in the range of -0.4–0.6 V vs Ag/AgCl from lactate-chelated aqueous $CuSO_4$ solution at pH 9.0, with ZnO NRs film serving as a sacrificial template [76].

The nano-cube-structures of Cu_2O were prepared on Cu foil by electrodeposition from alkaline lactate-stabilized copper sulfate solution (0.2 M $CuSO_4 \cdot 5H_2O$ + 2 M lactic acid in 0.5 M K_2HPO_4 buffer, pH $\sim$ 12 using 4 M KOH) in a two-electrode configuration at 30°C. The deposition of Cu_2O on Cu was accomplished in a galvanostatic mode at a current density of -0.3 mA cm^{-2} for 90 minutes [77]. Large-scale and highly oriented single-crystalline hexagonal Cu_2O nanotube arrays were synthesized using a two-step solution approach that involved the electrodeposition of oriented Cu_2O NRs, and a subsequent dissolution technique using NH_4Cl additive to convert Cu_2O nanorods to nanotubes during electrodeposition (Figures 4.11a, b). Cu_2O NRs were electrodeposited from a solution mixture of 3 mM $Cu(NO_3)_2$, 0.1 M NH_4Cl + 0.05 M KCl at a current density of 2.0 mA cm^{-2} for 120 minutes at 70°C in galvanostatic mode [78].

Various hierarchical nanostructures of ZnO were reported by a two-step electrochemical deposition process that involved the electrodeposition of ZnO crystals with different morphologies as a seed layer, followed by the electrochemical epitaxial growth of oriented nanorods on the surfaces of the primary ZnO nanostructures.

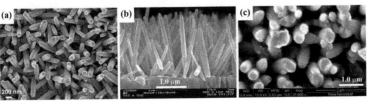

Figure 4.10. SEM images of ZnO NRs prepared on FTO (a) with (top view) or (b) without (cross-section) ZnO seed layer using electrodeposition. (c) SEM image of electrochemical grown Cu_2O NRs on FTO using sacrificial ZnO NRs templates.

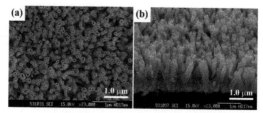

Figure 4.11. SEM images of oriented Cu₂O nanotubes converted from Cu₂O NRs on Cu plate prepared by electrodeposition: (a) top and (b) cross-section view.

A variety of hierarchical ZnO nanostructures, including 2-fold nanorod arrays on nanosheets, 6-fold nanorods on nanorods, and 6-fold nanoneedles on nanoneedles, were synthesized on ITO. The electrochemical depositions were performed in a three-electrode setup at 70°C. ZnO nanostructures with great control over morphology can be achieved by tuning the electrodeposition parameters. The great example includes the electrochemical synthesis of hexagonal nanosheets, nanorods, and nanoneedles from 0.05 M $Zn(NO_3)_2$ solutions mixed with 0.06 M KCl, 0.06 M KCl with 0.01 M ethylenediamine (EDA), and 0.01 M EDA, respectively [79]. The electrolyte for the second electrodeposition of ZnO nanorods was prepared by adding ammonia dropwise to the 0.05 M $Zn(NO_3)_2$ aqueous solution at 70°C under continuous stirring to get a clear solution. The second electrodeposition of ZnO nanoneedles was carried out in 0.05 M $Zn(NO_3)_2$/EDA solution by controlling the amount of EDA. All the electrodepositions were performed at −1.1 V vs SCE for 1.5 hours [80]. Figure 4.12 shows different hierarchical ZnO nanostructures.

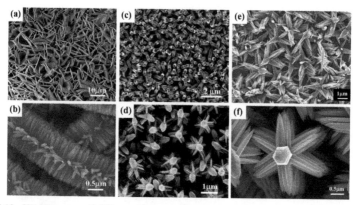

Figure 4.12. SEM images of different ZnO nanostructures electrodeposited on ITO: (a) primary hexagonal ZnO nanosheets, and (b) hierarchical ZnO NRs on nanosheets; (c) primary ZnO nanorods, and (d) 6-fold hierarchical ZnO nanorods; (e) hierarchical ZnO nanoneedles on nanoneedles, and (f) 6-fold hierarchical ZnO nanoneedles.

3.3b Semiconductor Nanowires

In photovoltaics and solar water splitting, the nanowire-type semiconductor nanostructures play a crucial role because NW's large aspect ratio provides a long optical path length for improved light absorption along the vertical axis, while its much smaller diameter facilitates the radial collection of minority carriers (holes for p-type and electrons for n-type semiconductors) over a shorter distance, thereby minimizing the probability of recombination within the absorber material.

Moreover, electrodeposition has been extensively used to obtain such nanostructures with or without the use of sacrificial templates or the structure-directing agents. The $Cu(OH)_2$ NWs were prepared by anodizing Cu-coated FTO at a constant current density of 10 mA cm^{-2} at room temperature from 3 M KOH solution for 3 minutes (Figure 4.13a). The resulting $Cu(OH)_2$ NWs were later transformed into Cu_2O NWs upon thermal annealing at 600°C in Ar atmosphere. The thin Cu_2O blocking layer was further electrodeposited on the synthesized Cu_2O NWs from an alkaline solution of lactate stabilized copper sulfate electrolyte in galvanostatic mode at a constant current density of -0.1 mA cm^{-2} for 30 minutes at a bath temperature of 30°C (Figure 4.13b) [81]. Highly crystalline, continuous parallel arrays of cadmium selenide (CdSe) NWs were obtained using electrodeposition at -0.59 V for 30 minutes using photoresist (NOA 60)-patterned ITO substrates from an electrolyte containing 0.1 M $CdCl_2$, 0.001 M SeO_2, and 1 M HCl, followed by subsequent removal of resist in an alkaline (0.5 M NaOH) developer solution for 45 minutes. The resulting CdSe nanowires (Figure 4.13c) were 100 nm in thickness, 300–500 nm in width, and several centimeters in length [82]. Using an LPNE technique described in section 3.2b, the single semiconductor CdSe NWs can uniformly be synthesized by the reaction of Cd^{2+} and H_2SeO_3 under controlled activation conditions using potentiostatic deposition at -0.60 V vs SCE [83]. In another approach, the photoconductive hemicylindrical shell NWs (HS-NWs) composed of nanocrystalline cadmium sulfide (CdS) were prepared. The CdS HS-NWs were synthesized in two steps: first by electrodepositing microcrystalline cadmium (Cd) NWs by the electrochemical step-edge decoration on the graphite electrode surface, and then converting these Cd nanowires into CdS by exposure to H_2S at an elevated temperature [84].

In addition to metallic and binary oxide materials, the electrodeposition technique can also be efficiently used to synthesize nanostructures of ternary compounds, such as copper indium diselenide ($CuInSe_2$) [85, 86] and quaternary compounds, such as copper thiocyanate (CuSCN) [87, 88, 89, 90, 91].

The p-type ternary semiconductor (I–III–VI) material, $CuInSe_2$ (CIS) NWs were pulse electrodeposited inside the AAO template pores. The electrolyte consisted of a mixture of 1.5 M copper sulfate hydrate, 2 mM indium sulfate hydrate, 3.5 mM selenous acid and lithium chloride, dissolved in 100 ml aqueous buffer solution

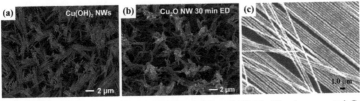

Figure 4.13. SEM images of (a) anodized $Cu(OH)_2$ NWs and (b) blocking layer-coated Cu_2O NWs synthesized on FTO using galvanostatic anodization/electrodeposition. (c) CdSe NWs electrodeposited at -0.59 V for 30 min using photoresist-patterned ITO.

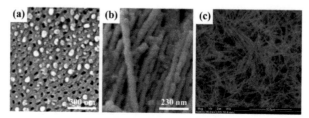

Figure 4.14. (a) SEM top view of $CuInSe_2$ (CIS) NWs embedded in a 2 μm long AAO/Al template anodized at 40 V. (b) SEM side-view of free-standing CIS NWs with average diameter of 30 nm after removal of the host AAO template. (c) Typical CIS nanowires prepared by one-step electrodeposition using AAO templates.

(pH 2.8) containing potassium hydrogen phthalate and hydrochloric acid. The dense and compact CIS NWs (Figures 4.14a, b) were obtained at optimized pulse waveforms with pulsing time $t_c = 0.3$ s at potential $V_c = -1$ V (vs Pt) and discharging time $t_r = 2$ s at $V_r = 0$ V yielded. Such electrodeposited CIS NWs can be annealed at 220°C in a vacuum (10^{-2} bar) to improve stoichiometry and crystallinity [86].

Alternatively, both p-type and n-type single-phase chalcopyrite CIS NW arrays were fabricated by a one-step electrodeposition route from aqueous solutions of copper sulfate, indium sulfate, selenium dioxide, and citric acid, using AAO as templates (Figure 4.14c). The conductivity type of synthesized CIS NWs is altered just by slightly varying the electrodeposition potential, for instance, p-type (slightly Cu-rich) and n-type (slightly In-rich) CIS NWs were obtained at -0.7 V and -0.75 V, respectively [85].

The CuSCN NWs were grown electrochemically from an aqueous electrolyte with the help of a chelating agent (EDTA) at room temperature in a two-step reaction, wherein Cu^{2+} is first reduced to Cu^+ and then in the second step Cu^+ chemically precipitates with SCN^- to form CuSCN NWs [88]. The formation of CuSCN NWs depends on the chelating agent in the electrolyte, and can be precisely controlled by the applied potential or the current density on different substrates. From CVs (Figure 4.15a), the cathodic current starts at around -0.1 V for the FTO substrate, and 0.2 V for the gold one, reaches the maximum, and then slightly decreases and increases again at -0.55 and -0.4 V for FTO and gold substrates, respectively. Au-substrates initiate the formation of CuSCN NWs at low current density (low applied

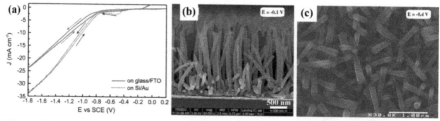

Figure 4.15. (a) CVs of glass/FTO and $Si/SiO_2/Au$ substrates in a mixture of electrolyte containing 12 mM Cu^{2+} ($Cu^{2+}/EDTA/SCN^- = 1:1:0.25$). SEM images of electrodeposited CuSCN NWs on (b) Si/SiO_2/Au substrate (E = -0.1 V, Q = 150 mC cm^{-2}), cross-section; (c) glass/ITO substrate (E = -0.4 V, Q = 50 mC cm^{-2}), top view.

potential) as compared to the FTO. Moreover, NWs on Au are vertically aligned compared to those on FTO (Figures 4.15b, c).

3.4 Nanostructured Materials using Template-Guided Electrodeposition

Electrodeposition is also better suited to obtain different nanostructures using *in situ* templates, such as membranes or surfactants/structure-directing agents. Both metal alloys and semiconductor nanostructures, especially NWs, have been synthesized using template-assisted electrodeposition methods. The electrodeposition of Au NWs from cyanide baths using PC templates with a pore diameter of 80 nm was also reported [92]. Well-defined Au NWs with uniform ends and a length equal to template thickness were synthesized via direct electrodeposition of gold through the pores of the PC template [93], as shown in Figure 4.16a. Porous Au NWs arrays were fabricated through a one-step galvanostatic (10 mA cm^{-2} current) electrodeposition method, utilizing 65 nm pore-diameter nanochannel alumina (NCA) templates [94], as shown in Figure 4.16b. Although silver (Ag) plating is a simple process, synthesis of discrete nanostructures is not a simple task as silver tries to agglomerate, often forming dendritic growth. Highly ordered Ag NWs arrays were prepared by pulsed electrodeposition method using self-ordered porous alumina templates [95]. Segmented all-Pt NWs were synthesized by a template-assisted pulse-reverse electrodeposition (PRED) method (Figure 4.16c) using ion track-etched nanoporous PC-membrane. PRED into the nanochannels was performed from an alkaline platinum bath (Platinum-OH, Metakem) at 65°C in a two-electrode configuration with a cathodic pulse (Uc = −1.3 V) for different pulse durations (t_c = 1 to 20 s) and anodic pulse (Ua = +0.4 V) for t_a = 1 s. The morphology is controlled by the electrokinetic effects on the local electrolyte distribution inside the nanochannels during the nanowire growth [96]. The poly(3,4-ethylene dioxythiophene) (PEDOT) NWs and thin films can also be synthesized by electropolymerization of 3,4-ethylene dioxythiophene (EDOT) in aqueous LiClO$_4$ within a template prepared using the LPNE process. These PEDOT NWs were 40−90 nm in thickness, 150−580 nm in width, and 200 μm in length [97].

Porous nanoballs of iron oxide can be synthesized from iron precursor by utilizing appropriate amount of CTAB surfactant and pulse potential conditions during the PRED synthesis of metallic iron nanoballs on FTO, which then can be transformed into porous Fe$_2$O$_3$ nanoballs array (Figures 4.17a, b) upon annealing after getting rid of surfactants [98].

In another study, branched core/shell nanoarrays were synthesized, in which porous nanosheets of metal hydroxides, such as Co(OH)$_2$ and Mn(OH)$_2$ were

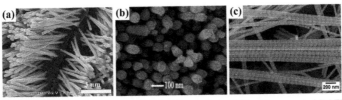

Figure 4.16. SEM images of (a) Au NWs, (b) porous Au NWs, (c) segmented Pt NWs.

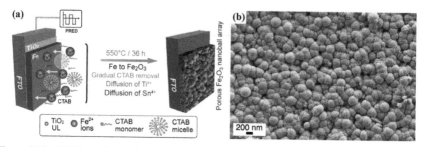

Figure 4.17. (a) Schematic showing formation of porous Fe_2O_3 nanoballs on FTO. (b) SEM image of porous Fe_2O_3 nanoballs.

electrodeposited on pre-formed Co_3O_4 nanowire scaffold as a core on various conductive substrates, such as Ni mesh, Ti foil, and stainless steel (SS) (Figure 4.18). The cathodic electrodeposition of $Co(OH)_2$ nanosheet shell growth on self-supported hydrothermally grown Co_3O_4 nanowire arrays scaffold was accomplished through a simple electrodeposition method in a three-electrode configuration at 25°C from aqueous $Co(NO_3)_2$ electrolyte. To deposit $Co(OH)_2$ shell nanosheet, four cycles of CVs were run in the potential range of −0.5 to −1.1 V vs SCE at the scan rate of 10 mV s^{-1} and the samples were annealed at 200°C in air for 1.5 hours to form homogeneous mesoporous Co_3O_4 core/shell NR arrays. The $Co_3O_4/Mn(OH)_2$ and Co_3O_4/Mn_3O_4 NW arrays were also fabricated under similar synthesis conditions [99].

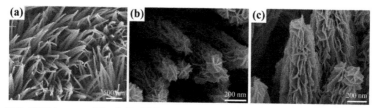

Figure 4.18. SEM images of $Co_3O_4/Co(OH)_2$ core/shell nanowire arrays grown on (a) Ni mesh, (b) Ti foil, and (c) SS substrates.

4. Factors Influencing the Electrochemical Growth of Nanostructures

In this section, we will briefly discuss various parameters that determine the fate of the electrochemical synthesis of nanostructured materials. By adjusting the composition, pH, solution concentration, bath temperature, applied potential, and current density, the final film characteristics, such as nanostructure morphology and crystallography can be controlled. Among the parameters mentioned above, a proper choice of the applied potential regime is one of the essential things in the electrodeposition of nanostructures from aqueous media. Electrodeposition in an aqueous system is thermodynamically hindered due to the competing hydrogen evolution reaction, and thus the growth can usually be restricted to a few monolayers. Therefore, one needs to account for the relationship between the applied potential

(V vs SHE) and the pH of the electrolyte before starting the electrochemical synthesis reactions. The thermodynamics of nanostructured growth of any desired material to be electrodeposited from the aqueous medium can be studied by constructing the potential-pH diagram, famously known as the Pourbaix diagram. A Pourbaix diagram represents thermodynamic and electrochemical equilibria in an aqueous solution, which plots the equilibrium potential (V_e) between a metal/semimetal and its various oxidized species as a function of pH at room temperature [100]. The applied potential and pH of the solution need to be carefully selected in order to have a stable and uniform electrodeposit of the desired material. The precursor solution concentration has a significant role in obtaining the nanostructured growth of the desired material. Generally, low concentrations (0.1 to 1 mM) are preferred. The electrolyte bath temperature needs to be lowered or elevated, as some reactions tend to be exothermic or endothermic, and do not occur at room temperature to precipitate the electrodeposits on the substrate during the reaction. During constant current or galvanostatic electrodeposition, current density flowing through the substrate electrode also decides the morphology of the nanostructures and generally, lower currents for longer electrodeposition durations are preferred. However, in the case of templated electrodeposition through PC or AAO membranes, large currents need to be used. Thus, the electrodeposition technique can easily be customized with proper electrochemistry knowledge about the material according to our needs to achieve the desired nanostructures by controlling the various parameters with ease. It has been widely used in electroplating industries to electrodeposit various materials on sophisticated devices where physical techniques cannot be employed. Thus, this chapter outlines the versatility of the less-energy intensive and simple electrodeposition technique.

5. Conclusions

Electrodeposition is gaining popularity in recent years due to its capability to prepare a range of high-quality low-dimensional multifunctional nanostructures via a simple yet elegant physicochemical route. Contrary to the energy-intensive synthesis processes, electrodeposition is a high-throughput and low-cost manufacturing process that offers an inexpensive, scalable, and straightforward bench-top approach for the growth of the thin film and nanostructures. Precise control of the potential or the current on a conductive substrate in a solution during electrodeposition allows for electrochemical oxidation or reduction of chemical species to form the desired nanostructures. This bottom-up fabrication technique is capable of fabricating one-dimensional nanostructures, such as nanoparticles, nanorods, nanowires, nanotubes, nanosheets, nanoclusters, and several other nanostructures (including the ones yet to be discovered) of various metals, metal oxides, metallic alloys, chalcogenides, and composites for a wide range of potential applications in photovoltaics, solar cells, catalysis, sensors, etc. This chapter gives brief insights into various facets of the electrodeposition technique to synthesize technologically important nanostructures of various materials. Different types of templates, patterns, or structure-directing agents utilized during the electrodeposition will contribute to the continued development of this technique in becoming popular among the research community.

Acknowledgments

Authors acknowledge the support of National Science Foundation (NSF) under award numbers OIA-1539035 and CHE-1508192. The authors are also thankful to the University of Alabama for the support through the RGC-level 2 award.

References

[1] Aricò, A.S., P. Bruce, B. Scrosati, J.-M. Tarascon and W. van Schalkwijk. 2005. Nanostructured materials for advanced energy conversion and storage devices. Nat. Mater., 4: 366–377.

[2] Gleiterr, H. 2000. Nanostructured materials: Basic concepts and microstructure. Acta Mater., 48: 1–29.

[3] Siegel, R.W. 1994. Nanostructured materials—mind over matter. Nanostruct. Mater., 4: 121–138.

[4] Siegel, R.W. 1993. Nanophase materials assembled from atom clusters. Mater. Sci. Eng.: B, 19: 37–43.

[5] Gleiter, H. 1995. Nanostructured materials: State of the art and perspectives. Nanostruct. Mater., 6: 3–14.

[6] Niederberger, M. 2007. Nonaqueous sol–gel routes to metal oxide nanoparticles. Acc. Chem. Res., 40: 793–800.

[7] Xu, P., X. Han, C. Wang, B. Zhang, X. Wang and H.-L. Wang. 2008. Facile synthesis of polyaniline-polypyrrole nanofibers for application in chemical deposition of metal nanoparticles. Macromol. Rapid Commun., 29: 1392–1397.

[8] Chiu, H.-C. and C.-S. Yeh. 2007. Hydrothermal synthesis of SnO_2 nanoparticles and their gas-sensing of alcohol. J. Phys. Chem. C, 111: 7256–7259.

[9] Strobel, R. and S.E. Pratsinis. 2009. Direct synthesis of maghemite, magnetite and wustite nanoparticles by flame spray pyrolysis. Adv. Powder Technol., 20: 190–194.

[10] Sang-Woo Kim, Shizuo Fujita and Shigeo Fujita. 2005. ZnO nanowires with high aspect ratios grown by metalorganic chemical vapor deposition using gold nanoparticles. Appl. Phys. Lett., 86: 153119.

[11] Shinde, P.S., G.H. Go and W.J. Lee. 2012. Facile growth of hierarchical hematite (a-Fe_2O_3) nanopetals on FTO by pulse reverse electrodeposition for photoelectrochemical water splitting. J. Mater. Chem., 22: 10469–10471.

[12] Li, G.-R., H. Xu, X.-F. Lu, J.-X. Feng, Y.-X. Tong and C.-Y. Su. 2013. Electrochemical synthesis of nanostructured materials for electrochemical energy conversion and storage. Nanoscale, 5: 4056–4069.

[13] Mohamad, S.A. and U.M.J. Fizik. 2007. Characterization of Electrodeposited Zn (Se, Te) Thin Films/polymer (PEO–Chitosan Blend) Junction for Solar Cells Applications. Jabatan Fizik, Fakulti Sains, Universiti Malaya.

[14] Choi, K.S. 2008. Shape control of inorganic materials via electrodeposition. Dalton Trans., 5432–5438.

[15] Gurrappa, I. and L. Binder. 2008. Electrodeposition of nanostructured coatings and their characterization—a review. Sci. Technol. Adv. Mater., 9: 043001.

[16] Ichinose, K., T. Mizuno, M.S. White and T. Yoshida. 2014. Control of nanostructure and crystallographic orientation in electrodeposited ZnO thin films via structure directing agents. J. Electrochem. Soc., 161: D195–D201.

[17] Wang, J., A. Gupta and S. Pan. 2017. A facile template-free electrodeposition method for vertically standing nanorods on conductive substrates and their applications for photoelectrochemical catalysis. Int. J. Hydrogen Energy, 42: 8462–8474.

[18] Wei, W.F., X.H. Mao, L.A. Ortiz and D.R. Sadoway. 2011. Oriented silver oxide nanostructures synthesized through a template-free electrochemical route. J. Mater. Chem., 21: 432–438.

[19] Xia, X., J. Tu, Y. Zhang, J. Chen, X. Wang, C. Gu, C. Guan, J. Luo and H.J. Fan. 2012a. Porous hydroxide nanosheets on preformed nanowires by electrodeposition: branched nanoarrays for electrochemical energy storage. Chem. Mater., 24: 3793–3799.

[20] Phuan, Y.W., W.J. Ong, M.N. Chong and J.D. Ocon. 2017. Prospects of electrochemically synthesized hematite photoanodes for photoelectrochemical water splitting: A review. J. Photochem. Photobiol. C, 33: 54–82.

[21] Martinez, L., D. Leinen, F. Martín, M. Gabas, J.R. Ramos-Barrado, E. Quagliata and E.A. Dalchiele. 2007. Electrochemical growth of diverse iron oxide (Fe_3O_4, α-FeOOH , and γ-FeOOH) thin films by electrodeposition potential tuning. J. Electrochem. Soc., 154: D126–D133.

[22] Spray, R.L. and K.S. Choi. 2009. Photoactivity of transparent nanocrystalline Fe_2O_3 electrodes prepared via anodic electrodeposition. Chem. Mater., 21: 3701–3709.

[23] Rangaraju, R., A. Panday, K. Raja and M. Misra. 2009. Nanostructured anodic iron oxide film as photoanode for water oxidation. J. Phys. D: Appl. Phys., 42: 135303–135312.

[24] LaTempa, T.J., X.J. Feng, M. Paulose and C.A. Grimes. 2009. Temperature-dependent growth of self-assembled hematite (alpha-Fe_2O_3) nanotube arrays: Rapid electrochemical synthesis and photoelectrochemical properties. J. Phys. Chem. C, 113: 16293–16298.

[25] Lucas-Granados, B., R. Sanchez-Tovar, R.M. Fernandez-Domene and J. Garcia-Anton. 2016. Study of the annealing conditions and photoelectrochemical characterization of a new iron oxide bi-layered nanostructure for water splitting. Sol. Energy Mater. Sol. Cells, 153: 68–77.

[26] Schrebler, R., C. Llewelyn, F. Vera, P. Cury, E. Munoz, R. del Rio, H.G. Meier, R. Cordova and E.A. Dalchiele. 2007. An electrochemical deposition route for obtaining alpha-Fe_2O_3 thin films-II. EQCM study and semiconductor properties. Electrochem. Solid State Lett., 10: D95–D99.

[27] Schrebler, R., K. Bello, F. Vera, P. Cury, E. Munoz, R. del Rio, H.G. Meier, R. Cordova and E.A. Dalchiele. 2006. An electrochemical deposition route for obtaining alpha-Fe_2O_3 thin films. Electrochem. Solid State Lett., 9: C110–C113.

[28] Schrebler, R.S., H. Altamirano, P. Grez, F.V. Herrera, E.C. Muñoz, L.A. Ballesteros, R.A. Córdova, H. Gómez and E.A. Dalchiele. 2010. The influence of different electrodeposition E/t programs on the photoelectrochemical properties of α-Fe_2O_3 thin films. Thin Solid Films, 518: 6844–6852.

[29] Shinde, P.S., H.H. Lee, S.Y. Lee, Y.M. Lee and J.S. Jang. 2015b. PRED treatment mediated stable and efficient water oxidation performance of the Fe_2O_3 nano-coral structure. Nanoscale, 7: 14906-14913.

[30] Riveros, G., D. Ramirez, E.A. Dalchiele, R. Marotti, P. Grez, F. Martin and J.R. Ramos-Barrado. 2014. Electrodeposition and characterization of hematite films obtained from DMSO solution. J. Electrochem. Soc., 161: D353–D361.

[31] Shinde, P.S., A. Annamalai, J.Y. Kim, S.H. Choi, J.S. Lee and J.S. Jang. 2015a. Fine-tuning pulse reverse electrodeposition for enhanced photoelectrochemical water oxidation performance of a-Fe_2O_3 photoanodes. J. Phys. Chem. C, 119: 5281–5292.

[32] Mohanty, U. 2011. Electrodeposition: A versatile and inexpensive tool for the synthesis of nanoparticles, nanorods, nanowires, and nanoclusters of metals. J. Appl. Electrochem., 41: 257–270.

[33] Kashchiev, D. 2000. Nucleation: Basic Theory with Applications (Butterworths-Heinemann). Oxford.

[34] Budevski, E.B., G.T. Staikov and W.J. Lorenz. 2008. Electrochemical Phase Formation and Growth: An Introduction to the Initial Stages of Metal Deposition. John Wiley & Sons.

[35] Yevtushenko, O., H. Natter and R. Hempelmann. 2007. Influence of bath composition and deposition parameters on nanostructure and thermal stability of gold. J. Solid State Electrochem., 11: 138–143.

[36] El-Deab, M.S. 2009. On the preferential crystallographic orientation of Au nanoparticles: Effect of electrodeposition time. Electrochim. Acta, 54: 3720–3725.

[37] Huang, C.-J., P.-H. Chiu, Y.-H. Wang and C.-F. Yang. 2006. Synthesis of the gold nanodumbbells by electrochemical method. J. Colloids Interface Sci., 303: 430–436.

[38] Dai, X. and R.G. Compton. 2006. Direct electrodeposition of gold nanoparticles onto indium tin oxide film coated glass: Application to the detection of arsenic(III). Anal. Sci., 22: 567–570.

[39] Ma, Y., J. Di, X. Yan, M. Zhao, Z. Lu and Y. Tu. 2009. Direct electrodeposition of gold nanoparticles on indium tin oxide surface and its application. Biosens. Bioelectron., 24: 1480–1483.

[40] Ma, Y., A.L. Highsmith, C.M. Hill and S. Pan. 2018. Dark-field scattering spectroelectrochemistry analysis of hydrazine oxidation at Au nanoparticle-modified transparent electrodes. J. Phys. Chem. C, 122: 18603–18614.

[41] Yang, B., S. Wang, S. Tian and L. Liu. 2009. Determination of hydrogen sulfide in gasoline by Au nanoclusters modified glassy carbon electrode. Electrochem. Commun., 11: 1230–1233.

[42] Wusimanjiang, Y., Y. Ma, M. Lee and S. Pan. 2018. Single gold nanoparticle electrode for electrogenerated chemiluminescence and dark field scattering spectroelectrochemistry. Electrochim. Acta, 269: 291–298.
[43] Sheridan, E., J. Hjelm and R.J. Forster. 2007. Electrodeposition of gold nanoparticles on fluorine-doped tin oxide: Control of particle density and size distribution. J. Electroanal. Chem., 608: 1–7.
[44] Liu, H. and R.M. Penner. 2000. Size-selective electrodeposition of mesoscale metal particles in the uncoupled limit. J. Phys. Chem. B, 104: 9131–9139.
[45] Penner, R.M. 2002. Mesoscopic metal particles and wires by electrodeposition. J. Phys. Chem. B, 106: 3339–3353.
[46] Tsai, M.-C., T.-K. Yeh and C.-H. Tsai. 2006. An improved electrodeposition technique for preparing platinum and platinum–ruthenium nanoparticles on carbon nanotubes directly grown on carbon cloth for methanol oxidation. Electrochem. Commun., 8: 1445–1452.
[47] Ourari, A., R. Zerdoumi, R. Ruiz-Rosas and E. Morallon. 2019. Synthesis and catalytic properties of modified electrodes by pulsed electrodeposition of Pt/PANI nanocomposite. Materials, 12: 723.
[48] Yu, P., J. Yan, J. Zhang and L. Mao. 2007. Cost-effective electrodeposition of platinum nanoparticles with ionic liquid droplet confined onto electrode surface as micro-media. Electrochem. Commun., 9: 1139–1144.
[49] Lu, G. and G. Zangari. 2006. Electrodeposition of platinum nanoparticles on highly oriented pyrolitic graphite: Part II: Morphological characterization by atomic force microscopy. Electrochim. Acta, 51: 2531–2538.
[50] Saminathan, K., V. Kamavaram, V. Veedu and A.M. Kannan. 2009. Preparation and evaluation of electrodeposited platinum nanoparticles on *in situ* carbon nanotubes grown carbon paper for proton exchange membrane fuel cells. Int. J. Hydrogen Energ., 34: 3838–3844.
[51] Sultana, S., N. Tachikawa, K. Yoshii, K. Toshima, L. Magagnin and Y. Katayama. 2017. Electrochemical preparation of platinum nanoparticles from bis(acetylacetonato)platinum(II) in some aprotic amide-type ionic liquids. Electrochim. Acta, 249: 263–270.
[52] Ustarroz, J., B. Geboes, H. Vanrompay, K. Sentosun, S. Bals, T. Breugelmans and A. Hubin. 2017. Electrodeposition of highly porous Pt nanoparticles studied by quantitative 3D electron tomography: Influence of growth mechanisms and potential cycling on the active surface area. ACS Appl. Mater. Interfaces, 9: 16168–16177.
[53] Zhu, W., G. Wang, X. Hong, X. Shen, D. Li and X. Xie. 2009. Metal nanoparticle chains embedded in TiO$_2$ nanotubes prepared by one-step electrodeposition. Electrochim. Acta, 55: 480–484.
[54] Tao, S., F. Yang, J. Schuch, W. Jaegermann and B. Kaiser. 2018. Electrodeposition of nickel nanoparticles for the alkaline hydrogen evolution reaction: Correlating electrocatalytic behavior and chemical composition. ChemSusChem., 11: 948–958.
[55] Wu, B. and J.J. Boland. 2006. Synthesis and dispersion of isolated high aspect ratio gold nanowires. J. Colloids Interface Sci., 303: 611–616.
[56] Wang, H.-J., C.-W. Zou, B. Yang, H.-B. Lu, C.-X. Tian, H.-J. Yang, M. Li, C.-S. Liu, D.-J. Fu and J.-R. Liu. 2009. Electrodeposition of tubular-rod structure gold nanowires using nanoporous anodic alumina oxide as template. Electrochem. Commun., 11: 2019–2022.
[57] Lu, Y., M. Yang, F. Qu, G. Shen and R. Yu. 2007. Enzyme-functionalized gold nanowires for the fabrication of biosensors. Bioelectrochem., 71: 211–216.
[58] Liu, Y.-C. and T.C. Chuang. 2003. Synthesis and Characterization of gold/polypyrrole core–shell nanocomposites and elemental gold nanoparticles based on the gold-containing nanocomplexes prepared by electrochemical methods in aqueous solutions. J. Phys. Chem. B, 107: 12383–12386.
[59] Rapecki, T., M. Donten and Z. Stojek. 2010. Electrodeposition of polypyrrole–Au nanoparticles composite from one solution containing gold salt and monomer. Electrochem. Commun., 12: 624–627.
[60] Dutta, R., B. Albee, W.E. van der Veer, T. Harville, K.C. Donovan, D. Papamoschou and R.M. Penner. 2014. Gold nanowire thermophones. J. Phys. Chem. C, 118: 29101–29107.
[61] Penner, R.M. 2014. Electrodeposited nanophotonics. J. Phys. Chem. C, 118: 17179–17192.
[62] Xiang, C., S.-C. Kung, D.K. Taggart, F. Yang, M.A. Thompson, A.G. Güell, Y. Yang and R.M. Penner. 2008. Lithographically patterned nanowire electrodeposition: A method for patterning electrically continuous metal nanowires on dielectrics. ACS Nano, 2: 1939–1949.

[63] Halpern, A.R., K.C. Donavan, R.M. Penner and R.M. Corn. 2012. Wafer-scale fabrication of nanofluidic arrays and networks using nanoimprint lithography and lithographically patterned nanowire electrodeposition gold nanowire masters. Anal. Chem., 84: 5053–5058.

[64] Chang, S.-S., C.-W. Shih, C.-D. Chen, W.-C. Lai and C.R.C. Wang. 1999. The shape transition of gold nanorods. Langmuir, 15: 701–709.

[65] Foss, C.A., G.L. Hornyak, J.A. Stockert and C.R. Martin. 1992. Optical properties of composite membranes containing arrays of nanoscopic gold cylinders. J. Phys. Chem., 96: 7497–7499.

[66] Martin, C.R. 1994. Nanomaterials: A membrane-based synthetic approach. Science, 266: 1961–1966.

[67] Martin, C.R. 1996. Membrane-based synthesis of nanomaterials. Chem. Mater., 8: 1739–1746.

[68] Lin, C.-C., T.-J. Juo, Y.-J. Chen, C.-H. Chiou, H.-W. Wang and Y.-L. Liu. 2008. Enhanced cyclic voltammetry using 1-D gold nanorods synthesized via AAO template electrochemical deposition. Desalination, 233: 113–119.

[69] Huang, C.-J., P.-H. Chiu, Y.-H. Wang, C.-F. Yang and S.-W. Feng. 2007. Electrochemical formation of crooked gold nanorods and gold networked structures by the additive organic solvent. J. Colloids Interface Sci., 306: 56–65.

[70] Wu, Q.Y., P. Diao, J. Sun, T. Jin, D. Xu and M. Xiang. 2015. Electrodeposition of vertically aligned silver nanoplate arrays on indium tin oxide substrates. J. Phys. Chem. C, 119: 20709–20720.

[71] Peulon, S. and D. Lincot. 1996. Cathodic electrodeposition from aqueous solution of dense or open-structured zinc oxide films. Adv. Mater., 8: 166–170.

[72] Elias, J., R. Tena-Zaera and C. Lévy-Clément. 2007. Electrodeposition of ZnO nanowires with controlled dimensions for photovoltaic applications: Role of buffer layer. Thin Solid Films, 515: 8553–8557.

[73] Pauporté, T., G. Bataille, L. Joulaud and F.J. Vermersch. 2010. Well-aligned ZnO nanowire arrays prepared by seed-layer-free electrodeposition and their cassie–wenzel transition after hydrophobization. J. Phys. Chem. C, 114: 194–202.

[74] Zong, X.L., R. Zhu and X.L. Guo. 2015. Nanostructured gold microelectrodes for SERS and EIS measurements by incorporating ZnO nanorod growth with electroplating. Sci. Rep., 5: 16454 (16451–16410).

[75] Izaki, M. and T. Omi. 1996. Transparent zinc oxide films prepared by electrochemical reaction. Appl. Phys. Lett., 68: 2439–2440.

[76] Haynes, K.M., C.M. Perry, M. Rivas, T.D. Golden, A. Bazan, M. Quintana, V.N. Nesterov, S.A. Berhe, J. Rodríguez, W. Estrada and W.J. Youngblood. 2015. Templated electrodeposition and photocatalytic activity of cuprous oxide nanorod arrays. ACS Appl. Mater. Interfaces, 7: 830–837.

[77] Xu, Q., X. Qian, Y.Q. Qu, T. Hang, P. Zhang, M. Li and L. Gao. 2016. Electrodeposition of Cu_2O nanostructure on 3D Cu micro-cone arrays as photocathode for photoelectrochemical water reduction. J. Electrochem. Soc., 163: H976–H981.

[78] Zhong, J.-H., G.-R. Li, Z.-L. Wang, Y.-N. Ou and Y.-X. Tong. 2011. Facile electrochemical synthesis of hexagonal Cu_2O nanotube arrays and their application. Inorg. Chem., 50: 757–763.

[79] Xu, L., Y. Guo, Q. Liao, J. Zhang and D. Xu. 2005. Morphological control of ZnO nanostructures by electrodeposition. J. Phys. Chem. B, 109: 13519–13522.

[80] Xu, L., Q. Chen and D. Xu. 2007. Hierarchical ZnO nanostructures obtained by electrodeposition. J. Phys. Chem. C, 111: 11560–11565.

[81] Luo, J., L. Steier, M.-K. Son, M. Schreier, M.T. Mayer and M. Grätzel. 2016. Cu_2O nanowire photocathodes for efficient and durable solar water splitting. Nano Lett., 16: 1848–1857.

[82] Erenturk, B., S. Gurbuz, R.E. Corbett, S.-A.M. Claiborne, J. Krizan, D. Venkataraman and K.R. Carter. 2011. Formation of crystalline cadmium selenide nanowires. Chem. Mater., 23: 3371–3376.

[83] Qiao, S., Q. Xu, R.K. Dutta, M. Le Thai, X. Li and R.M. Penner. 2016. Electrodeposited, transverse nanowire electroluminescent junctions. ACS Nano, 10: 8233–8242.

[84] Li, Q. and R.M. Penner. 2005. Photoconductive cadmium sulfide hemicylindrical shell nanowire ensembles. Nano Lett., 5: 1720–1725.

[85] Hernández-Pagán, E.A., W. Wang and T.E. Mallouk. 2011. Template electrodeposition of single-phase p- and n-type copper indium diselenide ($CuInSe_2$) nanowire arrays. ACS Nano, 5: 3237–3241.

[86] Phok, S., S. Rajaputra and V.P. Singh. 2007. Copper indium diselenide nanowire arrays by electrodeposition in porous alumina templates. Nanotechnol. 18: 475601–475608.

[87] Aldakov, D., C. Chappaz-Gillot, R. Salazar, V. Delaye, K.A. Welsby, V. Ivanova and P.R. Dunstan. 2014. Properties of electrodeposited CuSCN 2D layers and nanowires influenced by their mixed domain structure. J. Phys. Chem. C, 118: 16095–16103.

[88] Chappaz-Gillot, C., R. Salazar, S. Berson and V. Ivanova. 2012. Room temperature template-free electrodeposition of CuSCN nanowires. Electrochem. Commun., 24: 1–4.

[89] Chappaz-Gillot, C., R. Salazar, S. Berson and V. Ivanova. 2013. Insights into CuSCN nanowire electrodeposition on flexible substrates. Electrochim. Acta, 110: 375–381.

[90] Ramírez, D., K. Álvarez, G. Riveros, B. González and E.A. Dalchiele. 2017. Electrodeposition of CuSCN seed layers and nanowires: A microelectrogravimetric approach. Electrochim. Acta, 228: 308–318.

[91] Sanchez, S., C. Chappaz-Gillot, R. Salazar, H. Muguerra, E. Arbaoui, S. Berson, C. Lévy-Clément and V. Ivanova. 2013. Comparative study of ZnO and CuSCN semiconducting nanowire electrodeposition on different substrates. J. Solid State Electrochem., 17: 391–398.

[92] Soleimany, L., A. Dolati and M. Ghorbani. 2010. A study on the kinetics of gold nanowire electrodeposition in polycarbonate templates. J. Electroanal. Chem., 645: 28–34.

[93] Bahari Mollamahale, Y., M. Ghorbani, A. Dolati and D. Hosseini. 2018. Electrodeposition of well-defined gold nanowires with uniform ends for developing 3D nanoelectrode ensembles with enhanced sensitivity. Mater. Chem. Phys., 213: 67–75.

[94] Zhang, X., D. Li, L. Bourgeois, H. Wang and P.A. Webley. 2009. Direct electrodeposition of porous gold nanowire arrays for biosensing applications. ChemPhysChem., 10: 436–441.

[95] Sauer, G., G. Brehm, S. Schneider, K. Nielsch, R.B. Wehrspohn, J. Choi, H. Hofmeister and U. Gösele. 2002. Highly ordered monocrystalline silver nanowire arrays. J. Appl. Phys., 91: 3243–3247.

[96] Rauber, M., J. Brötz, J. Duan, J. Liu, S. Müller, R. Neumann, O. Picht, M.E. Toimil-Molares and W. Ensinger. 2010. Segmented all-platinum nanowires with controlled morphology through manipulation of the local electrolyte distribution in fluidic nanochannels during electrodeposition. J. Phys. Chem. C, 114: 22502–22507.

[97] Taggart, D.K., Y. Yang, S.-C. Kung, T.M. McIntire and R.M. Penner. 2011. Enhanced thermoelectric metrics in ultra-long electrodeposited PEDOT nanowires. Nano Lett., 11: 125–131.

[98] Shinde, P.S., M.A. Mahadik, S.Y. Lee, J. Ryu, S.H. Choi and J.S. Jang. 2017. Surfactant and TiO$_2$ underlayer derived porous hematite nanoball array photoanode for enhanced photoelectrochemical water oxidation. Chem. Eng. J., 320: 81–92.

[99] Xia, X., J. Tu, Y. Zhang, J. Chen, X. Wang, C. Gu, C. Guan, J. Luo and H.J. Fan. 2012b. Porous hydroxide nanosheets on preformed nanowires by electrodeposition: Branched nanoarrays for electrochemical energy storage. Chem. Mater., 24: 3793–3799.

[100] Gaskell, D.R. and D.E. Laughlin. 2017. Introduction to the Thermodynamics of Materials. CRC press.

CHAPTER 5

Nanostructured Materials Using Microemulsions

Sonalika Vaidya

1. Introduction

Ever since the term 'microemulsions' has been first coined [1] by Schulman, the use of microemulsions for carrying out the synthesis of nanostructures has been emerging at a potential rate. These microemulsions are homogeneous mixtures of water and oil stabilized by surfactants. Surfactants help in the stabilization of microemulsions by lowering the interfacial tension between the oil and water. Depending on the proportion of the components used and the hydrophilic-lipophilic balance of the surfactant, these microemulsions can be either water-in-oil type or oil-in-water type.

Synthesis in microemulsions is mostly carried out in a water-in-oil type of microemulsion. These contain reverse micelles, which are a type of surfactant aggregate, wherein the core of the reverse micelles is composed of an aqueous phase (Figure 5.1). Typically, a water-in-oil type of microemulsion is formed by a surfactant, co-surfactant, an oil phase, and an aqueous phase.

There have been reviews on this topic by eminent scientists, viz. Pileni [2], Lopez-Quintela [3], Capek [4], Eastoe [5], Holmberg [6], Uskoković [7], and Ganguli

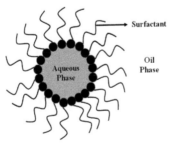

Figure 5.1. Schematic showing the structure of reverse micelle.

Institute of Nano Science and Technology, Habitat Centre, Phase-X, Sector-64, Mohali, Punjab-160062, India.

[8, 9]. This chapter focuses on highlighting the developments that have occurred in this field since 2015.

2. Synthesis of Nanostructured Materials

Microemulsion acts as a nanoreactor. The reaction takes place in a confined environment in reverse micelles. The main advantage of carrying out the reaction in microemulsions is that the size of the particles can be easily controlled as their growth is inhibited after nucleation. This ensures that the particles are formed with monodisperse size distribution. Moreover, one can easily control the morphology by varying the parameters, such as the nature of the solvent, nature of surfactant, nature of co-surfactant, water to surfactant ratio (Wo), that are involved in the formation of microemulsions. Various kinds of surfactants are being used, such as cationic, anionic, non-ionic, and ionic liquid to form microemulsions. These have been used for the synthesis of a variety of nanostructured materials. Looking at the literature, this method has been used for the synthesis of a variety of materials such as metal, metal alloy, oxides, chalcogenides, core-shell structures, etc.

For the synthesis of material with formula AB, two different microemulsions are prepared—one containing an aqueous solution of metal cation A, and the other containing the anion B. The two microemulsions are then mixed to form AB. The product is formed by an exchange of reactants present in different reverse micelles by the process of coalescence and de-coalescence of the reverse micelles. The resultant product can be separated through centrifugation, or magnetic separation (in case of magnetic materials). Figure 5.2 shows the mechanism of the formation of the product in microemulsions. The number of microemulsion systems and the ratio of the volume of the microemulsion involved can be varied depending on the composition and the stoichiometry of the product.

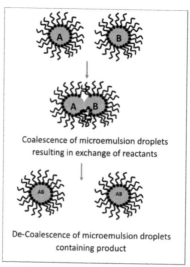

Coalescence of microemulsion droplets
resulting in exchange of reactants

De-Coalescence of microemulsion droplets
containing product

Figure 5.2. Schematic showing the mechanism of formation of nanostructured material AB in reverse micelles.

We cite some examples to showcase the versatility of the microemulsion-based method.

Wu et al. [10] synthesized $Mg(OH)_2$ in microemulsion containing CTAB as the surfactant, iso-propanol as the co-surfactant, cyclohexane as the oil and aqueous solution. The authors have shown through surface performance analysis that CTAB adsorbs on the surface of $Mg(OH)_2$, making it easier to be miscible with the organic phase. Najafi et al. [11] prepared magnetic FeCo alloy with varying Fe/Co ratio using CTAB/1-butanol/iso-octane. They found that a ratio of Fe:Co as 70:30 led to the formation of spherical particles of FeCo, with an average particle size of less than 10 nm. In another example, Nacy et al. [12] showcased that rod-shaped $La_2NiO_{4+\delta}$ (a layered nickelate oxide), synthesized using microemulsions based on CTAB/1-butanol/hexane, exhibited promising electrochemical performance as cathodes for $Li-O_2$ batteries.

$CoFe_2O_4$ nanoparticles with size ranging from 10–15 nm were synthesized by Laokul et al. [13] using AOT/n-hexane. AOT is an anionic surfactant. The authors showcased the ferromagnetic behavior of the ferrites.

Various nonionic surfactants, such as TX-100, Tergitol, Igepal, Brij, etc. have been used to synthesize a variety of materials. For instance, Pal et al. [14] used TX-100/1-hexanol/cyclohexane-based microemulsion to synthesize bimetallic alloy of Au-Pt with a size of 2.5 nm. NiO@ZnO core-shell nanoparticles were synthesized by Han et al. [15] using TX-100/1-hexanol/cyclohexane as the water-in-oil microemulsion. In another example, white light-emitting nanorods of $Sr_2SiO_4:Eu^{2+}$ were synthesized using the reverse micellar route using a Tergitol/1-octanol/cyclohexane system [16]. The authors have studied EPR, photoluminescence properties, and thermo-luminescence of the rods. Emsaki et al. [17] have synthesized $ZnO-ZnWO_4$ with different ratios using the microemulsion method containing Brij 35 as the surfactant, 1-butanol as the co-surfactant, and cyclohexane as the oil phase. The particles possessed irregular morphology with a particle size of 40–70 nm. The authors used these composites for photodegradation of methylene blue. In a study by Liu et al. [18], pH activable PEG-modified Fe-PDA nanoparticles for MRI guided photothermal therapy were synthesized using Igepal CO-520 and cyclohexane as the microemulsion system. The sensitive and efficient photothermal therapy of tumors has been demonstrated for both the cellular level *in vitro* and for the tumor-bearing small animals *in vivo*. Water-in-oil microemulsion based on benzyl alcohol and water was also demonstrated to synthesize materials. Xue et al. [19] used this system to prepare 2D silica-based nanomaterials. Igepal CO-520/aqueous solution/cyclohexane-based microemulsion was used by Biausque et al. [20] to form a bimetallic alloy with a composition of Ni_9Pt. These alloy nanostructures possessed cuboctahedra shape with a size of 2 nm. The authors have shown that these small-sized particles can be used for the dry reforming of methane. Richard et al. [21] used Igepal-based microemulsion to synthesize nanocrystals of $(H_3O)Y_3F_{10}\cdot xH_2O$. The authors showed that by varying various parameters, such as microemulsion composition, micelle hydrodynamic radius, nanoparticles with varied sizes can be formed. The authors in their study established a relation between the parameters and the size of nanoparticles. Xu et al. [22] have used this methodology to synthesize layered gadolinium hydroxide using oleylamine, which performed the function of surfactant, oil, and base. 1-butanol was used as a co-surfactant. The size of the hydroxide varied from 200–300 nm, depending on the amount of 1-butanol used. In another interesting study by Menaka et al. [23] anisotropic nanostructures of $NiMnO_3$ were obtained by thermal decomposition of the oxalate precursors of nickel and manganese having composition $Ni_{0.5}Mn_{0.5}(C_2O_4)\cdot 2H_2O$, which crystallized as nanorods. The authors used microemulsion method and used two different systems: Tergitol/co-surfactant/cyclohexane and CTAB/co-surfactant/cyclohexane. The authors used three different co-surfactants, viz., 1-butanol, 1-hexanol, and 1-octanol to see the effect of chain length of the co-surfactant on the product. In the case of cationic surfactant, rod-shaped structures were formed, whereas cubes were formed with Tergitol as a surfactant that eventually led to rod-shaped on increasing the chain length of the co-surfactant. The anisotropy of the precursor was maintained for the mixed metal oxide. Tojo et al. [24], in their review, highlighted the segregation of bimetallic nanocatalyst using microemulsions. The authors concluded that to obtain a core-shell structure, the minimum difference between the reduction potential of two metals should be 0.2 V. Ganguli et al. have highlighted the use of microemulsion method in synthesizing nanostructures as electrocatalysts for water splitting [9].

Synperonic series of surfactants is another kind of non-ionic surfactant that has been used to synthesize a variety of nanostructured materials. This surfactant has been used to form an oil-in-water kind of microemulsion. Sanchez et al. [25] used a microemulsion system based on water/Synperonic 91/5/iso-octane to synthesize Zn-doped TiO_2 for the degradation of phenol. The authors observed that the degradation was maximum at an optimal amount of Zn doping at 5%. Pineda et al. [26] formed oil-in-water based microemulsion using water/Synperonic 91/5/iso-octane. They used this system in combination with hydrothermal treatment to form the TiO_2-B (Bronze) phase of TiO_2. This is a polymorphic phase of TiO_2 with space group C2/m and possesses low density. Oil-in-water microemulsions formed using Synperonic 91/5/iso-octane/water were used by Pemartin-Biernath et al. [27] to synthesize cerium oxide and mixed Cu-Ce oxide nanoparticles.

There are some interesting uses of microemulsions. For instance, Petcu et al. [28] have showcased that microemulsions can be used for photodegradation of dyes without the use of solid catalysts. Here, ethyl acetate was used as an UV initiator and the photocatalytic effect of the water/oil interface was the basis of the entire process. The author showcased that a microemulsion system formed using a water/non-ionic surfactant/ethylacetate chain radical process resulted in complete photodegradation of dyes in ~ 25 minutes. Varol et al. [29] in their study demonstrated the use of the oil-water interface of the droplets formed in a miniemulsion to synthesize hollow nanostructures of metal oxides. The authors used this methodology to synthesize hollow nanostructures of ceria and iron oxide. Shang et al. [30] synthesized reverse bumpy ball structure (RBBS) nanoreactor containing Pt-γ Fe_2O_3 dimers. These were shown to act as an efficient catalyst for the reduction of 4-nitrophenol with hydrogen. RBBS nanoreactors are porous hollow nanostructures. These contain a large number of accessible nanoparticles in their interior. They formed RBBS nanoreactors based on m-SiO$_2$ using a modified emulsion-based bottom-up self-assembly process with high boiling point organics as a soft template. The water-in-oil microemulsion was used to encapsulate particles that had potential to be used as nanocarriers. Zhang et al. [31] demonstrated that when gold nanorods conjugated with porous silicon nanoparticles were encapsulated by nano hydrogel of calcium alginate, using water-in-oil microemulsions, they acted as an excellent nanocarriers for co-delivering therapeutics for biomedical applications.

Another class of microemulsion that has been used widely is based on ionic liquid. We cite here a few examples wherein ionic liquid-based microemulsions have been used. Yan et al. [32] in their review have highlighted the use of ionic liquid-based microemulsion for polymerization. They have cited examples in their review showcasing the use of ionic liquid-based microemulsion for the synthesis of polymers. In their review, they have stated that in comparison to conventional aqueous microemulsion, these ionic liquid-based microemulsions have an advantage for carrying out polymerization. After polymerization, the mixture which contains the catalyst, ligands, ionic liquid, and surfactant ionic liquids can be recovered and reused. They have also highlighted the fact that owing to high ionic conductivity and wide potential window, these ionic liquids act as a suitable media for electro-polymerization. These ionic liquids act as either immiscible liquids, immiscible oil, or surfactants. Wang et al. [33] have synthesized ZnInS$_4$ in ionic liquid microemulsion mediated hydrothermal method containing [BMIM][PF6] as ionic liquid and TX-100 as the surfactant.

3. New Techniques to Characterize Dynamics of Reverse Micelles

In this section, we discuss the use of Small Angle X-ray scattering (SAXS), Small-Angle Neutron Scattering (SANS), and Fluorescence Correlation Spectroscopy to understand the microemulsion system. These techniques were also used for understanding the formation of nanoparticle inside the core of reverse micelles.

Zhang et al. [34] investigated the use of nanoemulsions, prepared from a combination of emulsification with various pairs of co-surfactants, to form core-shell and multi core-shell structure using SANS, DLS, and Cryo-TEM. The authors carried out a detailed investigation to see the effect on molecular geometry and interfacial curvature driving the formation of various morphologies in the nanoemulsions.

SAXS studies have been used effectively to understand the microemulsion system and the formation of nanostructured material using this method. Ganguli et al. [35] studied the growth mechanism for the formation of nanorods of zinc oxalate formed in the microemulsion system. The system consisted of CTAB/1-butanol/iso-octane. The authors observed that the zinc oxalate nanostructures were formed within half an hour of the start of the reaction. From the studies, the authors concluded that the system consists of two different kinds of moieties, viz., spherical sized moiety of size 3–4 nm, which were assumed to be microemulsion droplets, and the second were larger aggregates assumed to be oxalate encapsulated by surfactant. Another study using SAXS by the authors [36] showed the variation in size and shape of the microemulsion droplets on introducing a hydrotrope (sodium salicylate) to the system consisting of CTAB as the surfactant, 1-butanol as the co-surfactant, and iso-octane as the oil phase. The shape of these droplets was observed to be ellipsoidal, and the size of these ellipsoids increased from 7 to 40 nm on increasing the concentration of sodium salicylate. Recently, our group has tried to understand the effect of co-surfactant on the growth of copper oxalate nanostructures using SAXS [37]. We used both the model-independent and model-dependent approach. The inferences obtained, by analyzing the data using the model-independent approach, were used to select the model used for fitting the SAXS data. The microemulsion system that was studied was CTAB/co-surfactant/iso-octane. The two co-surfactants studied were 1-butanol and 1-octanol. We observed the shape of the reverse micelles was ellipsoid with 1-butanol and spherical with 1-octanol. We concluded that apart from changes in the rigidity of the surfactant film, it is the shape of the reverse micelles that affect the shape of the final product. Sakuragi et al. [38] carried out SAXS measurement to study the structure of microemulsions formed from deep eutectic solvents (DES)/water/mixture of Tween 80 and Span 20. They observed a change in the morphology from spherical to cylindrical in compositions containing a large amounts of Tween 80. The authors also observed that the size of the aggregates increased with increasing the amount of DES. The scattering profiles were fitted using the core-corona model.

Sen et al. [39] used fluorescence correlation spectroscopy to study the size, size distribution, polydispersity, and diffusion of the microemulsion droplets for AOT/oil/water system and CTAB/1-butanol/oil/water system. The study was carried out by varying the oil phase. The authors observed that the size of microemulsion droplets for the system formed using AOT did not vary with the oil phase in contrast to the microemulsions formed using CTAB, wherein a substantial effect was observed on the size of the droplets and their distribution. Another study carried out by Ganguli et al. was effective in studying the growth kinetics for the formation and growth of iron [40] and nickel oxalates [41] using CTAB-based microemulsion. The formation of iron oxalate takes eight days, while nickel oxalates were formed in just fifteen hours. For iron oxalates, Ganguli et al. monitored the fluctuations in fluorescence of sulphorhodamine B dye, embedded in reverse micelles for eight consecutive days, and found that the growth was isotropic up to four days, with a critical size of 53 nm. Beyond this, the nanoparticle growth was anisotropic. For nickel oxalate, the authors studied the effects of two different parameters—one was the effect of the surfactant, and the other was the effect of water to surfactant ratio (Wo). Three different surfactants, viz., CTAB, CPB, and TX-100 were chosen to see the effect of the surfactant. The size of microemulsion droplet was largest for TX-100, which was found to be 11 nm, and it was 3–4 nm for CTAB. The size of the droplets increased linearly with Wo.

4. Conclusions

In this chapter, we have tried to highlight the use of the microemulsion method for the synthesis of a variety of nanostructured material. This aspect has been highlighted in this chapter by citing numerous examples which showcased the versatility of this methodology. We have also showcased the use of different techniques, such as FCS and SAXS to understand nanoparticle formation using microemulsion.

References

[1] Hoar, T.P. and J.H. Schulman. 1943. Transparent water-in-oil dispersions: the oleopathic hydro-micelle. Nature, 152: 102.

[2] Pileni, M.P. 2007. Control of the size and shape of inorganic nanocrystals at various scales from nano to macrodomains. J. Phys. Chem. C, 111: 9019.

[3] López-Quintela, M.A., C. Tojo, M.C. Blanco, L. García Rio and J.R. Leis. 2004. Microemulsion dynamics and reactions in microemulsions. Curr. Opin. Colloid Interface Sci., 9: 264.

[4] Capek, I. 2004. Preparation of metal nanoparticles in water-in-oil (w/o) microemulsions. Adv. Colloid Interface Sci., 110: 49.

[5] Eastoe, J., M.J. Hollamby and L. Hudson. 2006. Recent advances in nanoparticle synthesis with reversed micelles. Adv. Colloid Interface Sci., 128-130: 5.

[6] Holmberg, K. 2004. Surfactant-templated nanomaterials synthesis. J. Colloid Interface Sci., 274: 355.

[7] Uskoković, V. and M. Drofenik. 2007. Reverse micelles: Inert nano-reactors or physico-chemically active guides of the capped reactions. Adv. Colloid Interface Sci., 133: 23.

[8] Ganguli, A.K., A. Ganguly and S. Vaidya. 2010. Microemulsion-based synthesis of nanocrystalline materials. Chem. Soc. Rev., 39: 474.

[9] Das, A. and A.K. Ganguli. 2018. Design of diverse nanostructures by hydrothermal and microemulsion routes for electrochemical water splitting. RSC Adv., 8: 25065.

[10] Wu, H.H., X. Zhao, L.C. Wang, L.B. Ma and X.P. Huang. 2018. Preparation and surface modification of magnesium hydroxide in a CTAB/isopropanol/cyclohexane/water microemulsion. Micro Nano Lett., 13: 1642.

[11] Najafi, A. and K. Nematipour. 2017. Synthesis and magnetic properties evaluation of monosized FeCo alloy nanoparticles through microemulsion method. J. Supercond. Nov. Magn., 30: 2647.

[12] Nacy, A., X. Ma and E. Nikolla. 2015. Nanostructured nickelate oxides as efficient and stable cathode electrocatalysts for Li–O$_2$ batteries. Top Catal., 58: 513.

[13] Laokul, P., S. Arthan, S. Maensiri and E. Swatsitang. 2015. Magnetic and optical properties of CoFe$_2$O$_4$ nanoparticles synthesized by reverse micelle microemulsion method. J. Supercond. Nov. Magn., 28: 2483.

[14] Pal, A. 2015. Gold–platinum alloy nanoparticles through water-in-oil microemulsion. J. Nanostruct. Chem., 5: 65.

[15] Han, D.Y., C.Q. Wang, D.D. Li and Z.B. Cao. 2016. NiO/ZnO core-shell nanoparticles *in situ* synthesis via microemulsion method. Synth. React. Inorg. Metal-Org. Nano-Metal Chem., 46: 794.

[16] Gupta, S.K., M.K. Bhide, R.M. Kadam, V. Natarajan and S.V. Godbole. 2015. Nanorods of white light emitting Sr$_2$SiO$_4$:Eu^{2+}: microemulsion-based synthesis, EPR, photoluminescence, and thermoluminescence studies. J. Expet. Nanosci., 10: 610.

[17] Emsaki, M., S.A. Hassanzadeh-Tabrizi and A. Saffar-Teluri. 2018. Microemulsion synthesis of ZnO–ZnWO$_4$ nanoparticles for superior photodegradation of organic dyes in water. J. Mater. Sci.: Mater. Electron., 29: 2384.

[18] Liu, F., X. He, J. Zhang, H. Chen, H. Zhang and Z. Wang. 2015. Controllable synthesis of polydopamine nanoparticles in microemulsions with pH-activatable properties for cancer detection and treatment. J. Mater. Chem. B, 3: 6731.

[19] Xue, Y., Y.-S. Ye, F.-Y. Chen, H. Wang, C. Chen, Z.-G. Xue, X.-P. Zhou, X.-L. Xie and Y.-W. Mai. 2016. A simple and controllable graphene-templated approach to synthesise 2D silica-based nanomaterials using water-in-oil microemulsions. Chem. Commun., 52: 575.

[20] Biausque, G.M., P.V. Laveille, D.H. Anjum, B. Zhang, X. Zhang, V. Caps and J.-M. Basset. 2017. One-pot synthesis of size- and composition-controlled Ni-Rich NiPt alloy nanoparticles in a reverse microemulsion system and their application. ACS Appl. Mater. Interfaces, 9: 30643.

[21] Richard, B., J.-L. Lemyre and A.M. Ritcey. 2017. Nanoparticle size control in microemulsion synthesis. Langmuir, 33: 4748.

[22] Xu, Y., J. Suthar, R. Egbu, A.J. Weston, A.M. Fogg and G.R. Williams. 2018. Reverse microemulsion synthesis of layered gadolinium hydroxide nanoparticles. J. Solid State Chem., 258: 320.

[23] Jha, M., S. Kumar, N. Garg, K.V. Ramanujachary, S.E. Lofland and A.K. Ganguli. 2018. Microemulsion based approach for nanospheres assembly into anisotropic nanostructures of $NiMnO_3$ and their magnetic properties. J. Solid State Chem., 258: 722.

[24] Tojo, C., D. Buceta and M.A. López-Quintela. 2017. On metal segregation of bimetallic nanocatalysts prepared by a one-pot method in microemulsions. Catalysts, 7: 68.

[25] Sanchez-Dominguez, M., G. Morales-Mendoza, M.J. Rodriguez-Vargas, C.C. Ibarra-Malo, A.A. Rodriguez-Rodriguez, A.V. Vela-Gonzalez, S.A. Perez-Garcia and R. Gomez. 2015. Synthesis of Zn-doped TiO_2 nanoparticles by the novel oil-in-water (O/W) microemulsion method and their use for the photocatalytic degradation of phenol. J. Environ. Chem. Eng., 3: 3037.

[26] Pineda-Aguilar, N., L.L. Garza-Tovar, E.M. Sanchez-Cervantes and M. Sanchez-Dominguez. 2018. Preparation of TiO_2–(B) by microemulsion mediated hydrothermal method: effect of the precursor and its electrochemical performance. J. Mater. Sci.: Mater. Electronics, 29: 15464.

[27] Pemartin-Biernath, K., A.V. Vela-González, M.B. Moreno-Trejo, C. Leyva-Porras, I.E. Castañeda-Reyna, I. Juárez-Ramírez, C. Solans and M. Sánchez-Domínguez. 2016. Synthesis of mixed Cu/Ce oxide nanoparticles by the oil-in-water microemulsion reaction method. Materials, 9: 480.

[28] Petcu, A.R., A. Meghea, E.A. Rogozea, N.L. Olteanu, C.A. Lazar, D. Cadar, A.V. Crisciu and M. Mihaly. 2017. No catalyst dye photodegradation in a microemulsion template. ACS Sustainable Chem. Eng., 5: 5273.

[29] Varol, H.S., O. Álvarez-Bermúdez, P. Dolcet, B. Kuerbanjiang, S. Gross, K. Landfester and R. Muñoz-Espí. 2016. Crystallization at nanodroplet interfaces in emulsion systems: A soft-template strategy for preparing porous and hollow nanoparticles. Langmuir, 32: 13116.

[30] Shang, L., R. Shi, G.I.N. Waterhouse, L.-Z. Wu, C.-H. Tung, Y. Yin and T. Zhang. 2018. Nanocrystals@hollow mesoporous silica reverse-bumpy-ball structure nanoreactors by a versatile microemulsion-templated approach. Small Meth., 2: 1800105.

[31] Zhang, H., Y. Zhu, L. Qu, H. Wu, H. Kong, Z. Yang, D. Chen, E. Mäkilä, J. Salonen, H.A. Santos, M. Hai and D.A. Weitz. 2018. Gold nanorods conjugated porous silicon nanoparticles encapsulated in calcium alginate nano hydrogels using microemulsion template. Nano Lett., 18: 1448.

[32] Yuan, C., J. Guo, Z. Si and F. Yan. 2015. Polymerization in ionic liquid-based microemulsions. Polym. Chem., 6: 4059.

[33] Wang, A.-L., L. Chen, J.-X. Zhang, W.-C. Sun, P. Guol and C.-Y. Ren. 2017. Ionic liquid microemulsion-assisted synthesis and improved photocatalytic activity of $ZnIn_2S_4$. J. Mater. Sci., 52: 2413.

[34] Zhang, M., P.T. Corona, N. Ruocco, D. Alvarez, P. Malo de Molina, S. Mitragotri and M.E. Helgeson. 2018. controlling complex nanoemulsion morphology using asymmetric cosurfactants for the preparation of polymer nanocapsules. Langmuir, 34: 978.

[35] Sharma, S. and, A.K. Ganguli. 2014. Spherical-to-cylindrical transformation of reverse micelles and their templating effect on the growth of nanostructures. J. Phys. Chem. B, 118: 4122.

[36] Sethi, V., J. Mishra, A. Bhattacharyya, D. Sen and A.K. Ganguli. 2017. Hydrotrope induced structural modifications in CTAB/butanol/water/isooctane reverse micellar systems. Phys. Chem. Chem. Phys., 19: 22033.

[37] Sunanina, V., Sethi, S.K. Mehta, A.K. Ganguli and S. Vaidya. 2019. Understanding the role of co-surfactants in microemulsions on the growth of copper oxalate using SAXS. Phys. Chem. Chem. Phys., 21: 336.

[38] Sakuragi, M., S. Tsutsumi and K. Kusakabe. 2018. Deep eutectic solvent-induced structural transition of microemulsions explored with small-angle x-ray scattering. Langmuir, 34: 12635.

[39] Khan, M.F., M.K. Singh and S. Sen. 2016. Measuring size, size distribution, and polydispersity of water-in-oil microemulsion droplets using fluorescence correlation spectroscopy: Comparison to dynamic light scattering. J. Phys. Chem. B, 120: 1008.

[40] Sharma, S., N. Pal, P.K. Chowdhury, S. Sen and A.K. Ganguli. 2012. Understanding growth kinetics of nanorods in microemulsion: A combined fluorescence correlation spectroscopy, dynamic light scattering, and electron microscopy study. J. Am. Chem. Soc., 134: 19677.

[41] Sharma, S., N. Yadav, P.K. Chowdhury and A.K. Ganguli. 2015. Controlling the microstructure of reverse micelles and their templating effect on shaping nanostructures. J. Phys. Chem. B, 119: 11295.

CHAPTER 6

Methods of Manufacturing Composite Materials

Anton Yegorov,[1,2,*] *Marina Bogdanovskaya,*[1,2] *Vitaly Ivanov*[1,2]
and *Daria Aleksandrova*[1,2]

1. Introduction

The widespread use of composite materials, especially polymer composite materials (PCM), has made tremendous progress in the field of polymer chemistry and physics. Today, the most popular are PCM based on continuous fibers, rovings, and fabrics. Such fillers of polymeric materials as carbide fibers, carbon nanotubes (CNT), fullerenes, and other nanoobjects, are not far behind in terms of the growth rates of technologies.

The increased interest in obtaining nanocomposites is associated with the unique effect of nanoscale filler on the bulk properties of polymer composites [1, 2]. In addition, the introduction of nanoscale filler in the polymer matrix results in materials with high rigidity, toughness, and tribological properties [3].

2. Brief Description of Polymer Composite Materials

The complex of properties of PCM [4] is determined by the properties of the components (matrix, filler), such as their micro- and macrostructure, phase boundary, the response of these structures to external influences. PCM are heterophase materials in which a continuous matrix interacting with the filler, perceives external loads and redistributes them to the filler.

The interfacial layer is one of the most important parameters of the PCM, which characterizes the contact area of the matrix-filler. For example, in the volume of PCM in 1 mm³ with the degree of filling 50% vol., the contact area is 450–600 mm². You can get an unlimited amount of PCM and in a very wide range

[1] The Federal State Unitary Enterprise—Institute of Chemical Reagents and High Purity Chemical Substances of National Research Centre—Kurchatov Institute.
[2] National Research Centre—Kurchatov Institute.
* Corresponding author: egorov@irea.org.ru

to change their properties by combining components of different nature, shape, and size in one material.

The boundaries of PCM characteristics change are mainly determined by the upper and lower values of the properties characteristic of the main classes of materials (metals, ceramics, polymers), and the state of aggregation of substances. The main advantage of PCM is the production of materials with properties that significantly exceed the upper and lower bounds of the properties of the original components (Table 6.1).

The use of light elements (carbon, silicon, boron) is most promising for the production of materials with high mechanical properties, since the theoretical strength of a material depends on the radius of the atom forming a chemical bond. The heat resistance of PCM when filled with high-strength and stable objects (carbide fibers, CNT, fullerenes) will be determined mainly by the choice of the polymer matrix, as the less thermostable PCM component.

Table 6.1. The main mechanical properties of polymers and PCM based on them.

Characteristic	Polymer	PCM	The range of changes in the properties of the PCM
Density, kg/m³	760–1800	5–22000	104
Tensile strength, MPa	8–210	0,1–4000	104
Elastic modulus, GPa	0.1–10	0.01–1000	105
Relative extension, %	0.5–1000	0.1–1000	104
Resistivity	108–1020	105–1020	1025
Thermal conductivity, W/m*K	0.12–2.9	0.02–400	104
Coefficients of Linear Thermal Expansion, 1/°C	$(2–30) \times 10^{-5}$	$104–5 \times 10^{-5}$	10
Poisson's ratio	0.3–0.5	0.1–0.5	5

A key factor determining the final properties of the resulting hybrid materials is the dispersion of inorganic materials in the polymer matrix. Dispersion of inorganic filler is associated with a number of difficulties arising from the need to combine organic and inorganic material. These difficulties can be solved by adding a crosslinking agent through which both phases can be connected by a covalent bond [5, 6].

The mixing of the organic polymer matrix and the inorganic filler can be carried out by two main mixing mechanisms, such as simple and dispersing.

By simple mixing, we mean a process that results in a statistically random distribution of particles of the initial components in the volume of the mixture without changing their initial sizes.

Dispersing mixing is a process of mixing which is accompanied by a change (decrease) of the initial sizes of the particles of the components associated with their fragmentation, destruction of aggregates, deformation, and disintegration of the dispersed phase, etc. The main task of the dispersing mixing is to destroy aggregates of solid particles and distribute them in the bulk of the liquid polymer [7].

An elementary act of destruction of an aggregate is to overcome the adhesion forces between solid particles as a result of the action of an inhomogeneous field of mechanical stresses during deformation of the medium. Calculations show that the irreversible decomposition of aggregates and their dispersion in the liquid dispersion phase occurs with the ratio:

$6\pi R\tau/F \geq 4,$

where R is the radius of the particle, τ is the shear stress, F is the cohesive force of the particles in the aggregate.

It follows that the higher the viscosity of the medium, the more effective the dispersion. Reducing the temperature, increasing the frequency of rotation of the working bodies of the mixer, and reducing the gap where the flow of the dispersed system is, all contributes to dispersion.

A decrease in the adhesion forces of particles (F) in an aggregate also leads to its destruction and an improvement in dispersion. The introduction of surfactants, low molecular weight liquids, and other substances that are adsorbed on the surface and weaken the interaction of particles improves dispersion.

Mixing highly viscous polymers and liquids with solid fillers are the most common and effective processes for producing polymer composites of a dispersed-filled structure. These are mainly polymer melts into which various amounts of solid fillers are introduced. For their mixing, considerable efforts and high values of the minimum shear strain γ_{min} are required. It should be noted that the initial viscosity of the melts and the liquid components of the mixtures may change during the mixing process as a result of an increase in temperature and shear stress.

The mixing of such systems in the laminar mode is carried out with a roller, blade, screw, and rotary mixers of the closed type.

In modern plastics processing technology, continuous blending is the main method for producing thermoplastic-based composite materials. Laminar mixing of highly viscous polymer melts with dispersed fillers is carried out in continuous mode on worm, disk, and rotor-worm extruders in the process of obtaining PCM or products. Using the dependences of shear deformation on the technological parameters of processing and design characteristics of the equipment, it is possible to optimize the technology of obtaining dispersed-filled material and products.

The dependencies of statistical quality criteria for mixing on the shear strain are described by one characteristic curve. This allows you to apply the model of laminar mixing for the organization of the process of obtaining dispersion-filled PCM according to various schemes and using equipment of different designs.

3. Approaches for the Formation of Polymer Composite Materials

According to the type of material formation, the synthesis of polymer nanocomposites can be divided according to two main principles, among which are the so-called bottom-up and top-down principles (Figure 6.1).

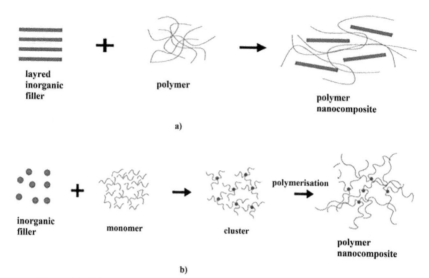

Figure 6.1. Schematic illustration of (a) bottom-up principle, (b) top-down principle.

In the case of a bottom-up principle, the formation of the structure of the final composite begins at the nanoscale level.

In most cases, there can be three basic ways of incorporating inorganic nanoparticles into a polymer matrix. The first uses the top-down principle and consists of direct mixing of the filler and the polymer matrix, carried out in solution or melt. The second is based on *in situ* polymerization in the presence of previously prepared inorganic nanoparticles, or vice versa, *in situ* preparation of nanoparticles in the presence of a polymer. In the third case, both inorganic and organic components are prepared *in situ*. The last two methods use the bottom-up principle [8–10].

The formation of composite materials *in situ*, based on the bottom-up principle, allows the creation of well-defined multi-level structures with properties that are fundamentally different from the properties of the original components. In most cases, the organic component acts as a reaction medium in which inorganic particles are formed. In the case of using metals and their oxides as an inorganic component, given nanoparticles are obtained by chemical transformation of the initial elements [8, 11]. Metals and metal oxides are good sorbents, catalysts, sensing elements, reducing agents. On the other hand, polymers are durable and chemically stable organic materials. Thus, the final hybrid material has a set of properties that cannot be achieved by each of the components separately (Figure 6.2).

The most available methods for producing nanocomposites based on polymers and inorganic fillers are such methods as the sol-gel method, mixing the polymer or prepolymer in solution or melt, and *in situ* polymerization.

The introduction of the nanofiller into the polymer matrix can also be carried out in two main ways, such as *ex situ* or *in situ* (Figure 6.3). In the first case, inorganic nanoparticles are prepared beforehand and then introduced into the polymer in the melt or solution. The method is based on the physical trapping of nanoparticles with a polymer matrix, in which it is difficult to achieve uniform distribution.

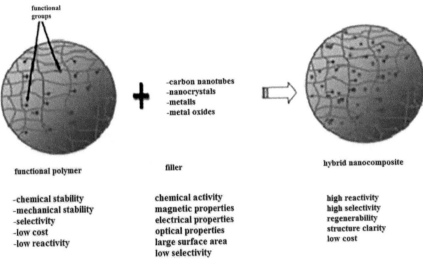

functional
groups

-carbon nanotubes
-nanocrystals
-metals
-metal oxides

functional polymer   filler   hybrid nanocomposite

-chemical stability   chemical activity   high reactivity
-mechanical stability   magnetic properties   high selectivity
-selectivity   electrical properties   regenerability
-low cost   optical properties   structure clarity
-low reactivity   large surface area   low cost
                             low selectivity

Figure 6.2. Synergistic effect of hybrid polymer-metal composite.

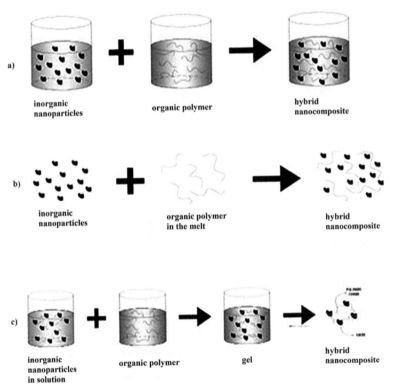

a)

inorganic   organic polymer   hybrid
nanoparticles   nanocomposite

b)

inorganic   organic polymer   hybrid
nanoparticles   in the melt   nanocomposite

c)

inorganic   organic polymer   gel   hybrid
nanoparticles   nanocomposite
in solution

Figure 6.3. Methods for producing nanocomposites: *ex situ* (a) mixing in the solution, (b) mixing in the melt, (c) *in situ* sol-gel method.

To overcome this drawback, an *in situ* method is used. This method allows you to control the size of particles and their morphology [11]. One of the most successful chemical processes that allows metal nanoparticles to be introduced into the organic phase is the sol-gel method.

Consider the above methods separately.

3.1 Sol-gel Polymerization

From an ecological point of view, the optimal methods for obtaining composite materials are drainless, in particular, the sol-gel method. Sols and gels are fluid and gelatinous colloidal systems; colloid systems are dispersed systems, intermediate between true solutions and coarse dispersed systems. The sol-gel method is a convenient way to obtain dispersed materials; it allows one to exclude numerous washing steps, since compounds that do not add impurities to the composition of the final product are used as starting materials [12]. This method is based on polymerization reactions of inorganic compounds, and includes the following stages, such as solution preparation (alcohols of different nature serve as solvents), gel formation, drying, heat treatment. Typically, the starting materials are alkoxides of metals with the general formula M (OR) (M = Si, Ti, Zr, V, Zn, Al, Sn, Ge, Mo, W, lanthanides, etc., R = Alk, Ar), which hydrolyze by adding water; the reaction is carried out in organic solvents. Subsequent polymerization (condensation) leads to the formation of a gel. For example, when n = 4,

$$M(OR)_4 + 4H_2O \rightarrow M(OH)_4 + 4ROH,$$

$$mM(OH)_4 \rightarrow (MO_2)_m + 2mH_2O.$$

The real process is much more complicated and proceeds through a multi-stage mechanism. At the same time, flow conditions, such as the use of catalysts, the nature of the metal, and the alkoxy group are essential.

Thus, the sol-gel process includes hydrolysis, polymerization, or chemically controlled condensation of a gel precursor, nucleation, and particle growth followed by agglomeration. Tetramethoxysilane (TMOS) or tetraethoxysilane (TEOS) are most often used as precursors, which form a host-silica gel-structure around the dopant guest and thereby create a trap. Nucleation proceeds through the formation of a polynuclear complex, the concentration of which increases until some supersaturation is reached, determined by its solubility. From this moment begins the growth of the nucleating seed and new nucleating seeds are not formed. At the gel formation stage, the gels can be impregnated with ions of various metals.

The obtained oxopolymers have an ultrathin porous mesh structure with pore sizes from 1 to 10 nm, similar to the structure of zeolites. Depending on the synthesis conditions, their specific surface SD is from 130 to 1260 m^2/g, bulk density is from 0.05 to 0.10 g/cm^3. Drying conditions, during which the removal of volatile components occurs, determine the texture of the product [13]. The formation of the structure and texture of the product is completed at the stage of heat treatment.

Template synthesis is one of the most successful techniques of sol-gel technology, which allows the formation of nanocomposite materials that meet the following requirements:

- a certain size and shape of crystallites or crystals;
- a narrow distribution of pore size in a given range;
- the formation at the molecular level of the specific structure of the nanocomposite, for example, a material with anisotropic organization at the meso-level (10–1000 nm).

The use of templates has long been known in metallurgy. One of the long-standing directions is the change in the shape of crystallites formed from solutions and melts of various natures. In the sol-gel technology, the most frequently used template term or templity, template synthesis has been consciously used relatively recently.

For example, the introduction of NaH_2PO_4 in nanoscale dispersions of iron hydroxide contributes to changing the shape of the growing particles α-Fe_2O_3 [14]. It was found that the shape of the formed crystals is influenced by both the nature of the anion of the introduced salt and its amount.

The use of the techniques of template sol-gel synthesis has become widely used to create silicate materials with controlled porosity. Silicate nanocomposites obtained by the sol-gel method are distinguished by chemical inertness, hardness, and a high degree of crosslinking of the silicate skeleton. Such materials also have the ability to incorporate organic matter, which can then be removed by chemical or thermal means.

The approach consists in the copolymerization of two precursors with the subsequent selective removal of one of them (Figure 6.4) [15]. This class of template agents is much broader because it generalizes the entire range of monomers

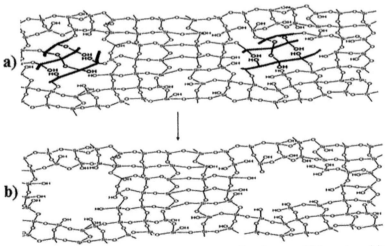

Figure 6.4. Scheme of the template sol-gel synthesis, including the step of (a) copolymerizing two precursors, and (b) selective removal of one of them.

with mutual affinity. In particular, the synthesis of macroporous materials by copolymerizing polyvinyl alcohol or polyethylene oxide with tetraethoxysilane is very successful. Phase segregation of the introduced second phase is polymerized and its removal by heat treatment leads to the formation of macroporous materials with a wide pore size distribution.

The sol-gel polymerization method can also be based on the hydrolytic polycondensation of tetraethoxysilane with an organic or organoxytrialkoxysilane. At the same time, the structure of the organic or organoxy group bound to the silicon atom (after its burning out or hydrolysis) determines the size and shape of the pores [15]. Bornyloxytriethoxysilane or fenyloxytriethoxysilane can in particular be used as pore-forming agents. Under the influence of molecular templates, the surface morphology of the coatings can change. For example, the introduction of branched oligomers into silica sols causes a significant increase in the development of the surface topography (Figure 6.5) [16].

This method is effective and difficult to implement. It is advisable to use it in the manufacture of composites with the participation of metals and their oxides. To develop a method for producing a composite based on polyimide and carbon nanotubes, it is possible to use simpler as well as efficient synthesis methods.

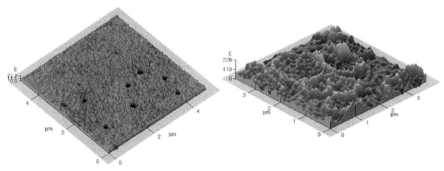

Figure 6.5. The state of the surface of silicate films was obtained from silica sol without organic modifiers (left) and with the addition of branched oligomer (right). Images were obtained with an atomic force microscope.

3.2 Melt Mixing Technique

Melt mixing technique is a standard method for producing composites, including calcining a polymer matrix at high temperatures and adding a filler followed by mixing to achieve the desired particle distribution. The advantage of this method is environmental friendliness, since there is no need to use large quantities of solvent. In addition, this method can be combined with such industrial stages as molding and extrusion, which makes it convenient and economical.

However, this method is mainly applicable in cases where layered silicate acts as an inorganic filler. The most common method for producing polymer-silicate nanocomposites is the mechanical mixing of a polymer melt with a modified silicate of organic cations. In this case, intercalation of polymer particles (intercalated systems) is achieved, and only a fraction of the particles of layered silicates are stratified into single layers of nanoscale thickness. As a result, the physicomechanical characteristics are improved, as, for example, in the case of polystyrene, polyethylene oxide, polypropylene compositions [17–22]. When polyolefin composites are obtained by this method, the filler is modified with maleic anhydride [23], or the olefin is copolymerized with a polar comonomer [20, 21, 24]. Modification increases the compatibility of the polymer with the filler.

3.3 Mixing in Solution

This approach is based on the use of a common dissolution system, suitable for both polymer dissolution and swelling of inorganic fillers [25]. Using the example of a layered silicate, the process consists of several stages. First, the silicate layers swell in the solvent (for example, in chloroform). After mixing the solutions of the polymer and layered silicate, the polymer chains intercalate into the interlayer space of the silicate, replacing the solvent. After removal of the solvent, the intercalated structure remains, which is a nanocomposite [26, 27].

The process includes a solvent in which the polymer and prepolymer are soluble to affect the exfoliation of the layered silicate into separate layers (Figure 6.6). When selecting a suitable solvent, weak bonds are destroyed holding the layers together, after which the polymer is easily adsorbed on a separate plate.

The leading factor of polymer intercalation into layered silicate from solution is entropy. Entropy arises due to the desorption of solvent molecules, which compensates for the decrease in entropy of bound intercalated polymer chains [28]. With this approach, intercalation occurs only in the case of certain polymer/solvent pairs.

This method is effective for intercalating polymers with low or neutral polarity into layered structures, and is convenient for producing thin films with intercalated clay layers oriented along polymer chains (and film surface) [29]. When used in industry, this method requires large amounts of organic solvents, which are usually environmentally harmful and uneconomical [30]. In addition, this method is difficult to implement when using carbon nanotubes.

In the work of Ku et al., a homogeneous distribution of single-layer and multi-walled CNTs in a polyimide matrix has been reported [31]. This result

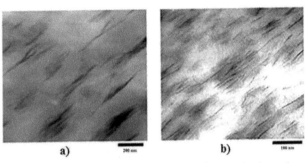

Figure 6.6. Example of (a) intercalated, and (b) exfoliated organoclay in a polymer matrix.

was achieved by prolonged (48 hours) ultrasonic dispersion of a polyimide solution containing terminal amino groups and a mixture of CNTs with 1-ethyl-3-(3-dimethylaminopropyl) carbodimide. The resulting compound was a CNT functionalized with polyimide groups. Further mixing of the modified CNT solution with a solution of a pure polyimide matrix and the subsequent removal of the solvent gave a composite material with a uniform distribution of the filler.

In [32], a polyimide/multi-layer CNT nanocomposite was obtained by ultrasonic dispersion of a mixture of modified CNTs and a solution of polyamic acid in N, N-dimethylacetamide, followed by thermal imidization. The resulting material had an increased tensile strength compared to pure polyimide. When the nanocomposite is filled with multi-walled CNTs, up to 10% by weight of its dielectric constant reached a value of 60, which is about 17 times greater than the analogous parameter of pure polyimide.

3.4 In Situ Polymerization

The next method of obtaining nanofilled polymers is direct synthesis of the material by *in situ* polymerization, i.e., matrix polymer synthesis in the presence of an inorganic nanofiller. In this case, the monomer or oligomer undergoes polymerization. The method allows you to create well-defined multi-level structures with properties fundamentally different from the properties of the original components. For example, the modulus of elasticity, strength, heat resistance, barrier properties of nylon-6 compositions with montmorillonite are doubled compared with the initial polymer [33, 34]. Note that the process of obtaining such a nanocomposite developed by Japanese scientists has been implemented on an industrial scale.

When *in situ* polymerization is used to obtain polyimide/CNT composites, preliminary ultrasonic dispersion of CNTs is carried out in a solution of one of the monomers (usually a diamine), and the polymerization reaction takes place in the presence of a filler.

For the first time, polymerization *in situ* was reported by Imai et al. [35]. In this work, the polyimide/carbon black composite was obtained directly by polycondensation of a mixture of carbon black, and ammonium salt obtained from nonamethylenediamine and pyromellitic acid. Later, Ounies et al. published a study of the effect of increasing the strength and conductive characteristics of composite materials PI/single-walled CNT [36]. In both papers, a jump in conductivity was noted by more than 8 orders of magnitude to a value of 10^{-6}–10^{-7} S/cm when a certain critical concentration was reached. When using multi-layer CNTs as a filler, it was possible to achieve an increase in conductivity by 11 orders of magnitude, to a value of 10^{-4} S/cm at a critical concentration of approximately 0.15% by volume [37], and an increase in concentration to 3.7% by volume gave electrical conductivity of 10^{-1} S/cm. In general, based on literature data, it is difficult to estimate the required number of CNTs for the manifestation of the electrically conductive properties of the composite, since the reported percolation threshold values vary within very wide limits: from less than 1% to values greater than 10%. The reasons for such large differences can be both the uneven distribution of CNTs (due to their poor dispersion in the polymer matrix) and experimental difficulties, which lead to errors in determining the threshold values of the electronic conductivity.

The *in situ* polymerization method is also effective in the polymerization of polar monomers, in particular, to obtain nanocomposites by emulsion polymerization. So, for example, when achieving the complete dispersion of the sodium form of montmorillonite in water, nanocomposites based on polymethyl methacrylate, polystyrene, styrene copolymer, and acrylonitrile were obtained [38]. Another approach to the synthesis of polystyrene–montmorillonite nanocomposites has been proposed in the source [39]. The radical polymerization initiator was fixed in the interlayer space of the silicate lattice by cation exchange with sodium ions. This allowed the polymerization of styrene directly in the interlayer space of silicate, followed by exfoliation of particles of this filler under the action of the resulting polymer.

This method is also used for the synthesis of nanocomposites based on polyethylene terephthalate, polyimide [40], as well as thermosetting polymeric matrices. Thus, in [24], the effect of the type of layered silicates and their modifiers, curing agents, and polymerization conditions on the structure and properties of nanocomposites based on epoxy resins was studied.

As you know, epoxy resins or oligomeric products of small molecular weight are widely used for the production of composite materials. Based on them, functional nanocomposites can be obtained. Thus, the modification of epoxy resins with other resins, in particular, vinyl esters, made it possible to create a binder with the structure of interpenetrating networks and nanoscale phases. Fiberglass based on such a modified binder has a complex of properties inherent in both epoxy and vinyl ester binder [41]. The introduction of copper nanoparticles into composites based on epoxy binders reduces their flammability [42, 43].

In situ polymerization can be carried out both in the presence of a solvent and in the presence of a catalyst.

Source [44] described the synthesis of nanocomposites based on graphene and epoxy resin by *in situ* polymerization. The synthesis is carried out first by dispersing the filler in acetone using ultrasound. The resulting mixture is then added to the epoxy matrix before being placed in a vacuum oven. After evaporation of about 80% of the solvent, m-phenylenediamine is added with vigorous stirring. After stirring, the mixture is poured into a stainless-steel mold, dried at 60°C for 5 hours to remove residual solvent, pre-cured in an oven at 80°C for 2 hours, and confirmed at 120°C for another 2 hours with the composite. Figure 6.7 shows that better dispersion was achieved in materials with epoxide/graphene and epoxy/modified graphene oxide composites compared to epoxy/graphene oxide composites.

Nevertheless, a good dispersion of graphene oxide was observed in the polypropylene matrix [45]. In order to achieve it, the Ziegler-Natta catalyst was used. The catalyst was incorporated into the graphene oxide layers, as shown in Figure 6.8. The Grignard reagent was used before the addition of titanium tetrachloride to obtain a Ziegler-Natta graphene-oxide catalyst system. Then this catalyst is added at 60°C in a hexane-propylene solution with vigorous stirring. Triethylaluminum ($AlEt_3$) and dimethoxydiphenylsilane (DDS) as initiators were added to the mixture to initiate

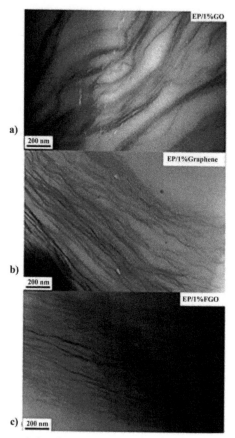

Figure 6.7. Results of transmission electron microscopy of samples (a) epoxide/graphene, (b) epoxide/ modified graphene oxide, (c) epoxide/graphene oxide.

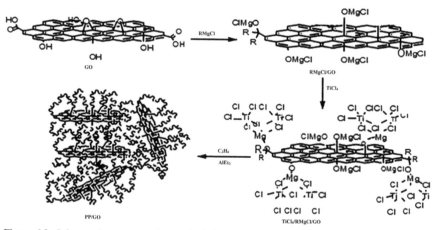

Figure 6.8. Scheme of nanocomposite synthesis based on polypropylene and graphene oxide in the presence of a catalyst.

the polymerization reaction. The final composite is obtained by filtration, washing, and drying.

The *in situ* polymerization method has several advantages. First of all, thermoplastics and thermosetting composites can be synthesized by this method [46, 9]. It allows the polymer to be "stitched" to the surface of the filler, which improves the properties of the final material. Exfoliated structures are achieved using this method due to the improved distribution and intercalation of the filler in the polymer matrix.

4. Production of Composites Based on High-Temperature Thermoset Plastics

As noted above, the production of composite materials, by mixing the matrix and filler in solution or melt, is compatible with existing industrial technologies, and therefore, is well known and well developed. However, these methods are unsuitable for the production of composites of high-temperature polymers or thermoset composite. In turn, methods of producing composite materials based on such matrices are not as detailed in the literature as those obtained from solutions or melts. In this regard, we would like to focus on obtaining polymer composites by the *in situ* method using the example of a polyimide matrix, which is a typical example of a high-temperature polymer, since most polyimide matrices are insoluble in organic solvents, and their softening temperature exceeds 300°C.

Polyimides are produced by polycondensation of carboxylic acid dianhydrides with diamines. There are single-stage and two-stage polymerization reactions. In the one-step method, the stages of acylation and cyclization proceed simultaneously in a high-boiling solvent at 180–200°C. In the case of a two-stage process, the reaction proceeds with the formation of an intermediate product of polyamic acid, and in the second stage, cyclodehydration (imidization) is carried out with the formation of polyimide. However, in both cases, the initial monomers are dissolved in a suitable solvent, and at this stage an inorganic filler may be added to the reaction mixture. Depending on the method of synthesis, the target polyimide composites can be obtained in the form of powders or films.

4.1 Composite Materials Based on Polyimides

One of the interesting fillers for polyimides is single-walled (SWCNT) or multi-walled (MCNT) carbon nanotubes. The combination of CNTs and polyimides will play an important role in the development of new highly efficient nanocomposites [47]. There are two general methods for producing such composites. One is to mix the CNT with the polymer matrix in a molten form to produce a composite. Another is to disperse the CNT in the polymer solution, solidify the resulting solution, and remove the solvent.

Park and his colleagues reported on the method of efficiently dispersing SWCNTs in a polyimide matrix [48]. The resulting SWCNT polyimide films are electrically conductive and optically transparent. A sharp increase in conductivity was observed between 0.02% and 0.1% vol. SWCNT, and during this process, the nanocomposite was transformed from a capacitor to a conductor. The introduction of 0.1% vol. SWCNTs increased conductivity by 10 orders of magnitude, which exceeds the antistatic criterion for thin films for use in astronautics (1×10^{-8} S cm^{-1}). Polyimide film containing 1.0% vol. The SWCNT still transmitted 32% of the visible light at 500 nm, while the film produced by direct blending passed less than 1%. Dynamic mechanical test showed that by the addition of 1.0% vol., the SWCNT increases the elastic modulus by 60%, and the thermal stability of the polyimide is improved in the presence of the SWCNT.

Connell et al. [49] reported on the synthesis of alkoxysilane polyamic acids, and SWCNTs were added to a previously prepared polyamic acid solution. When loading 0.05% wt., the SWCNT achieved a percolation barrier, which can be seen from the sharp decrease in the surface resistance of the material. Surface resistance and volume resistance indicates that the SWCNT polyimide composite is conductive. However, the presence of SWCNT in polyimide has very little effect on the temperature at which the fracture starts (Tg) and the tensile strength of the polymer. An increase in the ionic strength of the polyimide matrix by the addition of an inorganic salt ($CuSO_4$) led to the formation of a SWNT network sufficient for conductivity, for example, adding 0.014% wt. $CuSO_4$ in a composite containing 0.03% wt.

The SWCNT led to films reduced by 4 orders of magnitude by surface and volume resistance [50, 51]. An increased electrical conductivity of nanocomposite films is observed; however, electrical percolation occurs at larger loads than those commonly used in SWCNT polyimide nanocomposites. The modulus of the films slightly increases with increasing content of single-walled carbon nanotubes. Electrospun fibers were obtained from the same SWCNT polyamide suspensions used to make films. High resolution scanning electron microscopy images have shown that the SWCNT is located inside the fibers and may have a direction parallel to the fiber axis [52].

San and his colleagues [53] reported on the production of functionalized CNTs using polyimides with pendant hydroxyl groups. It was found that the resulting polyimide-functionalized CNTs are soluble in the same solvents as the original polyimide. A significant advantage of this method is that these functionalized nanotubes can be used directly to produce polyimide-CNT composites with a relatively high content of nanotubes.

Bean and colleagues [54] obtained polyimide-CNT composites by carrying out *in situ* polymerization in the presence of MCNT. The percolation barrier of the electrical conductivity of the resulting PIMUNT composites is about 0.15% by volume. The electrical conductivity increases by more than 11 orders of magnitude to 10^{-4} S cm^{-1} when the percolation barrier is reached, and subsequently increases to 10^{-1} S cm^{-1} with an increase in the concentration of MCNT to 3.7% by volume.

Nakashima [47] reported on the synthesis of fully aromatic polyimides containing disulfonic acid triethylammonium salts (Figure 6.9). The polyimides obtained have an increased ability to dissolve MWCNTs in themselves. The main driving force for solubilization of MWCNTs are the π-π-interactions between the condensed aromatic part of the polyimide and the MWCNT surface. A high concentration of MWCNTs

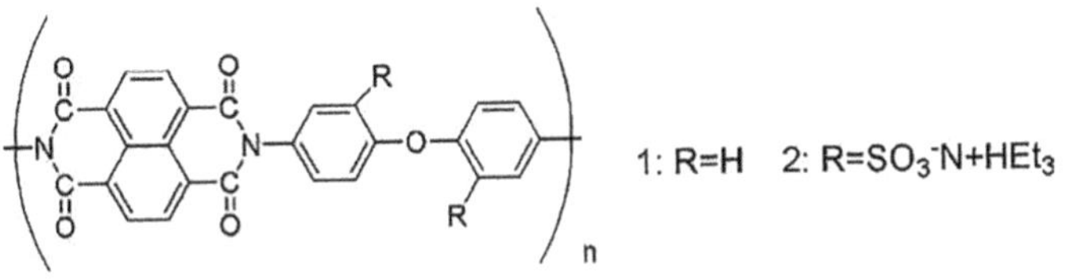

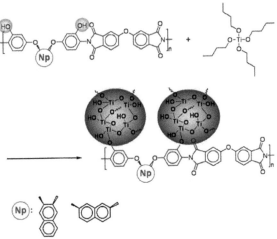

Figure 6.9. Polyimides containing disulfonic acid salts [43].

1: R=H   2: R=SO$_3$⁻N+HEt$_3$

in polyimide solutions leads to the formation of gels consisting of individually dissolved MWCNTs.

Ando and his colleagues [55] obtained new nano-ZnO/hyperbranched polyimide hybrid films by *in situ* sol-gel polymerization. Films obtained from colorless, fluorinated polyimides and homogeneously dispersed ZnO nanoparticles show good optical transparency. Later on, two types of model compounds (with and without ZnO) and a hyperbranched polyimide (HBPI) film with ZnO microparticles were obtained. These materials were used to study the mechanism of fluorescence of the original HBPI and *in situ* hybrid films. In *in situ* hybrid films, an effective energy transfer from ZnO nanoparticles to the main HBPI chain was observed, while in ordinary HBPI films only energy transfer from local excited states was observed. This shows that HBPI are terminally associated with ZnO particles through a monoethanolamine group, which is an effective way to transfer energy, which leads to fluorescence.

Liu and his colleagues [56] obtained hybrid optical films based on PI-nanocrystalline titanium with a relatively high titanium content and a large thickness of soluble polyimides containing hydroxyl groups (Figure 6.10). Two types of new soluble polyimides were synthesized from hydroxy-substituted diamines and various commercially available tetracarboxylic dianhydrides. The hydroxyl groups in the main chain of the polyimide bind the organic and inorganic parts and also control the molar ratio of titanium butyloxide to hydroxyl groups. This leads to

Figure 6.10. Hybrids based on PI nanocrystalline titanium [56].

homogeneous hybrid solutions. Flexible hybrid films were obtained, and analysis showed that these films have relatively good surface flatness, thermal stability, variable refractive index, and high optical transparency. Three-layer anti-reflection coatings based on these hybrid films were obtained, and the reflectivity was less than 0.5% in the visible region. These characteristics suggest that these films can be applied in optics.

Polyimide conductive composites are made from appropriate polyimides and conductive fillers, such as carbon nanotubes, graphite, and acetylene black. The polyimide precursor (polyamic acid) was synthesized from 3,4,3',4'biphenyl tetracarboxylic dianhydride and 4,4'diaminodiphenyl ether with the help of intensive mechanical mixing at −5°C. Experimental results showed that the electrically conductive composite based on carbon nanotubes and polyimide possess better electrical, mechanical, and adhesive properties than the other two composites [57].

A new highly porous composite based on polyimide and silicon with high flexibility, mechanical strength, and heat resistance has been developed. The composite material was prepared using a new process consisting of phase separation of a mixture of a polyimide precursor (polyamic acid), solvent, and alkoxide of silicon using CO_2 at high pressure (40°C, 20 MPa), the formation of silicate by sol-gel reaction, and solvent extraction with supercritical CO_2. The composite has a bimodal porous structure with micropores of 10–30 μm and nanopores of 50 nm. Silicon nanoparticles with a diameter of less than 100 nm are dispersed in the polyimide matrix. The porosity of this composite is 78%, which is higher than the porosity of the polyimide produced by the foaming method.

The relative dielectric constant of the composite is below 1.4 at 1 MHz. The porous sheet of a polyimide silicon composite material proved to be quite flexible, and does not collapse even when bent. It should be noted that the Young's modulus (0.80 GPa) and the decomposition temperature (600°C) of this composite are higher than those of ordinary porous polyimide with the same porosity. These properties make a composite material based on polyimide and silicon suitable for use as a flexible thermally insulating material [58].

4.2 Composites with Nanostructured Silicon Carbide

Over the past decade, many research papers have been devoted to combining polymers with nanoparticles to obtain materials with increased rigidity, toughness, and tribological properties [59]. By adding nanoscale fillers to the polymer matrix, the material acquires new chemical and physical properties. This is due to the influence of the unique nature of the nanoscale filler on the bulk properties of polymer-based nanocomposites [60, 61]. Polymer nanocomposites are intensively used in various fields, due to their ease of processing, low production cost, good adhesion to the substrate, and unique physicochemical properties. Dispersion of inorganic materials in a polyimide matrix is a complex task and a key factor influencing the final properties of hybrid materials. Adding a crosslinking agent is a solution to a number of difficulties associated with dispersing. By adding a crosslinking agent, organic and inorganic materials can be covalently bonded, and the compatibility between these two phases is improved [62, 63].

Silicon carbide nanoparticles (SiC) are chosen for their unique physical properties, such as excellent chemical resistance, heat resistance, high electron mobility, excellent thermal conductivity, and outstanding mechanical properties. They are used to produce highly efficient composites [64–67], and are used in electronics [68, 69]. These properties make SiC nanoparticles a suitable material for the production of polymer nanocomposites with an enhanced structure [70].

The properties of nanocomposite films obtained from the new polyimide and silicon carbide nanoparticles SiC using two simple methods are reported. In the first, SiC nanoparticles were first functionalized with epoxy (γ-glycidoxypropyltriethoxy silane) terminal groups (mSiC), and then the solution was mixed with a polytriazolimide. A homogeneous solution for the preparation of a polytriazolimide/mSiC film was heated under vacuum. In the second method, a new diamine containing the 1,2,4-triazole cycle - 4,4'-(4-(2,3-diphenylphenoxal-6-yl)-4H-1,2,4-triazole-3,5-diyl), and a commercially available dianhydride (4,4'-(hexafluoroisopropylidene) diphthal dianhydride) react *in situ* in the presence of SiC nanoparticles to form a homogeneous mixture of polyamic acid and silicon carbide (PAA/SiC), which is then transformed into a polytriazolimide in a vacuum in a high-temperature process/SiC film. The results of the study showed that strong chemical bonding between SiC nanoparticles and the polymer matrix leads to an increase in glass transition temperature Tg from 300°C to more than 350°C, tensile strength from 108 MPa to 165 MPa, and a temperature of 5% mass loss (T5%) from 380°C to 500°C. The photoluminescence intensity has increased, and the spectrum shows a shift to the blue region with increasing content of SiC [71].

A highly efficient composite material based on silicon carbide (SiC) and bismaleimide modified with allyl novolac for abrasive tools and wear-resistant elements was developed and characterized. The results showed that the residual strength at 440°C for 1 hour decreased to 64% of the strength without heat treatment, and the thermal-oxidative stability is better than for SiC/polyimide composites made in a similar way. The proportion of polymer in the composition of the composite affects the mechanical properties. The flexural strength of a composite increases with an increase in the proportion of bismaleimide, but its excess leads to the formation of bubbles in the composite. The best composite with a bending strength of 82.4 MPa was obtained using 13% wt. bismaleimide. After treatment at 280°C for 1 hour, the bending strength increased by 34% due to the further crosslinking of the polymer at a higher temperature [72].

It is expected that the combination of polyimides and other organic/inorganic compounds will play an important role in the development of innovative high-performance nanocomposites for various applications.

4.3 Preparation of Powder Polyimide Composites

The process of obtaining a powder composite material can be divided into four stages:

- modification of the surface of the inorganic filler (if necessary);
- dispersion of inorganic filler in a solution of high-boiling solvent and diamine;

- carrying out polymerization and imidization reactions;
- isolation of the resulting composite material in the form of a powder.

4.3.1 Modification of the Surface of the Inorganic Filler

Modification of the surface of the inorganic component before introducing into the composite, on the one hand, allows an increase in the affinity between the organic and inorganic phases of the material, on the other hand, makes it possible to vary the properties of the materials obtained. For example, it is well known that doping of carbon nanotubes leads to a sharp increase in the conductivity of nanotubes, due to a change in their electronic structures caused by charge transfer [73–76]. The authors [77–79] reported on the production and electrical properties of air-stable polymer composites filled with CNTs (doped with iodine). It is reported that conductivity increases 2–5 times as compared to composites with unalloyed CNTs.

4.3.2 Dispersion of the Inorganic Filler in a Solution of High-Boiling Solvent and Diamine

As noted earlier, the key stage in obtaining a composite with an inorganic nanosized filler is dispersion of the filler in solution. As high-boiling solvents, choose those that are capable of dissolving both the initial monomers and the polymerization product. In the case of precipitation of the product from the solution in the process of growth of the polymer chain, a product with a low molecular weight will be obtained. Suitable solvents can be, for example, high-boiling aprotic polar solvents, such as N, N-dimethylformamide, dimethyl sulfoxide, etc.

The processing time and frequency of acoustic waves during ultrasonic dispersion are the determining parameters of the process. The effect of dispersion time and ultrasound frequency on the quality of the distribution of carbon nanotubes and nanoscale boron carbide in the final polyimide composite is considered in [80]. It was shown that when using a frequency of 10–15 kHz for both fillers, agglomerates of nanoparticles are observed on the surface of the composite, regardless of the time of dispersion. Increasing the frequency to 20 kHz at small (5–10 minutes) exposure times leads to an uneven distribution of the filler in the final composite with agglomerates of 2–5 μm in size. And only with an increase in the dispersion time to 15–20 minutes, a uniform distribution of the inorganic filler over the surface and in the structure of the composite is observed.

After obtaining a uniform dispersion of the filler and the diamine in the solvent, the usual mechanical stirring can be used to further carry out the process. Dianhydride is added to the resulting dispersion with stirring, and then proceed to the next stage.

4.3.3 Polymerization and Imidization

The polymerization reaction is usually carried out at elevated temperatures for a long time. If necessary, a polymerization initiator, for example, benzoic acid, may be added to the reaction mixture. The parameters of this stage of the process will depend on the starting monomers. For polyimide composites, the process is carried out in steps: they hold the reaction mass at 90°C for 4 hours, and then, at a temperature of 180–200°C for 12–16 hours.

4.3.4 Selection of the Resulting Powder Composite Material

Isolation of the composite is carried out by precipitating the polymer from the solvent of the reaction medium. To do this, the reaction mass is poured into a large volume of solvent that does not dissolve the final polymer. In the case of polyimides, ethyl alcohol is usually used as such a solvent. The precipitation is filtered, washed, and dried to constant weight.

4.3.5 Production of Film Polyimide Composites

The process of obtaining film polyimide composite material can be divided into five stages [81–84]:

1. Modification of the surface of the nanoscale filler (if necessary);
2. Dispersing inorganic filler in a solution of high-boiling solvent and diamine;
3. Conducting the polymerization reaction to obtain a solution of the precursor (polyamic acid);
4. Applying the resulting solution of polyamic acid on a substrate;
5. Drying the solvent followed by stepwise imidization in a vacuum oven.

The last stage is the drying of the solvent and stepwise imidization is carried out in a vacuum drying oven. An extremely important step is pre-drying, at a temperature of 70–80°C for 8–12 hours, allowing you to remove the solvent. It is important to avoid rapid heating of the plate in order to avoid the formation of bubbles on the surface of the film due to the rapid evaporation of the solvent. Further gradual increase in temperature provides the most complete and uniform imidization. Step polymerization begins at a temperature of 150°C, keeping the plate under these conditions for 20–30 minutes, then the temperature is increased by 50°C, and maintained for the same amount of time. Thus, the imidization temperature is increased in increments of 50°C, and the imidization is completed at a temperature of 300°C.

The most important criterion in obtaining polymer composite films is uniformity and uniform thickness.

The simplest way to apply a polyamic acid solution to a substrate is to immerse the substrate in the solution or pour the solution onto the substrate. However, this method does not allow the control of the thickness of the resulting film. At present, for the manufacture of polyimide films, the method of spin-coating is used, which appeared relatively recently, but gained most popularity in the manufacture of polyimide films. Using this method, it is possible to adjust the thickness of the resulting film by changing the speed of rotation of the plate.

Samples of films can be obtained on various laboratory instruments designed to apply liquids to coatings. Concomitant evaporation of the solvent, which is induced by rapid rotation, leads to the formation of a semi-solid film. On the previously cleaned and dried plate of glass or metal, put the prepared polymer solution. Place the plate on the Spin Coater vacuum chuck and rotate at a speed ranging from 100 to 6000 revolutions per minute, depending on the desired film thickness.

The study was carried out with the financial support of the Russian Foundation for Basic Research, within the framework of the RFBR research project No. 18-29-18087\20.

References

[1] Wang, P.J., C.H. Lin, S.L. Chang and S.J. Shih. 2012. Facile, efficient synthesis of a phosphinated hydroxyl diamine and properties of its high-performance poly(hydroxylimides) and polyimide–SiO$_2$ hybrids. Polym. Chem., 3: 2867–2874.

[2] Agag, T., T. Koga and T. Takeichi. 2001. Studies on thermal and mechanical properties of polyimide–clay nanocomposites. Polymer., 42: 3399–3408.

[3] Zou, H., S. Wu and J. Shen. 2008. Polymer/silica nanocomposites: preparation, characterization, properties, and applications. Chem. Rev., 108: 3893–3957.

[4] Nikolaev, A.F., V.K. Kryzhanovsky and V.V. Burlov. 2011. Technology of polymer materials/ educational aid, SPb.: "Profession", 4: 393–465. https://www.studmed.ru/nikolaev-a-f-kryzhanovskiy-v-k-burlov-v-v-tehnologiya-polimernyh-materialov_6564d7f8baf.html.

[5] Zhu, J., S. Wei, N. Haldolaarachchige, D.P. Young and Z. Guo. 2011. Electromagnetic field shielding polyurethane nanocomposites reinforced with core–shell Fe–silica nanoparticles. J. Phys. Chem. C, 115: 15304–15310.

[6] Kobayashi, Y., H. Katakami, E. Mine, D. Nagao, M. Konno and L.M. Liz-Marzan. 2005. Silica coating of silver nanoparticles using a modified Stöber method. J. Colloid. Interface. Sci., 283: 392–396.

[7] Polymeric composite materials: structure, properties, technology: educational aid. edited by A.A. Berlin - SPb.: "Profession". 2014. 592 p. https://plastinfo.ru/information/literature/136_2018/.

[8] Ruiz-Hitzky, E., P. Aranda, M. Darder and M. Ogawa. 2011. Hybrid and biohybrid silicate based materials: molecular vs block-assembling bottom-up processes. Chem. Soc. Rev., 40: 801–828.

[9] Kickelbick, G. 2007. Hybrid Materials: Synthesis, Characterization, and Applications. Wiley-VCH, 18: 117–131.

[10] Ariga, K. and H.S. Nalwa. 2009. Bottom-up Nanofabrication: Supramolecules-I. American Scientific Pub., 26: 307–324.

[11] Sarkar, S., E. Guibal, F. Quignard and A. SenGupta. 2012. Polymer-supported metals and metal oxide nanoparticles: synthesis, characterization, and applications. J. Nanopart. Res., 14: 1–24.

[12] Sharygin, L.M. 2011. Sol-gel technology for producing nanomaterials. Ekaterinburg: Ural Branch of the Russian Academy of Sciences, 46 p.

[13] Hench, L.L. and J.K. West. 1990. The sol-gel process. Chem. Rev., 90: 33–72.

[14] Sugimoto, T., Y. Wang, H. Itoh and A. Muramatsu. 1998. Systematic control of size, shape and internal structure of monodisperse α-Fe$_2$O$_3$ particles. Colloids Surf. A: Physicochemical and Engineering Aspects, 134: 265–279.

[15] Srebnik, S. and O. Lev. 2003. Theoretical investigation of imprinted crosslinked silicates. J. Sol-Gel Sci. Tech., 26(1-3): 107–113.

[16] Maksimov, A.I., V.A. Moshnikov, Yu.M. Tairov and O.A. Shilova. 2008. Basics of sol-gel nanocomposite technology. SPb.: "Elmore", 255 p.

[17] Vaia, R.A., R.K. Teukolsky and E.P. Giannelis. 1994. Interlayer structure and molecular environment of alkylammonium layered silicates. Chem. Mater., 6: 1017–1022.

[18] Alexandre, M. and P. Dubois. 2000. Polymer-layered silicate nanocomposites: preparation, properties and uses of a new class of materials. Mater. Sci. Engng. R, 28: 1–62.

[19] Sikka, M., L.N. Cerini, S.S. Ghosh and K.I. Winey. 1996. Melt intercalation of polystyrene in layered silicates. J. Polym. Sci. B: Polymer Physics, 34: 1443–1449.

[20] Manias, E., A. Touny, L. Wu, K. Strawhecker, B. Lu and T.C. Chung. 2001. Polypropylene/ montmorillonite nanocomposites. Review of the synthetic routes and materials properties. Chem. Mater., 13(10): 3516–3523.

[21] Kawasumi, M., N. Hasegawa, M. Kato, A. Usuki and A. Okada. 1997. Preparation and mechanical properties of polypropylene–clay hybrids. Macromoleculs, 30: 6333–6338.

[22] Vaia, R.A., B.B. Sauer, O.K. Tse and E.P. Giannelis. 1997. Relaxations of confined chains in polymer nanocomposites: glass transition properties of poly (ethylene oxide) intercalated in montmorillonite. J. Polym. Sci. B, 35: 59–67.

[23] Tjong, S.C., Y. Meng and A.S. Hay. 2002. Novel preparation and properties of polypropylene–vermiculite nanocomposites. Chem. Mater., 14: 44–51.

[24] Wang, K.H., I.J. Chung, M.C. Jang, J.K. Keum and H.H. Song. 2002. Deformation behavior of polyethylene/silicate nanocomposites as studied by real-time wide-angle X-ray scattering. Macromolecules, 35: 5529–5535.

[25] Bonner, J.G. and P.S. Hope. 1993. Polymer blends and alloys. J. Folkers and Blackie Akad. and Prof., pp. 46–74.

[26] Hussain, F., M. Hojjati, M. Okamoto and R.E. Gorga. 2006. Polymer-matrix nanocomposites, processing, manufacturing, and application: an overview. J. Compos. Mater., 40: 1511–1575.

[27] Laoutid, F., L. Bonnaud, M. Alexandre, J.M. Lopez-Cuesta and P. Dubois. 2009. New prospects in flame retardant polymer materials: from fundamentals to nanocomposites. Mater. Sci. Eng. R Rep., 63(3): 100–125.

[28] Hu, G.H., Y. Sun and M. Lambla. 1996. Devolatilization: A critical sequential operation for *in situ* compatibilization of immiscible polymer blends by one-step reactive extrusion. Polym. Eng. and Sci., 36: 676–684.

[29] Sanchez, C., H. Arribart and M.M. Giraud Guille. 2005. Biomimetism and bioinspiration as tools for the design of innovative materials and systems. Nat. Mater., 4: 277–288.

[30] Gao, F. 2004. Clay/polymer composites: the story. Mater. Today, 7: 50–55.

[31] Qu, L., Y. Lin, D.E. Hill, B. Zhou, W. Wang, X. Sun, A. Kitaygorodskiy, M. Suarez, J.W. Connell, L.F. Allard and Y.-P. Sun. 2004. Polyimide-functionalized carbon nanotubes: synthesis and dispersion in nanocomposite films. Macromolecules, 37(16): 6055–6060.

[32] Zhu, B.K., S.H. Xie, Z.K. Xu and Y.Y. Xu. 2006. Preparation and properties of the polyimide/multi-walled carbon nanotubes (MWNTs). Compos. Sci. Technol., 66: 548–554.

[33] Kojima, Y., A. Usuki, M. Kawasumi, A. Okada, T. Kurauchi and O. Kamigaito. 1993. One-pot synthesis of nylon 6–clay hybrid. J. Polym. Sci., A, 31: 1755–1758.

[34] Kojima, Y., A. Usuki, M. Kawasumi, A. Okada, T. Kurauchi and O. Kamigaito. 1993. Sorption of water in nylon 6-clay hybrid. J. Appl. Polym. Sci., 49: 1259–1264.

[35] Imai, Y., T. Fueki, T. Inoue and M. Kakimoto. 1998. A new direct preparation of electroconductive polyimide/carbon black composite via polycondensation of nylon–salt-type monomer/carbon black mixture. J. Polym. Sci., Part A: Polym. Chem., 36: 1031–1034.

[36] Ounaies, Z., C. Park, K.E. Wise, E.J. Siochi and J.S. Harrison. 2003. Electrical properties of single wall carbon nanotube reinforced polyimide composites. Compos. Sci. Technol., 63: 1637–1646.

[37] Hill, D.E., Y. Lin, A.M. Rao, L.F. Allard and Y.P. Sun. 2002. Functionalization of carbon nanotubes with polystyrene. Macromolecules, 35(25): 9466–9471.

[38] Choi, Y.S., M.Z. Xu and I.J. Chung. 2005. Synthesis of exfoliated acrylonitrile–butadiene–styrene copolymer (ABS) clay nanocomposites: role of clay as a colloidal stabilizer. Polymer., 46: 531–538.

[39] Weimer, M., H. Chen, E. Giannelis and D. Sogah. 1999. Direct synthesis of dispersed nanocomposites by *in situ* living free radical polymerization using a silicate-anchored initiator. J. Am. Chem. Soc., 121: 1615–1616.

[40] Davis, C.H., L.J. Mathias, J.W. Gilman, D.A. Schiraldi, J.R. Shields, P. Trulove and H.C. Delong. 2002. Effects of melt-processing conditions on the quality of poly (ethylene terephthalate) montmorillonite clay nanocomposites. J. Polym. Sci. B: Polymer Physics, 40(23): 2661–2666.

[41] Sorina, T.G., A.Kh. Khayretdinov, T.V. Penskaya, Yu.G. Klenin and A.P. Korobko. 2005. Study of physicomechanical properties of modified epoxy vinyl ester resin and fiberglass based on it. Plastics, 5: 28–31.

[42] Sorina, T.G., D.K. Polyakov, A.P. Korobko and T.V. Penskaya. 2001. Third Intern. conf. Theory and practice of production technologies for products made of composite materials and new metal alloys—21st century/Moscow, 374 p.

[43] Korobko, A.P., T.V. Penskaya, D.K. Polyakov and T.G. Sorina. 2002. Proc. of The 9-th Int. Conf. on Composites Engineering (ICCE-9)/Cal. USA, P. 411.

[44] Guo, Y., C. Bao, L. Song, B. Yuan and Y. Hu. 2011. *In situ* polymerization of graphene, graphite oxide, and functionalized graphite oxide into epoxy resin and comparison study of on-the-flame behavior. Ind. Eng. Chem. Res., 50: 7772–7783.

[45] Huang, Y., Y. Qin, Y. Zhou, H. Niu, Z.-Z. Yu and J.-Y. Dong. 2010. Polypropy-lene/graphene oxide nanocomposites prepared by *in situ* ziegler-natta polymerization. Chem. Mater., 22: 4096–4102.

[46] Pavlidou, S. and C.D. Papaspyrides. 2008. A review on polymer-layered silicate nanocomposites. Prog. Polym. Sci., 33: 1119–1198.

[47] Shigeta, M., M. Komatsu and N. Nakashima. 2006. Individual solubilization of single-walled carbon nanotubes using totally aromatic polyimide. Chem. Phys. Lett., 418: 115–118.

[48] Park, C., Z. Ounaies, K.A. Watson, R.E. Crooks, J. Smith, S.E. Lowther, J.W. Connell, E.J. Siochi, J.S. Harrison and T.L.S. Clair. 2002. Dispersion of single wall carbon nanotubes by *in situ* polymerization under sonication. Chem. Phys. Lett., 364: 303–308.

[49] Smith, J.G., J.W. Connell, D.M. Delozier, P.T. Lillehei, K.A. Watson, Y. Lin, B. Zhou and Y.P. Sun. 2004. Space durable polymer/carbon nanotube films for electrostatic charge mitigation. Polymer., 45: 825–836.

[50] Smith, Jr. J.G., D.M. Delozier, J.W. Connell and K.A. Watson. 2004. Carbon nanotube–conductive additive-space durable polymer nanocomposite films for electrostatic charge dissipation. Polymer., 45: 6133–6142.

[51] Watson, K.A., S. Ghose, D.M. Delozier, J.J.G. Smith and J.W. Connell. 2005. Transparent, flexible, conductive carbon nanotube coatings for electrostatic charge mitigation. Polymer., 46: 2076–2085.

[52] Delozier, D.M., K.A. Watson, J.G. Smith Jr., T.C. Clancy and J.W. Connell. 2006. Investigation of aromatic/aliphatic polyimides as dispersants for single wall carbon nanotubes. Macromolecules, 39: 1731–1739.

[53] Hill, D., Y. Lin, L. Qu, A. Kitaygorodskiy, J.W. Connell, L.F. Allard and Y.P. Sun. 2005. Functionalization of carbon nanotubes with derivatized polyimide. Macromolecules, 38: 7670–7675.

[54] Jiang, X., Y. Bin and M. Matsuo. 2005. Electrical and mechanical properties of polyimide–carbon nanotubes composites fabricated by *in situ* polymerization. Polymer., 46: 7418–7424.

[55] Gao, H., D. Yorifuji, J. Wakita, Z.H. Jiang and S. Ando. 2010. *In situ* preparation of nano ZnO/ hyperbranched polyimide hybrid film and their optical properties. Polymer., 51: 3173–3180.

[56] Liou, G.S., P.O.H. Lin, H.J.U. Yen, Y.Y. Yu and W.C. Chen. 2010. Flexible nanocrystalline-titania/ polyimide hybrids with high refractive index and excellent thermal dimensional stability. J. Polym. Sci. Part A Polym. Chem., 48: 1433–1440.

[57] Qiao-Hui, G., Z. Xiao-Ping, W. Su-Qin, F. Hong-Wei, L. Yong-Hong and H. Hao-Qing. 2009. Heat-resistant polyimide electrical conductive composites. Polym. Mat. Sci. Eng., 25(2): 52–54.

[58] Fukubayashi, Y. and Y. Satoshi. 2014. Porous polyimide-silica composite: A new thermal resistant flexible material. Mat. Res. Soc. Symposium Proceedings, pp. 1645.

[59] Zou, H., S. Wu and J. Shen. 2008. Polymer/silica nanocomposites: preparation, characterization, properties, and applications. Chem. Rev., 108: 3893–3957.

[60] Wang, P.J., C.H. Lin, S.L. Chang and S.J. Shih. 2012. Facile, efficient synthesis of a phosphinated hydroxyl diamine and properties of its high-performance poly(hydroxylimides) and polyimide–SiO$_2$ hybrids. Polym. Chem., 3: 2867–2874.

[61] Agag, T., T. Koga and T. Takeichi. 2001. Studies on thermal and mechanical properties of polyimide–clay nanocomposites. Polymer., 42: 3399–3408.

[62] Zhu, J., S. Wei, N. Haldolaarachchige, D.P. Young and Z. Guo. 2011. Electromagnetic field shielding polyurethane nanocomposites reinforced with core–shell Fe–silica nanoparticles. J. Phys. Chem. C, 115: 15304–15310.

[63] Kobayashi, Y., H. Katakami, E. Mine, D. Nagao, M. Konno and L.M. Liz-Marzan. 2005. Silica coating of silver nanoparticles using a modified Stöber method. J. Colloid. Interface. Sci., 283: 392–396.

[64] Mavinakuli, P., S. Wei, Q. Wang, A.B. Karki, S. Dhage, Z. Wang, D.P. Yuong and Z. Guo. 2010. Polypyrrole/silicon carbide nanocomposites with tunable electrical conductivity. J. Phys. Chem. Part C, 114: 3874–3882.

[65] Guo, Z., T.K. Kim, K. Lei, T. Pereira, J.G. Sugar and H.T. Hahn. 2008. Strengthening and thermal stabilization of polyurethane nanocomposites with silicon carbide nanoparticles by a surface-initiated-polymerization approach. Compos. Sci. Technol., 68: 164–170.

[66] Majewski, P., N.R. Choudhury, D. Spori, E. Wohlfahrt and M. Wohlschloegel. 2006. Synthesis and characterization of star polymer/silicon carbide nanocomposites. Mater. Sci. Eng. Part A, 434: 360–364.

[67] Bazzar, M., M. Ghaemy and R. Alizadeh. 2012. Novel fluorescent light-emitting polymer composites bearing 1,2,4-triazole and quinoxaline moieties: reinforcement and thermal stabilization with silicon carbide nanoparticles by epoxide functionalization. Polym. Deg. Stab., 97: 1690–1703.

[68] Rittenhouse, T.L., P.W. Bohn, T.K. Hossain, I. Adesida, J. Lindesay and A. Marcus. 2004. Surface state origin for the blue shifted emission in anodically etched porous silicon carbide. J. Appl. Phys., 95: 490–496.

[69] Fan, J.Y., X.L. Wu, F. Kong, T. Qiu and S.G. Huang. 2005. Luminescent silicon carbide nanocrystallites in 3C-SiC/polystyrene films. Appl. Phys. Lett., 86: 171903.

[70] Guo, Z., T.Y. Kim, K. Lei, T. Pereira, J.G. Sugar and H.T. Hahn. 2008. Strengthening and thermal stabilization of polyurethane nanocomposites with silicon carbide nanoparticles by a surface initiated-polymerization approach. Compos. Sci. Technol., 68: 164–170.

[71] Bazzar, M. and M. Ghaemy. 2013. 1,2,4-Triazole and quinoxaline based polyimide reinforced with neat and epoxide-end capped modified SiC nanoparticles: Study thermal, mechanical and photophysical properties. Compos. Sci. Technol., 86: 101–108.

[72] Zheng, H.F., Z. H. Li and Y.M. Zhu. 2007. High performance composite of silicon carbide/ bismaleimide. Key Eng. Mat., 336(2): 1377–1379.

[73] Xia, H., Q. Wang and G. Qiu. 2003. Polymer-encapsulated carbon nanotubes prepared through ultrasonically initiated *in situ* emulsion polymerization. Chem. Mater., 15: 3879–3886.

[74] Suslick, K.S. 1990. Sonochemistry. Science, 247: 1439–1445.

[75] Andrews, R., D. Jacques, M. Minot and T. Rantell. 2002. Fabrication of carbon multiwall nanotube/ polymer composites by shear mixing. Micromol. Mater. Eng., 287: 395–403.

[76] Lee, R.S., H.J. Kim, J.E. Fischer, A. Thess and R.E. Smalley. 1997. Conductivity enhancement in single-walled carbon nanotube bundles doped with K and Br. Nature, 388: 255–257.

[77] Chen, J., M.A. Hamon, H. Hu, Y. Chen., A.M. Rao, P.C. Eklund and R.C. Haddon. 1998. Solution properties of single-walled carbon nanotubes. Science, 282: 95–98.

[78] Eklund, P.C., L. Grigorian, K.A. Williams, G.U. Sumanasekera and S. Fang. 2000. Metallic nanoscale fibers from stable iodine-doped carbon nanotubes. U.S. Patent 6,139,919.

[79] Babasaheb, R. Sankapal, Kristina Setyowati, Jian Chen and Haiying Liu. 2007. Electrical properties of air-stable, iodine-doped carbon-nanotube–polymer composites. Appl. Phys. Lett., 91: 173103.

[80] Egorov, A.S. 2018. The development of technology for new composite materials modified by silicon carbide and carbon nanotubes. PhD Thesis Yegorov A.S. Published in Russian. http://www. irea.org.ru/upload/Dissertacion/Автореферат%20Егоров%20A_C.pdf.

[81] Wozniak, A.I., V.S. Ivanov, V.M. Retivov and A.S. Yegorov. 2015. Novel polymer nanocomposite with silicon carbide nanoparticles. Orient. J. Chem., 31(3): 1545–1550.

[82] Wozniak, A.I., V.S. Ivanov, O.V. Kosova and A.S. Yegorov. 2016. Thermal properties of polyimide composites with nanostructured silicon carbide. Orient. J. Chem., 32(6): 2967–2974.

[83] Ivanov, V.S., A.I. Wozniak, A.S. Yegorov. 2016. Heat-resistant composite materials based on polyimide matrix. Orient. J. Chem., 32(6): 3155–3164.

[84] Averina, E.A., A.S. Yegorov, M.V. Bogdanovskaya, A.I. Wozniak and O.A. Zhdanovich. 2018. Polyimide composite materials containing modified nanostructured boron carbide. Orient. J. Chem., 34(2): 743–749.

CHAPTER 7

Quantum Dots and Their Synthesis Processes

Prashant Ambekar[1],* and Jasmirkaur Randhawa[2]

1. Introduction

The possibility of zero-dimensional quantum confinement was realized in the year 1981, when Ekimov and co-workers at Ioffe Physical-Technical Institute, St. Petersburg observed unusual optical spectra for a sample of glass containing CdS and CdSe semiconductors [1]. The first explanation for the unusual optical behaviour has also been given by Ekimov, suggesting that nanocrystallites of the semiconductor got precipitated in glass due to heating and the quantum confinement of electrons in these nanocrystals, which were named quantum dots. A large amount of experimental and theoretical work has been done in the first half of the 1980s decade, studying the size-dependent development of bulk electronic properties in semiconductor crystallites of size ~ 15 to several hundred angstroms [2]. In some initial works, these crystallites have been termed as "clusters" because they were very small to have electronic wave function similar to bulk material despite exhibiting the same unit cell and bond length as the bulk semiconductors. For clusters, it has been concluded that complete delocalization of electrons has not yet occurred [3]. They are fluorescent semiconducting nanocrystals (NCs) with a radius that is comparable to that of the Bohr exciton radius of the material [4]. QDs possess various unique physical properties, such as quantum confinement effects, tunable electronic and optical properties, surface effects, high quantum yields, quantum tunneling effects, etc., due to their modified energy structure and increased surface to volume ratio. A/The visible change in color from lemon green to blood red, with a gradual decrease in the size of QDs is depicted in Figure 7.1. A typical example of emission wavelengths of different types and sizes of semiconductor QDs are listed in Table 7.1 by Reshma et al. [5]. QDs are among the most researched materials at

[1] Dharampeth M.P. Deo Memorial Science College, Nagpur, India-440033.
[2] Government College of Engineering, Nagpur, India-441108.
  Email: jbrandhawa2@gmail.com
* Corresponding author: pacific0701@gmail.com

Figure 7.1. The CdSe QD samples (A, C, E, G, and I) as seen under normal light and samples (B, D, F, H, and J) as seen under UV light (For interpretation of the references to color in this figure legend, the reader is referred to the web version of this article). [Reprinted from Synthesis, characterization, and application of CdSe quantum dots by Karan Surana, Pramod K. Singh, Hee-Woo Rhee and B. Bhattacharya. 2014. Journal of Industrial and Engineering Chemistry 20: 4188–4193. Copyright (2014) with the permission from Elsevier].

Table 7.1. Emission wavelength and size of the different QDs (Reprinted with the permission from ref. [5]).

Sr. No.	Quantum dots	Size/range/diameter (nm)	Emission range (nm)
1.	CdS	2.8–5.4	410–460
2.	CdTe	3.1–9.1	520–750
3.	CdSe	2–8	480–680
4.	CdTe/CdSe	4–9.2	650–840
5.	InP	2.5–4.5	610–710
6.	InAs	3.2–6	860–1270
7.	PbSe	3.2–4.1	1110–1310
8.	DT-Ag_2S [14]	5.4–10	1000–1300

present catering to the recent applications, such as light emitting diodes [6, 7], biolables [8, 9], medicine [10], lasers [11], and sensors [12, 13].

2. Types of QDs

On the basis of the nature of materials, configuration/structure, and applicability, the QDs developed so far could be listed in different types as—

a. Semiconductor Quantum Dots (SQDs)

b. Carbon-based Quantum Dots (CQDs)

c. Infrared Quantum Dots (IR QDs)

d. Dilute Magnetic Semiconductor Quantum Dots (DMS QDs)

e. Core-Shell Quantum Dots (CSQDs)

Semiconductor Quantum Dots (SQDs): The first in the genre are nanocrystals of group II–VI, III–V, and IV–VI binary, forming a large number of luminescent materials with unique optical, electrical, and physical properties [1]. They demonstrate size, shape, composition, and nanoscale interface-controlled fluorescence properties over a wide range of emission spectra, ranging from 450 to 1500 nm, making them a potential candidate for multiplex optical sensing, in long term *in vitro* and *in vivo* imaging [15].

Carbon-based Quantum Dots (CQDs): Xu et al. in 2004 discovered a new class of carbon nanomaterials, i.e., CQDs, while working on the purification of single-walled carbon nanotubes [16]. They show good conductivity, high chemical stability, environmental friendliness, broadband optical absorption, low toxicity, strong photoluminescence (PL) emission, optical properties, and can be synthesized easily at a large scale. Their physiochemical properties are seen to be controlled by surface passivation/functionalization with several polymeric, inorganic, organic, or biological materials [17]. The CQDs so far developed are of three kinds, viz., Polymer Dots (PDs), Carbon Nanodots (CNDs), and Graphene Quantum Dots (GQDs), of which GQDs and CQDs are fluorescent.

Infrared Quantum Dots (IR QDs): Due to the limited penetration depth of visible range photons, IR and near IR quantum dot-based devices are demanded by bio-imaging techniques looking at their performance in deep tissue imaging, wherein the absorption window of a spectral range of hemoglobin and water are blocked. Moreover, IRQDs are also in demand due to their importance in harvesting the Sun's infrared energy, which is available most of the time when direct sun rays are not available ($\sim$ 480 W/m^2). PbS, PbSe, InAs are a few examples of IRQDs developed so far [18].

Dilute Magnetic Semiconductor Quantum Dots (DMSQDs): The ferromagnetism in nanomaterials enhances its usefulness in opto-spintronics [19]. Ferromagnetic DMSs with an energy gap in the visible range are obtained by doping paramagnetic transition metal ions into a wide bandgap of semiconductors, such as ZnO doped with transition metals, viz., V^{2+}, Cr^{2+}, Mn^{2+}, Fe^{2+}, Co^{2+}, Ni^{2+}, of which V^{2+}, Fe^{2+}, and Co^{2+} doping exhibits ferromagnetism at room temperature [20]. This class of materials appears to be extremely sensitive to the conditions of sample preparation and post-synthetic treatment, and the ultimate source of the observed ferromagnetism remains controversial [21].

Core-Shell Quantum Dots (CSQDs): Core-shell QDs (CSQDs) are developed to improve the photoluminescence efficiency of single QDs, as well as to enhance their sensing applications. Achievement and sustenance of quantum confinement is generally obtained by encapsulation with organic surfactant. Organic encapsulation acts as a surface trap state, aiding non-radiative de-excitation of the charge generated by the photon, and hence reduces the fluorescent quantum yield. In CSQDs, epitaxial layers of inorganic material are grown over the core material, which improves the quantum efficiency due to increased confinement of electron-hole pair in the core and dangling bonds on the surface. The core and the shell are typically composed of groups II–VI, IV–VI, and III–V semiconductors, with configurations, such as (CdS) ZnS, (CdSe) ZnS, (CdSe) CdS, and (InAs) CdSe [22]. Different categories of CSQDs are formed depending on the position of the valence and conduction band, and the essential energy gap between them in the semiconductors, as depicted in Figure 7.2.

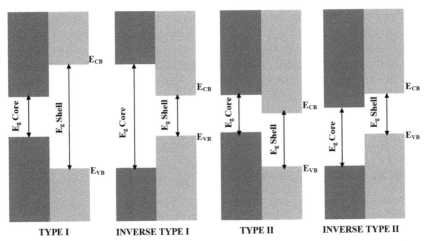

Figure 7.2. Types of core-shell quantum dots. [Reprinted from Core–shell quantum dots: Properties and applications by D. Vasudevan, Rohit Ranganathan Gaddam, Adrian Trinchi, Ivan Cole. 2015. Journal of Alloys and Compounds, 636: 395–404. Copyright (2015) with the permission from Elsevier].

3. Preparation Techniques

Quantum dots are prepared by both physical as well as chemical methods. The physical methods usually involve nucleation and growth of the particle in vapor phase or in solution. In the chemical methods, chemical reaction of formation of a substance is a necessary step. The steps occurring in chemical method are nucleation and crystal growth, which is followed by necessary supersaturation.

The two approaches used for the synthesis of QDs are popularly known as a top-down and bottom-up approach. In the top-down method, the bulk material is brought into smaller nanosized dimension by using different processes, tools, and particles. Typical examples could be given as optical lithography, chemical etching, use of electrons/ions/atoms particle beam, thin-film evaporation, molecular beam epitaxy, milling, and grinding.

In a/the bottom-up approach, the synthesis process is carried out either in the gas phase or in liquid state. In the gas phase process, the material to be synthesized is mixed/reacted in an atomic state in gas. The atoms of the material are produced *in situ* by evaporation. The synthesized bulk material is obtained by condensation with/without the addition of stabilizers. This method is used for the preparation of mono-dispersed metal particles and bulk powder of several oxides.

The synthesis process of nanoparticles carried out in solution phase is termed as arrested precipitation, in which further growth of synthesized particles is arrested at some stage by stabilizing its surface using surfactants (ionic and covalent both are used). Growth can also be arrested by using cavities of micelles, zeolites, membranes, etc., and also by quenching, i.e., sudden variation in temperature and pressure [23].

3.1 Electron Beam Lithography

Mark Reed and co-workers of Texas Instruments in Dallas made the first lithographic quantum dots in 1987 using electron-beam lithography [24]. An electron beam scans the semiconductor surface, which has been coated with a thin polymer layer called a resist. (Similar effects can also be achieved by means of x-rays or ion beams.) A series of process steps replace the resist with a thin layer of metal in areas where the beam was scanned at high intensity. A shower of reactive gas then etches away the unprotected quantum-well material, leaving the pillars behind. Using this technique, pillars or other features as small as 1,000 angstroms across can be quite easily constructed. However, the process becomes increasingly difficult as the scale falls to about 100 angstroms, which is the limit of the best-known resist.

3.2 Molecular Beam Epitaxy

Molecular beam epitaxy is considered the most valuable and pure method of preparation of SQDs [25]. The approach involves deposition of atomic or molecular beams produced in some sources on crystalline substrates with atomically smooth surface in ultrahigh vacuum. Spontaneous formation of ordered quantum dot arrays on crystalline surfaces represents an example of self-organization of matter in the course of stress relaxation in thin epitaxial films; these stresses originate from the mismatch between the lattice constants of the film and substrate. This unique method produces quantum dot arrays with ultra-small dispersion of the size distribution function, and helps to study the quantum size effects. Alchalabi et al. prepared PbSe quantum dots with very small dispersion of the size distribution function (only 2%) using this method [26]. However, the method requires complex equipment and high-purity materials.

3.3 High Temperature Colloidal Synthesis

Production of high optical quality monodisperse (< 5% RMS in diameter) CdS, CdSe, and CdTe nanocrystals by a simple synthetic route has been first reported by Murray et al. [27]. This technique is also called high temperature colloidal synthesis, namely pyrolysis of organometallic precursors at 300°C. In this technique, the organometallic reagent is rapidly injected into a hot coordinating solvent to produce temporally discrete homogeneous nucleation. The coordinating solvent facilitates slow growth and annealing, resulting in uniform surface derivatization and regular core structure. Powders of the desired size are precipitated to get monodispersed nanocrystallites (slightly prolate $1.1 \sim 1.3$) of average size tunable from 12 to 115 AU, which can be dispersed in a variety of solvents. Chemical processes involving the compounds, such as alkylcadmium, silylchalconides, phosphines, phosphine chalcogenides are carried out in an airless manner. For producing 1.0 M stock solutions of trioctylphosphine selenide [TOPSe] and trioctylphosphine telluride [TOPTe], required masses of selenium and tellurium shot have been dissolved directly in sufficient TOP [28].

The typical preparation of Tri-n-octylphosphine (TOP) or Tri-n-octylphosphine oxide (TOPO) capped CdSe nanocrystallites is reprinted here as "Method 1 and 2 with permission from J. Am. Chem. Soc. 1993, 115: 8706–8715. Copyright (1993) American Chemical Society".

Method 1: Fifty grams of TOPO is dried and degassed in the reaction vessel by heating to 200°C at 1 Torr for 20 minutes, flushing periodically with argon. The temperature of the reaction flask is then stabilized at 300°C under 1 atm of argon.

Solution A is prepared by adding 1.00 mL (13.35 mmol) of Me_2Cd to 25.0 mL of TOP in the dry-box. Solution B is prepared by adding 10.0 ml of the 1.0 M TOPSe stock solution (10.00 mmol) to 15.0 ml of TOP. Solutions A and B are combined and loaded into a 50-mL syringe in the dry-box.

The heat is removed from the reaction vessel. The syringe containing the reagent mixture is quickly removed from the dry-box, and its contents are delivered to the vigorously stirring reaction flask in a single injection through a rubber septum. The rapid introduction of the reagent mixture produces a deep yellow/orange solution with an absorption feature at 440–460 nm. This is also accompanied by a sudden decrease in temperature to 80°C. Heating is restored to the reaction flask, and the temperature is gradually raised to 230–260°C.

Aliquots of the reaction solution are removed at regular intervals (5–10 minutes), and absorption spectra were taken to monitor the growth of the crystallites. The best quality samples are prepared over a period of a few hours of steady growth by modulating the growth temperature in response to changes in the size distribution as estimated from the absorption spectra. The temperature is lowered in response to a spreading of the size distribution and increased when growth appears to stop. When the desired absorption characteristics are observed, a portion of the growth solution is transferred by cannula and stored in a vial. In this way, a series of sizes ranging from 15 to 115 AU in diameter can be isolated from a single preparation.

CdTe nanocrystallites are prepared by Method 1, with TOPTe as the chalcogen source, an injection temperature of 240°C, and growth temperatures between 190 and 220°C.

Method 2: A second route to the production of CdE (E = S, Se, Te) nanocrystallites replaces the phosphine chalcogenide precursors in Method 1 with $(TMS)_2S$, $(TMS)_2Se$, and $(BDMS)_2Te$, respectively. Growth temperatures between 290 and 320°C were found to provide the best CdS samples. The smallest (12 AU) CdS, CdSe, and CdTe species are produced under milder conditions with injection and growth carried out at 100°C.

The detailed procedures for Isolation and Purification of Crystallites, Size-Selective Precipitation, and Surface Exchange are described by Murray et al. [27].

High Temperature Colloidal Synthesis Method has also been used by Schwartz et al. [21] for the preparation of colloidal ZnO DMS-QDs by alkaline-activated hydrolysis and condensation of zinc acetate solutions in dimethyl sulfoxide (DMSO). The process carried out is "Reprinted (adapted) with permission from (Dana A. Schwartz, Nick S. Norberg, Quyen P. Nguyen, Jason M. Parker, and Daniel R. Gamelin. 2003. J. Am. Chem. Soc., 125: 13205–13218). Copyright (2003)

American Chemical Society". For the preparation of colloidal nanocrystalline ZnO, typically 30 mL of 0.552 M N(Me)$_4$OH in EtOH (1.8 equivalent of OH$^-$) has been added dropwise at the rate of 2 mL/min to a 90-mL solution of 0.101 M Zn(OAc)$_2$ in DMSO with constant stirring. Then the transition-metal doping has been achieved by the addition of Co(OAc)$_2$.4H$_2$O or Ni(ClO$_4$)$_2$.6H$_2$O to the Zn(OAc)$_2$.2H$_2$O precursor solution. The nanocrystals prepared have been precipitated from DMSO by the addition of ethyl acetate or heptane. They have been later washed with ethyl acetate and resuspended in DMSO or ethanol. Addition of 10 mg of Zn(OAc)$_2$ restores the optical clarity of the suspension. On the other hand, for preparation of Co^{2+}:ZnO nanocrystals, the solvent has been allowed to evaporate from a concentrated ethanol suspension of nanocrystals in the air at room temperature. For the treatment with trioctylphosphine oxide (TOPO), the nanocrystals have been precipitated from DMSO by addition of dodecylamine, and washed with ethanol to remove DMSO. The excess Zn^{2+} precursor from the DMSO solution has been carefully removed completely prior to the addition of the amine, since the amine causes rapid growth of the nanocrystals with variable size. The dodecylamine-capped nanocrystals have been then resuspended in toluene and precipitated with ethanol twice. The resulting powder has been heated in TOPO at 180°C for at least 30 minutes, and then cooled to below 80°C, precipitated, and washed with ethanol. The resulting powders may be re-suspended (with the addition of 1 mg of additional TOPO) in a variety of nonpolar solvents to form stable, high-optical-quality colloidal solutions.

3.4 Hydrothermal Method

The process of preparation of QDs at high vapor pressure and moderate temperature yields crystallization under a controlled atmosphere. The hydrothermal method uses comparatively low temperature under-water synthesis with an added advantage of a good reaction rate to prepare specific materials. The basic need of this method is the solubility of minerals/precursors in hot water and under high pressure. Therefore, it requires a bulky and strong-walled vessel called a hydrothermal bomb or autoclave. This method has been proposed in the middle of the 19th century, and has been used extensively due to its low cost, environment non-toxicity, and simple experimental facilities [29]. For the first time, by using this wet chemical route, a series of oxidize-stable CdTe nanoclusters with narrow size distributions and extremely small particle sizes ranging from 1.3 to 2.4 nm have been prepared in an aqueous solution using 2-mercaptoethanol and 1-thioglycerol as stabilizers in the year 1996 by Rogach et al. [30]. However, the process is a bit time consuming as it requires keeping materials in an/the autoclave for more time, and sometimes for days. Therefore, a new modified method called Microwave-assisted Hydrothermal method was developed. Wang et al. [31] have given the following modified one-step synthesis technique to prepare CdTe Quantum Dots. The excerpts are reproduced here with permission.

"The one-step synthesis of water-soluble cysteamine (CA), CA-CdTe QDs was performed using TeO$_2$, NaBH$_4$ (96%), and CdCl$_2$·2.5H$_2$O (99%) as precursors. CA (0.6816 g) was added to 100 mL of 2×10^{-2} mol/L CdCl$_2$·2.5H$_2$O solution under stirring. The pH of the solution was then adjusted to 5.85 by dropwise addition of 1 mol/L NaOH solution. The solution was deaerated under nitrogen flow for

30 minutes. While stirring, TeO_2 and $NaBH_4$ were added to the original oxygen-free solution. The typical molar ratio of $Cd^{2+}/Te^{2-}/CA$ was 1:0.05:2. The resulting mixture solution was heated to 100°C and refluxed at different times to control the size of the CdTe QDs."

Tian et al. [32] have published details about the different hydrothermal methods to prepare the graphene quantum dots. Dong et al. [33] described the comparatively simple method to prepare GQDS using citric acid at 200°C. In this method, citric acid is directly heated till the formation of graphene oxide, which is then reduced to graphene quantum dots using NaOH solution, maintaining a pH value of 7.0 repeatedly to produce a final product.

3.5 Solvothermal Method

This method has a small deviation from the hydrothermal method, wherein organic solvents are used as a solvent to synthesize water-based sensitive materials. It also evades air for air-sensitive precursors in a hermetically sealed system and prevents the volatilization of toxic substances. As this process needs to be performed at high temperatures and pressures, the costs involved are higher. This method shows ease in precise control over the size, shape, and crystallinity of metal oxide nanoparticles or nanostructure products by simply changing certain experimental parameters. The reaction temperature, time, solvent type, surfactant type, and precursor type can also be altered to have precise control over the particles, such as nanostructured to make nanostructured titanium dioxide [34], graphene [35], carbon spheres [36], etc.

3.6 Core-Shell Structured Quantum Dots

The CSQDs are prepared either by organometallic route or by aqueous phase synthesis [37]. Core metal nanoparticles are prepared by high temperature thermal decomposition of organometallic compounds in the absence of water and oxygen [38]. A typical example is the synthesis of cobalt metal from octacarbonyl dicobalt complex at 500°C in the presence of surfactant. The surfactant added controls the particle size, and can later be removed by repeated washing, and the core QDs are stored in toluene and ethanol. Contact with air leads to the formation of oxide, and hence this method is used for the fabrication of metal-metal oxide core shell nanoparticles.

In order to enhance the air stability of CSQDs, a number of inorganic and organic surface coatings have been developed (Figure 7.3) to protect the core–shell structure. These surface modifications could be achieved through many methods, such as covalent bond formation, passive adsorption, electrostatic forces, and multivalent chelation [39]. These organic/inorganic surfaces play a major role in tailoring the properties of CSQD. The first inorganic surface coating of a core shell QD to reduce lattice mismatch has been reported by Bleuse et al. [40]. The inorganic surface coating provides considerable stability against photooxidation, higher quantum yield, extraordinary photoluminescence, enhanced optical properties, and increased half lifetime. Besides semiconductors, other types of inorganic shell materials used include polymers, silica, metals (Au, Ni, Co, Fe) and metal oxides,

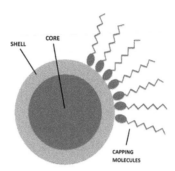

Figure 7.3. Surface modification of core-shell quantum dots. [Reprinted from Core–shell quantum dots: Properties and applications by D. Vasudevan, Rohit Ranganathan Gaddam, Adrian Trinchi, Ivan Cole. 2015. Journal of Alloys and Compounds, 636: 395–404. Copyright (2015) with the permission from Elsevier].

carbon, etc., to act as protection for chemically highly active core from oxidation or erosion, to provide bio-affinity through functionalization of amine/thiol terminal groups, for some specific applications, such as catalysis, solar energy absorption, and permanent magnetic properties, to enhance the adsorption capacity for environmental remediation applications [37]. The QDs having a semiconducting surface coating of band gap energy value higher than the shell are called a quantum dot quantum well (QDQW), for example, a QDQW with structure ((CdS) HgS) CdS [41].

Organic surface modification of CSQDs is achieved by attaching functional moieties either by a covalent bond or mere physical adsorption on the surface. In general, the usual method is to cap the synthesized QDs with thiolate ligands during the growth period. Different capping ligands that have been used are Thioglycerol, mercaptoacetate, 1,4-dithiothreitol, 2-mercaptoethanol, cysteine, methionine, and glutathione, amines such as hexadecylamine, n-butylamine, and n-hexylamine have also been used in conjunction with TOP and TOPO [42].

3.7 Chemical Ablation

In this process, organic molecules are carbonized to carbonaceous materials by using strong oxidizing acids. The materials can be converted into small sheets by controlled oxidation. One simple process of synthesizing CQDs in an aqueous solution has been reported, where carbonates are dehydrated with concentrated sulfuric acid, then the carbonaceous material is broken into QDs with nitric acid, and then finally passivated with amine-terminated compounds (4,7,10-trioxa-1,13-tridecanedi-amine) [43]. Surface passivation is an important criterion for the photo-luminescence application of CQDs, the emission wavelength of which can be tuned by selecting the proper material and controlling the duration of nitric acid treatment. One-pot preparation of photoluminescent CQDs has been reported using carbon as a source of QD and polyethyleneimine (PEI-cationic branched polyelectrolyte) oxidized with HNO_3, the PL response of which has been found tp be very much pH-sensitive, in sharp contrast to the generally reported pH-insensitivity of CQDs [17].

References

[1] Ekimov, A.I. and A.A. Onushchenko. 1981. Quantum size effect in three-dimensional microscopic semiconductor crystals. JETP Letr., 34: 346.

[2] Louis Brus. 1986. Electronic wave functions in semiconductor clusters: Experiment and theory. J. Phys. Chem., 90: 2555–2560.

[3] Rossetti, R., S. Nakahara and L.E. Brus. 1983. Quantum size effects in the redox potentials, resonance Raman spectra and electronic spectra of CDS crystallites in aqueous solutions. J. Chem. Phys., 79: 1086.

[4] Vasudevan, D., S.A. Ziaudeen, R.R. Gaddam, P.K. Pallapothu, M.K. Sugumar and J. Rangarajan. 2013. Supra gap excitation properties of differently confined PbS-nano structured materials studied with opto-impedance spectroscopy. J. Nanophotonics, 7: 073075.

[5] Reshma, V.G. and P.V. Mohanan. 2019. Quantum dots: Applications and safety consequences. J. Lumin., 205: 287–298.

[6] Park, N.M., T.S. Kim and S.J. Park. 2001. Band gap engineering of amorphous silicon quantum dots for light-emitting diodes. Appl. Phys. Lett., 78: 2575–2577.

[7] Jang, H.S., H. Yang, S.W. Kim, J.Y. Han, S.G. Lee and D.Y. Jeon. 2008. White light-emitting diodes with excellent color rendering based on organically capped CdSe quantum dots and $Sr3SiO_5$: Ce^{3+}, Li+ phosphors. Adv. Mater., 20: 2696–2702.

[8] Bruchez, M., M. Moronne, P. Gin, S. Weiss and A.P. Alivisatos. 1998. Semiconductor nanocrystals as fluorescent biological labels. Science, 281: 2013–2016.

[9] Chan, W.C. and S. Nie. 1998. Quantum dot bioconjugates for ultrasensitive nonisotopic detection. Science, 281: 2016–2018.

[10] Bagalkot, V., L. Zhang, E. Levy-Nissenbaum, S. Jon, P.W. Kantoff, R. Langer and O.C. Farokhzad. 2007. Quantum dot-aptamer conjugates for synchronous cancer imaging, therapy, and sensing of drug delivery based on bi-fluorescence resonance energy transfer. Nano Lett., 7: 3065–3070.

[11] Huffaker, D.L., G. Park, Z. Zou, O.B. Shchekin and D.G. Deppe. 1998. 1.3 µm room temperature GaAs-based quantum-dot laser. Appl. Phys. Lett., 73: 2564–2566.

[12] Zhang, C.Y., H.C. Yeh, M.T. Kuroki and T.H. Wang. 2005. Single-quantum-dot-based DNA nanosensor. Nat. Mater., 4: 826–831.

[13] Shi, L., V. De Paoli, N. Rosenzweig and Z. Rosenzweig. 2006. Synthesis and application of quantum dots FRET-based protease sensors. J. Am. Chem. Soc., 128: 10378–10379.

[14] Yan Zhang, Guosong Hong, Yejun Zhang, Guangcun Chen, Fengli, Hongjie Dai and Qiangbin Wang. 2012. ACS Nano, May 22, 6(5): 3695–3702. DOI 10.1021/nn301218z.

[15] Jie Zhoul, Yun Liu, Jian Tang and Weihua Tang. 2017. Materials Today, Volume 20, Number 7, September.

[16] Wang, Y. and A. Hu. 2014. Carbon quantum dots: synthesis, properties and applications. J. Mater. Chem. C, 2(34): 6921–6939.

[17] Namdari Pooria, Babak Negahdari and Ali Eatemadi. 2017. Synthesis, properties and biomedical applications of carbon-based quantum dots: an updated review. Biomed. Pharmacother., 87: 209–222.

[18] Sargen Edward. 2005. Infrared quantum dots. Adv. Mater., 17(5). March 8; DOI: 10. 1002/adma.200401552.

[19] Pearton, S.J., C.R. Abernathy, M.E. Overberg, G.T. Thaler, D.P. Norton, N. Theodoropoulou, A.F. Hebard, Y.D. Park, F. Ren, J. Kim and L.A. Boatner. 2003. Wide band gap ferromagnetic semiconductors and oxides. J. Appl. Phys., 93: 1.

[20] Pavle, V. Radovanovic and Daniel R. Gamelin. 2003. High-temperature ferromagnetism in Ni^{2+}-doped ZnO aggregates prepared from colloidal diluted magnetic semiconductor quantum dots. Phys. Rev. Lett., 91(15). 10 October 2003; DOI: 10.1103/PhysRevLett.91.157202.

[21] Schwartz Dana A., Nick S. Norberg, Quyen P. Nguyen, Jason M. Parker and Daniel R. Gamelin. 2003. Magnetic quantum dots: Synthesis, spectroscopy, and magnetism of Co^{2+}- and Ni^{2+}- doped ZnO nanocrystals. J. Am. Chem. Soc., 125: 13205–13218.

[22] Zeng, Z., C.S. Garoufalis, A.F. Terzis and S. Baskoutas. 2013. Linear and nonlinear optical properties of ZnO/ZnS and ZnS/ZnO core shell quantum dots: effects of shell thickness, impurity, and dielectric environment. J. Appl. Phys., 114: (023510–023510(9)).

[23] Sergey B. Brichkin and Vladimir F. Razumov. 2016. Colloidal quantum dots: synthesis, properties and applications. Russ. Chem. Rev., 85: 1297.

[24] Reed, Mark A. 1993. Quantum dots. Sci. Am. January 1993, pp. 118–123.

[25] Chang, L.L. and K. Ploog. 1985. Molecular Beam Epitaxy and Heterostructures (Nato Science Series E Vol. 87) (Netherlands: Springer).

[26] Alchalabi, K., D. Zimin, G. Kostorz and H. Zogg. 2003. Self-assembled semiconductor quantum dots with nearly uniform sizes. Phys. Rev. Lett., 90: 026104.

[27] Murray, C.B., D.J. Noms and M.G. Bawendi. 1993. Synthesis and characterization of nearly monodisperse CdE (E = S, Se, Te) semiconductor nanocrystallites. J. Am. Chem. Soc., 115: 8706–8715.

[28] Zingaro, R.A., S. teeves, B.H. Ir gorlic and K.J. Orgunomet. 1965. Phosphine tellurides. J. Organomet. Chem., 4: 320.

[29] Yang, W.H., W.W. Li, H.J. Dou and K. Sun. 2008. Hydrothermal synthesis for high-quality CdTe quantum dots capped by cysteamine. Mater. Lett., 62: 2564.

[30] Rogach, A.L., L. Katsikas, A. Kornowski, Dangsheng Su, A. Eychmuller and H. Weller. 1996. Synthesis and characterization of thiol stabilized CdTe nanocrystals. Ber. Bunsenges. Phys. Chem., 100: 1772–1778.

[31] Wang, R., Y. Wang, Q. Feng, L. Zhou, F. Gong and Y. Lan. 2012. Synthesis and characterization of cysteamine-CdTe quantum dots via one-step aqueous method. Mater. Lett., 66: 261–263.

[32] Tian, P., L. Tang, K.S. Teng and S.P. Lau. 2018. Graphene quantum dots from chemistry to applications. Mater. Today Chem., 10: 221e258.

[33] Yongqiang Dong, Jingwei Shao, Congqiang Chen, Hao Li, Ruixue Wang, Yuwu Chi, Xiaomei Lin and Guonan Chen. 2012. Blue luminescent graphene quantum dots and graphene oxide prepared by tuning the carbonization degree of citric acid. Carbon, 50: 4738–4743.

[34] Xie, Rong-Cai, Shang and Jian Ku. 2007. Morphological control in solvothermal synthesis of titanium oxide. J. Mater. Sci., 42(16): 6583–6589. DOI:10.1007/s10853-007-1506-0.

[35] Choucair, Mohammad, Thordarson, Pall and Stride, John A. 2008. Gram-scale production of graphene based on solvothermal synthesis and sonication. Nat. Nanotechnol., 4(1): 30–3. doi:10.1038/nnano.2008.365. PMID 19119279.

[36] Hu, Gang, Ma, Ding, Cheng, Mojie, Liu, Lin and Bao, Xinhe. 2002. Direct synthesis of uniform hollow carbon spheres by a self-assembly template approach. Chem. Commun., (17): 1948–1949. doi:10.1039/B205723A.

[37] Vasudevan, D., Rohit Ranganathan Gaddam, Adrian Trinchi and Ivan Cole. 2015. Core–shell quantum dots: Properties and applications. J. Alloy Compd., 636: 395–404.

[38] Nadagouda, M.N. and R.S. Varma. 2007. A greener synthesis of core (Fe, Cu)-shell (Au, Pt,Pd, and Ag) nanocrystals using aqueous vitamin C. Cryst. Growth Des., 7: 2582–2587.

[39] Gao, X., L. Yang, J.A. Petros, F.F. Marshall, J.W. Simons and S. Nie. 2005. *In vivo* molecular and cellular imaging with quantum dots. Curr. Opin. Biotechnol., 16: 63–72.

[40] Bleuse, J., S. Carayon and P. Reiss. 2004. Optical properties of core/multishell CdSe/Zn (S, Se) nanocrystals. Phys. E: Low-Dimensional Syst. Nanostruct., 21: 331–335.

[41] Lifshitz, E., H. Porteanu, A. Glozman, H. Weller, M. Pflughoefft and A. Echymüller. 1999. Optically detected magnetic resonance study of CdS/HgS/CdS quantum dot quantum wells. J. Phys. Chem. B, 103: 6870–6875.

[42] Chen, L.Y., H.L. Chou, C.H. Chen and C.H. Tseng. 2012. Surface modification of CdSe and CdS quantum dots-experimental and density function theory investigation. Nanocryst. – Synth., Charact. Appl., 8: 149.

[43] Tian, L., D. Ghosh, W. Chen, S. Pradhan, X. Chang and S. Chen. 2009. Nanosized carbon particles from natural gas soot. Chem. Mater., 21(13): 2803–2809.

CHAPTER 8

Chemical Vapor Deposition (CVD) Technique for Nanomaterials Deposition

Abhishek K. Arya,[2] Rahul Parmar,[1] K.S. Gour,[2] Decio B. de F.N.,[3] R. Gunnella,[1] J.M. Rosolen[3] and V.N. Singh[2,]*

1. Introduction

In recent years, nanoscience and nanotechnology have been assisting the researchers and scientists to understand the problems, such as deformation/distortion in the structure of materials, and this is one of the reasons behind the degradation in efficiency of energy conversion and storage devices. Nanoscience and nanotechnology have a wide range of applications at nanometer scale (10^{-9} meters), i.e., inorganic, organic, metallic, metal-oxides, ceramic, polymers, etc. Physical and chemical properties of any material play a very crucial role to understand crystal structure, surface morphology, electrical structure, chemical bonding, phases of crystals, interface between solid-liquid, energy capacity, power density, storage capacity, etc. Further, these structures need to be well understood at nanoscale and then improved in the devices at an industrial scale. At present, researchers have mechanized many efficient chemical and physical techniques to grow nanoparticles, such as spray coating, electrospinning, sputtering (DC/RF), plasma deposition, dip coating, spin coating, electro-chemical deposition (ECD), atomic layer deposition (ALD), chemical bath deposition (CBD), molecular beam epitaxy (MBE), ball milling, etching through chemical acids to deform the materials into nanomaterials, photolithography (Electron Beam Lithography, X-Ray Lithography), nanomasking, and chemical vapor deposition, etc. All these techniques are efficient at particular levels, and considered requisites. Among these techniques, some are economical and some are expensive.

[1] Sez. Fisica Scuola di Scienze e Tecnologie, Universita di Camerino, Via Madonna delle Carceri, I-62032 Camerino (MC), Italy.
[2] CSIR-National Physical Laboratory, Dr. K.S. Krishnan Road, New Delhi 110012, India.
[3] Departamento de Quimica-FFCLRP, Universidade de São Paulo, Ribeirào Preto-14040-930 SP, Brazil.
* Corresponding author: singhvn@nplindia.org

In this chapter, we are going to discuss the low cost, easy to operate, and efficient way to deposit the nanomaterials at both laboratory and industrial scale by using the chemical vapor deposition technique. The chemical vapor deposition (CVD) method is very popular among the researchers due to its flexibility in operation and working efficiency in the field of nanoscience and nanotechnology. In this chapter, a brief history, classification, working principle, applications in different nanomaterials, and its advantages and disadvantages are to be discussed.

2. Definition

There are many flexible definitions of the CVD method which have been given by researchers and scientists in their own words. In simple reported words, the CVD is a technique to deposit the nanostructure/nanomaterial in the solid-state form with some required sources on the specific target/sample with the help of vapor/gases in the inert atmosphere at a controlled temperature and pressure zone. In other words, CVD is a technique to deposit the desired solid nanostructures on the surface of substrate by decomposition or chemical reactions of specific precursors in the presence of particular catalysts by the flow of inert carrier gases at a controlled flow rate inside the reaction chamber/tube. It produces the high-quality thin films or nanostructures for energy conversion and storage devices at a commercial scale. Jan-Otto et al. have reported the definition of CVD in the 'Handbook of Deposition Technologies for Films and Coatings' as the process in which the substrate is exposed to one or more volatile precursors, which reacts on the substrate to get the desired thin film, or as the family of processes where a solid material is deposited from a vapor by a chemical reaction occurring on a normally heated substrate surface [1]. Another definition is that this method can be used to deposit good quality thin films, powder, single, and polycrystalline films, different kinds of oxide coatings, texture, or shape substrates. Guo et al. reported that CVD is the process in which a precursor is converted to nanoparticles, which is widely used for ceramic nanopowders depositions [2]. According to Wilson, CVD is the process of synthesizing one material on another material from a vapor precursor caused to react by heating. In this article, the CVD was used for surface functionalization of a textile, i.e., coating of nanoparticles on textiles for integrated sensors and actuators applications [3]. Makhlouf et al. reported that the CVD process to coat any ceramic, metallic alloys, and inter-metallic compounds in solid form from a gaseous phase, is by chemical reaction between volatile precursors (gases phases) and the surface to be coated (heated substrate) [4]. The CVD method has been used widely in material science for commercial production of thin film coating of metal oxides, sulfides, silicates, carbides, nitrides for photovoltaic, corrosive resistive coatings, electronics/microelectronic, magnetic properties, micro, catalysis, different kinds of mirror coatings, and nanostructures materials in the field of nanoscience and nanotechnology. So overall, CVD techniques are a bottom-up approach to synthesize or deposit the nanomaterials on the substrate at different conditions or parameters. This approach is very popular for homogeneous nanoparticles deposition on the large surface area as well as on any shape's substrate where the deposition of nanoparticles occurs from an atomic scale to a nano scale, which would change the properties of the materials in output devices.

2.1 A Brief History

In 1855, the CVD method was reported by Wohler for metal deposition in which tungsten (WCl6) was used as a precursor with hydrogen as carrier gas. After that the CVD system was used by John Howarth to produce black carbon and later it was used for burning of wooden wastes. In 1880, Sawyer and Mann filed the patent to make carbon fiber filament in electric lamp industry by the CVD method, and later this process of deposition was patented for metal deposition for lamp filaments applications. In 1890, the popular Mond Process was explained to deposit the pure nickel from nickel ores. In 1909, the deposition of silicon by hydrogen reduction of $SiCl_4$ for silicon thin films was used in electronics and photocells applications. Later in the year 1950, the Tri-iso-butyl-aluminum was used as a catalyst to deposit the highly pure aluminum by polymerization of olefins using the Ziegler-Natta method, and later this pure aluminum was used in large scale for VLSI applications [5]. So, after the second world war, researchers started to produce coated materials in various fields, and thus the CVD technique came into trend. In 1960, the terms chemical vapor deposition and physical vapor deposition (PVD) were used as technical aspects, and in this same year CVD was introduced for semiconductor materials fabrications for many electronics and photovoltaic applications. Later in the year 1960, the carbide of titanium (TiC) was deposited first, and CVD tungsten was developed. In 1963, the plasma-based CVD was introduced in electronics applications. In 1968 and 1980, the CVD for cemented carbide coating and diamond coating were introduced, respectively. Later in 1990, the Metal-Organic CVD for metal and ceramic deposition, and combined CVD, PVD, plasma tools were used to develop the semiconductor device fabrication in electronic and optoelectronic applications [6].

2.2 Deposition Parameters in CVD

2.2.1 Precursors

Precursors in the form of gas or liquid provide reactive species, and when these precursors come in contact with the heated substrate, they generate raw material (nanomaterial), so proper selection is very important, as it affects the other growth parameters. Types of precursors fall into several general groups, which are the halides, carbonyls, and hydrides. The general characteristics of a precursor which are taken in consideration can be summarized as follows [6].

- Good stability at room temperature.
- React cleanly in the reaction zone.
- Sufficient volatility at low temperature so that it can easily be transported to the reaction zone without condensing.
- High degree of purity and yield when produced.
- Able to react without producing side reactions.

2.2.2 Carrier Gas

Carrier gases in the CVD process directly affect its parameters, i.e., deposition time, growth time. Inert gases, such as argon, helium, and hydrogen have been largely studied as carrier gases [7, 8]. Carrier gases have different momentum and thermal diffusivity, i.e., mean free path and mass diffusivity of reactant molecules have a significant effect on the deposition rate, composition, and morphology of the structure. The general characteristics of a carrier gas can be summarized as follows:

- Transport of the reagents in gas phase (often with carrier gas) to the reaction zone.
- Diffusion (or convection) through the boundary layer.
- Adsorption of precursors on the substrate.
- Surface diffusion of the precursors to growth sites and reaction without diffusion is not needed, as this may lead to rough growth of film.
- Surface chemical reaction, formation of a solid film, and formation of by-products.
- Desorption of by-products.
- Diffusion of by-products through the boundary layer.
- Transport of gaseous by-products out of the reactor.

2.2.3 Substrate

The CVD method can be used to grow films on various types of substrates. Substrate selection depends upon the individual case of synthesis, but they have to obey some general characteristics, such as stability, good adhesion with film, growth temperature stability, and their inertness in the growth environment.

For example, the effect of substrate's pore structure with bulk phase reactant concentration, reactant diffusion, and deposition temperature are studied experimentally and explained qualitatively by a theoretical modeling analysis. In this report, results revealed that the reaction mechanism depends on water vapor and chloride vapor concentrations. Consequently, diffusivity, bulk phase reactant's concentration, and substrate's pore dimension are important in the CVD process. The effect of deposition temperature and narrow deposition zone as compared to the substrate thickness also suggested a mechanism named Langmuir-Hinshelwood Mechanism. This mechanism got involved in the CVD process for a very fast reaction rate. Further, gas permeation data indicated that deposition of solid in a substrate's pores could result in the pore size reduction, which strongly depends on the initial pore size distribution of the substrate [9].

2.2.4 Catalysts

Selection of the catalyst becomes very important for synthesis of particular nanomaterials. Some of the recent techniques, such as the catalyst enhanced chemical vapor deposition (CECVD) method have emerged as new enhanced techniques. It is particularly suitable for the deposition of metallic films on thermally sensitive substrates. Palladium, platinum, and nickel have been found to be very suitable catalysts for the deposition of metallic layers on polymers. For example, catalyst nanoparticles play a key role in carbon nanotubes growth by catalytic chemical vapor deposition (CCVD). CNTs growth via CCVD process includes the decomposition of carbon source near the catalyst surface through catalytic mechanism. Further, there is diffusion of carbon into catalyst particles and finally solid carbon structure due to its super saturation in the catalyst particles. In this described process, they are capable of decomposing the hydrocarbons used for CNTs growth. Overall, transition metals have been reported to be appropriate catalysts. In recent researches, alloys of these metals have proved to be better catalysts and produce CNTs of high quality [10].

2.2.5 Growth/Deposition Rate

Growth rate is a dependent parameter which depends on the physical parameters, such as the temperature of the substrate, operating pressure in the reactor, composition, and chemistry of different phases. Kinetics and mass transport can both play a significant role in the film deposition. At lower growth temperature, deposition rate is controlled by the kinetics of chemical reactions occurring either in the gas phase or on the substrate surface. In the case of film, growth rate increases exponentially with substrate temperature according to the Arrhenius equation:

$$\text{Growth rate} \propto \exp(\frac{E_A}{R \cdot T})$$

where E_A is the activation energy, R is the gas constant, and T is the temperature. As the temperature increases, the growth rate becomes nearly independent of temperature. Further, this growth rate is controlled by the mass transport of reagents through the boundary layer to the growing surface. The pressure of the CVD reactor also influences the growth rate. As the pressure falls, gas phase reactions tend to become less important, but layer growth is often controlled by surface reactions. At very low pressures (e.g., 10^{-4} Torr), mass transport is completely absent and layer growth is primarily controlled by the gas and substrate temperature and by desorption of precursor fragments and matrix elements from the growth surface.

Steps involved in CVD growth process

- Reactant molecules diffuse through the boundary layer near gas-solid interface.
- They adsorb on the surface.
- Get diffused on the surface.
- Further, react with each other and the solid product is formed. Any gaseous by-product formed may be adsorbed on the surface.
- Desorbs and diffuses outward into the gas stream and gets carried away.

2.3 Classification of Chemical Vapor Deposition (CVD)

2.3.1 Atmospheric Pressure Chemical Vapor Deposition (APCVD)

2.3.2 Metal-organic Chemical Vapor Deposition (MOCVD)

2.3.3 Low-pressure Chemical Vapor Deposition (LPCVD)

2.3.4 Plasma-enhanced Chemical Vapor Deposition (PECVD)

2.3.5 Microwave Plasma Enhanced Chemical Vapor Deposition (MW PECVD)

2.3.6 Aerosol-assisted Chemical Vapor Deposition (AACVD)

2.3.7 Photochemical Vapor Deposition (PCCVD)

2.3.8 Chemical Beam Epitaxy (CBE)

Type of CVD System: Formation of films on a substrate by chemical reaction of vapor phase precursor which is understood as chemical vapor deposition (CVD). To maintain the vapor phase of different precursors, various techniques are applied, such as direct heat (Thermal CVD), Higher frequency radiation (Photo-assisted CVD), Plasma (Plasma CVD), etc. Further, according to the reaction pressure and by specific types of precursors, they are named differently, such as Atmospheric Pressure CVD (APCVD), Metal Organic CVD (MOCVD), Atomic Layer Chemical Vapor Deposition (ALCVD), Chemical Beam Epitaxy (CBE), and a high vacuum CVD technique.

2.3.1 Atmospheric Pressure CVD (APCVD)

This CVD method is used for deposition of undoped and doped oxide thin films at atmospheric pressure (1 atmosphere = 101325 Pa or 760 Torr) with high deposition rate. Due to relatively low temperature, the deposited oxide has low density and the coverage is moderate. Low temperature APCVD is needed for many insulating films (SiO_2, BPSG glasses). On the other hand, high temperature APCVD is used to deposit epitaxial Si and compound films (cold wall reactors) or hard metallurgical coatings, such as TiC and TiN (hot wall reactors). Figure 8.1 shows the schematic of atmospheric pressure CVD technique. In this schematic, the precursor and catalyst are placed on the heater for controlled evaporation. The gaseous flow further goes to the reactor for the growth. Here in the reactor, the substrate is placed on the susceptor, and excess gas goes to the vent for further cleaning.

Advantages of APCVD

- Low equipment cost.
- Large area uniformity achieved through control of temperature and gas.
- Simple process control and source replenishment because the source gas generation is physically separated from the deposition chamber.

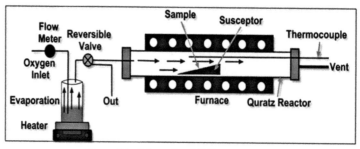

Figure 8.1. Schematic of APCVD technique.

APCVD Limitations

- Wafer throughput is low due to low deposition rate.
- Film thickness uniformity can be an issue.
- Step coverage is not very good.
- Contamination is a problem and maintaining stoichiometry can be hard.
- Large number of pinhole defects can occur.

2.3.2 Metal-Organic Chemical Vapor Deposition (MOCVD)

Organo Metallic Chemical Vapor Deposition (OM-CVD) is widely employed in solid state chemistry and electronics for the selective deposition of mono/poly-metallic film of high purity. Few reports have been reported on its application to the preparation of oxide-supported metal particles and films. In particular, potential advantages will be gained in the case of crystalline oxides, such as zeolites. Conventionally used ion exchange or wet impregnation techniques have no general applicability in heterogeneous catalysis for the preparation of high-purity and high-performance materials. Unconventional techniques, such as solution-phase metal impregnation of zeolites and *in situ* microwave decomposition of intrazeolite organometallics have also been used for the preparation of zeolite-encapsulated metal clusters [11].

2.3.3 Low-Pressure Chemical Vapor Deposition (LPCVD)

Due to mass transport velocity and speed of reaction on surface, LPCVD is used instead of APCVD. Pressure and gas diffusion are reciprocal to each other in LPCVD. Pressure in LPCVD is usually around 10–1000 Pa. In LPCVD procedure, it has a set of quartz tube inside a winding heater that begins with cylinder weight at low weight of around 0.1 Pa. The cylinder is then warmed to the ideal temperature and the vaporous species (working gas) is embedded into the cylinder at the weight foreordained between 10–1000 Pa. This working gas comprises of weakening gas and the receptive gas that will respond with the substrate and make a strong stage material on the substrate. After the working gas enters the cylinder, it spreads out around the hot substrates that are as of now in the cylinder at a similar temperature. The substrate temperature is critical and impacts what responses happen. This working gas responds with the substrates and structures the strong stage material, and the overabundance material is siphoned out of the cylinder [12]. Figure 8.2 shows the schematic of low-pressure CVD technique.

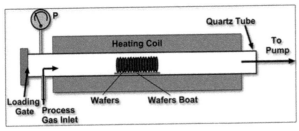

Figure 8.2. Schematic of low-pressure CVD technique.

2.3.4 Plasma-Enhanced Chemical Vapor Deposition (PECVD)

In this method, to create the desired solid surface, such as SiO_2, Si_3N_4 (Si_xN_y), $Si_xO_yN_z$, and amorphous Si film on the substrate, plasma is purged in the deposition chamber with reactive gases. Plasma is an ionized gas with high free electron content (about half). Plasmas are isolated into two conditions- chilly (likewise called non-warm) and warm. In warm plasma, electrons and particles in the gas are at a similar temperature in any case. In chilly plasmas, the electrons have a much higher temperature than the unbiased electrons and particles. In this way, cool plasma can use the vitality of the electrons by changing only the weight. This enables a PECVD framework to work at low temperature (somewhere in the range of 100 and 400°C). PECVD must contain two anodes (in a parallel plate arrangement), plasma gas, and receptive gas in a chamber. To start the PECVD procedure, a wafer is put on the base cathode, and responsive gas with the testimony components is brought into the chamber. Plasma is then brought into the chamber between the two cathodes, and voltage is connected to energize the plasma. The energized state of plasma at that point barrages the receptive gas, causing separation. This separation stores the ideal component onto the wafer. The schematic of plasma-enhanced CVD technique is shown in Figure 8.3 which gives a representation of system parts by their names.

Advantages of PECVD

- Low temperature synthesis.
- More compression and higher film density for higher dielectric.
- Ease of cleaning the chamber.

Limitations

- Stress of plasma bombardment.
- Initial expenses of the equipment.
- Small batch size.

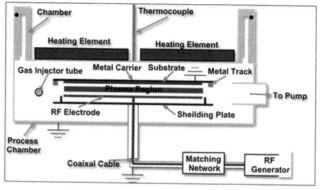

Figure 8.3. Schematic of plasma-enhanced CVD technique.

2.3.5 Microwave Plasma Enhanced Chemical Vapor Deposition (MW PECVD)

MW-PECVD reactors are broadly utilized for developing precious stone with grain sizes crossing the range from nanometers through microns to millimeters. Precious stones can be kept in microwave MW plasma-improved synthetic vapor statement PECVD reactors with a scope of grain sizes going from nanometers through microns to millimeters (Figure 8.4). Single gem materials rely on the selection of factors, such as gas blend, development conditions, substrate properties, and development time [13].

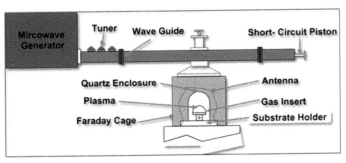

Figure 8.4. Microwave plasma enhanced CVD technique.

2.3.6 Aerosol-Assisted Chemical Vapor Deposition (AACVD)

This involves utilization of a fluid gas to transport dissolvable antecedents on a warmed substrate. The strategy has generally been utilized when an atmospheric pressure CVD demonstrates volatility or is thermally flimsy [14].

2.3.7 Photochemical Vapor Deposition (PCVD)

The procedure of photograph processing helps in improvement of CVD method, in which it includes association of light radiation with forerunner atoms either in the gas stage or on the growth stage on the surface. Forerunner atoms must assimilate energy, since customarily basic inorganic antecedents have been utilized, which require utilization of UV radiation. The utilization of organometallic antecedents (with p- and s-fortified moieties) opens up the conceivable outcomes for a more extensive scope of wavelengths. Yet, this can prompt an expanded potential for carbon fuse. Photochemical CVD has comparable potential of focal points to those of PECVD; to be specific, low temperature statement, changes in the properties of developed layers, i.e., dopant joining, free control of substrate temperature, and separation of forerunner. However, with concealing or laser actuation, it is conceivable to accomplish chosen region development [15].

2.3.8 Chemical Beam Epitaxy (CBE)

CBE is a high vacuum CVD method that utilizes unstable metal-natural antecedents and vaporous co-forerunners. Firmly, this method is a combined procedure of metal-natural antecedent, sub-atomic epitaxy (MOMBE) that utilizes unpredictable metal-natural antecedents and co-forerunner vapor. In CBE, MOMBE compound's response happens on the substrate, prompting single precious stone. Thus, gas-stage responses assume no huge job in film development [15].

2.4 Application of CVD in Nanomaterials

2.4.1 Transition Metal (TM) Oxides Nanoparticles

The TMs display specific properties and transport different outputs with different reactants in surrounding elements, such as sulphides, oxides, selenides, nitrides, chalcogenides, and some MOFs having specific organic compounds, etc. The TMs are the elements that are placed in the d-block in the periodic table from IV to VII groups. TMs have variable oxidation states because of partially-filled d-orbit. The TMs have been used as nanoparticle materials at the nanoscale in electronic device applications, i.e., semiconductors devices, energy storage, and conversion devices i.e., Li^+ and Na^+ ion batteries, capacitors, supercapacitors, gas sensing devices, photovoltaics, etc.

In this section, we have discussed about metal oxides deposited using CBD and their properties at a nano scale. Scandium (Sc) is a rare TM on earth, but still it is available in oxides and can be useful for many applications. Xu et al. reported optical and microstructural properties of Sc_2O_3 thin film deposited by MO-CVD, and studied the effect of deposition temperature on its properties [16]. Luo et al. reported the effect of Sc_2O_3 layer deposited by PAMBE-CVD at low a temperature of about 100°C for high electron-mobility (μ_e) transistors based on AlGaN/GaN material [17]. Putkonen et al. reported Sc_2O_3 from $Sc(thd)_3$, $Sc(tmod)_3$, and $Sc(mdh)_3$ at 450–600°C in oxygen atmosphere using flow-type hot-wall ALE-CV. Figures 8.5a–d show the comparative study of AFM image of Sc_2O_3 films [18]. Lee et al. reported the homogeneous and dense Sc_2O_3 and Ta-doped SnO_2 thin films on corning glass substrate as having high optical and mechanical properties using cold-wall, horizontal, and low-pressure type MOCVD for transparent conductive oxides (TCOs) [19, 20].

One of TM is TiO_2 which is widely used in energy related devices, such as Li-ion batteries (LIBs) as cathode material, capacitors, photocatalyst, solar cells, gas sensors, nanorods, nanowires, thin films, 1D, 2D, 3D film, and other shape and sized nanostructured materials. Puma et al. reported synthesis of TiO_2 for the photocatalyst on the activated carbon by MOCVD. In this case, the titanium tetra-isopropoxide (TTIP), tetrabutyltitanate (TBOT), Titanium tetrachloride/tetra-nitra-totitanium, and activated carbon were used as precursors for N_2 as a carrier gas [21]. Song et al. reported TiO_2 nanoparticles having particle size 1–5 nm on a silver substrate at 90 K using RLA-CVD [22]. Pradhan et al. synthesized TiO_2 nanorods (50–100 nm diameter with 0.5–2 µm length) using low pressure MOCVD, and they observed that in the presence of NH_3, the growth rate of TiO_2 nanorods was increased [23]. The Rutile phase of TiO_2 nanowires by using surface reaction limited pulsed CVD, as reported by Shi et al., as shown in Figure 8.6 [24]. Xie et al. reported mass production of TiO_2 nanoparticles (30–80 nm) synthesis by propane/air turbulent flame CVD by the oxidation of $TiCl_4$ in high strength propane/air turbulent flame [25].

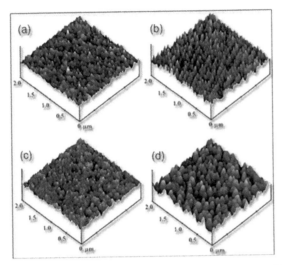

Figure 8.5. AFM images of Sc_2O_3 films deposited from $Sc(thd)_3/O_3$ at 350°C (a) and 400°C (b) as well as from $(C_5H_5)_3Sc/H_2O$ at 250°C (c) and 350°C (d). The thicknesses of the measured samples were 70 (a, b) and 150 (c, d) nm. The height axes were 40 (a–c) and 100 (d) nm. Reprint with permission [18], Copyright 2001, American Chemical Society.

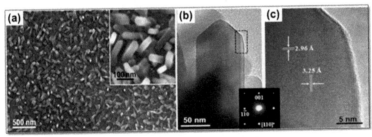

Figure 8.6. (a) NRS of TiO_2 grown on Si substrate. (b) TEM image of a TiO_2 NR. (c) HRTEM image of a NR acquired from the rectangle region in part (c). Reprint with permission [24], Copyright 2011, American Chemical Society.

The oxides of vanadium metal have been used in energy related materials for the last five decades. This material showed very interesting results due to its special properties, such as flexible oxidation states, and phase transition (semiconductors to metal and metal to insulator phase, and vice-versa) at critical temperature. These metal oxides have their many phases, and each of its phase is stable at a specific temperature and pressure conditions, i.e., VO, VO_2, V_2O_3, V_2O_5, V_3O_7, V_4O_7, V_5O_9, V_6O_{13}, V_7O_{13}, V_8O_{15}, and V_9O_{17}. They generally form vanadium oxide (VnO_{2n-1}) for their possible oxide forms. Nag et al. reported VO_2 thin film and its nanoparticle synthesis by various CVD methods, i.e., MO-CVD, AP-CVD, LP-CVD, AA-CVD [26]. Manning et al. reported the single phase W-doped VO_2 synthesized by AP-CVD on silicon-coated (50 nm) glass substrate [27]. Barreca et al. reported the highly oriented V_2O_5 nanocrystalline thin film using the PE-CVD method [28]. Nandakumar et al. reported the carbon-free V_2O_5 thin film at 180°C on Sb-doped n-type silicon using LT-CVD [29]. Based on these studies, we can summarize that the vanadium oxide phases and crystallinity depend on the deposition parameters, such as temperature, pressure, and technique used for depositing thin films. Vanadium oxides are the most widely used material in Li^+/Na^+ ion batteries, capacitors, supercapacitors, gas sensing devices, etc. Another TM is chromium oxide, and there are very few methods that have been reported to deposit chromium oxide by CVD. Gupta et al. reported Cr_2O_3 thin film on selective areas by using the AP-CVD method, in which CrO_3 precursor was used on the single crystal SiO_2 deposited-TiO_2 substrate for depositing Cr_2O_3 film [30]. Sousa et al. reported a single crystal chromia-Cr_2O_3 thin film on sapphire deposited by laser assisted-CVD method at low temperature and low pressure [31]. Zhong et al. reported the chromium oxide nanoparticles/thin film deposition using MO-CVD method for electro-magnetic devices applications [32]. Manganese (Mn) is the most popular element for energy storage related applications based on transition metal oxides. It has variable oxidation states, for example, Mn^{+2} (MnO), $Mn^{+2/+3}$ (Mn_3O_4), Mn^{+3} (Mn_2O_3), Mn^{+4} (MnO_2, or MnO_3), Mn^{+6} (MnO_3), and Mn^{+7} (Mn_2O_7), and some of its oxides are stable in the natural conditions, and also commercialized in the forms of MnO, Mn_3O_4, Mn_2O_3, and MnO_2 [Sigma-Aldrich Ltd., USA]. The MnO_2 nanoparticles are very popular in the energy storage/conversion devices, such as Li^+/Na^+ ion batteries, capacitors, or supercapacitors due to their structural stability during charge-discharge process for a long time. These nanoparticles can be in any shapes and sizes, such as nanorods, nanowires, nanoflowers, nanospheres, etc., and it depends on the deposition condition and phases of crystal structures (α, β, γ, λ, δ-Mn_xO_y). Table 8.1 shows the deposition parameters of CVD for CNT synthesis reported by various research groups.

Manganese oxides 3D crystal structures are most widely used as cathode materials in battery applications. It has low electrical conductivity, so various research groups are working on this material, and especially MnOx with particular conductive polymers, carbon materials (carbon nanotube, graphene, graphene oxides, etc.) to improve the mechanical and electrical properties. Table 8.2 shows the cobalt oxide thin film deposition by various CVD methods at different temperatures.

Le et al. reported the manganese oxides nanoparticles by CVD by using the manganese carbonyl $(Mn_2CO)_{10}$ as a precursor [50]. Matsumoto et al. has reported the manganese and manganese oxides thin films by CVD method for advanced silicon devices. They used bis (ethyl cyclopentadienyl) manganese, and $(EtCp)_2$-Mn was used as a precursor at 70°C–80°C temperature with 25 sccm H_2 gas flow rate [51]. The Li-manganese oxide ($Li_{0.47}Mn_{0.27}O_{0.26}$) thin film on soda lime glass substrate by MO-CVD process was reported by Oyedotun et al. In this process, the lithium manganese acetylacetonate was used a precursor, and deposition temperature and N_2 gas flow rate was kept at 420°C and 2.0 dm^3/min in MO-CVD system, respectively [52]. Iron oxides are the compounds that have Fe^{+2}, Fe^{+3} bonded with oxygen atoms, which make a stable ore or compound in nature. These metal oxides also show the tunable oxidation states, such as FeO (F^{+2}), Fe_2O_3 (Fe^{+3}), FeO_2 (Fe^{+4}), and some mixed valency (Fe^{+2}, Fe^{+3}) of Fe_3O_4 (Magnetite), Fe_4O_5, Fe_5O_7, $Fe_{13}O_{19}$, etc. Crystal phases, such as α-Fe_2O_3 (Hematite), β-Fe_2O_3, and γ-Fe_2O_3 (Maghemite) are also available

Table 8.1. Deposition parameters of CVD for CNT synthesis.

Deposition method	CNT type	Precursors	Catalyst	Carrier gas	Temperature range (°C)	Ref.
CVD (Temp membrane)	-	Ethylene, Pyrene	Ni,	Ar, 50 sccm	545, 900	[33]
PE-CVD	SWCNT, VA-CNT	Methane, C_2H_2:NH_3	Ferritin, Ni	Ar, 60 sccm	600	[34, 35]
CVD	SWCNT	Methane	Aerogel-Fe/Mo	Ar, 100 sccm	850–1000	[36]
Fast heating CVD	SWCNT	CH_4/H_2	Fe/Mo	H_2	900	[37]
Th-CVD	MWCNT	C_2H_2		N_2	675, 700, 850	[38]
CCVD	SWCNT	Methane	CO, Fe, Co-Fe	H_2, 75 ml/300 ml/min	1000	[39]
AE-CVD		C_2H_2	Ni	Ar	700–1000	[40]
Injection-CVD	MWCNt	Ferrocence, toluence	Fe	Ar^+H_2 (10%), 750 ml/min	590, 740, 850, 940	[40]
LT-CVD	SWCNT	C_2H_2.NH_3	Fe and Al/Fe/Al layer	NH_3/H_2	350	[41]
CVD	SWCNT	Ethanol	CoMn doped Mesoporus silica (SBA16)	Ar (50 sccm)	850	[42]
Th-CVD	SWCNT	Methane	Ni-Cu-Al	H_2 and N_2	700–750	[43]
Infusion-CVD	MWCNT	Ethanol	Ni	None	700	[44]
RFM_PECVD	SWCNT/ MWCNT	Methane	Ni and Zeolite	H_2	550, 850	[45]
Water-assisted CVD	MWCNT	Ethylene + water + H_2	Fe/Al_2O_3/ SiO_2/Si	Ar	750	[46]
Th-CVD	VA-CNT	C_2H_2	Co: Ni with Pd, Cr and Pt	Ar,	500–550	[47]
MPE-CVD	CNT	$CH_4 + H_2$	Ni/Si/TiN	-	520	[48]
rfPE-CVD	WACNT	$C_2H_2 + H_2$	Fe	-	600	[48]
ECR-CVD	HA-CNT	$CH_4 + H_2$	Co	-	600	[49]

in iron oxide compounds. Park et al. reported the γ-Fe_2O_3 and Fe_3O_4 nanoparticles and thin film using CVD [53]. Yang et al. reported Fe_2O_3-ZSM-5 catalyst preparation by MO-CVD [54]. Yubero et al. synthesized α-Fe_2O_3 hematite thin films deposition by IBI-CVD methods. In this research, the bombardment of accelerated ions of $O^{+2}/O^{+2}+Ar^+$ with volatile $Fe(CO)_5$ precursor on the substrate (silicon wafer, fused quartz, and KBr pallets) surface at a pressure of 3×10^{-5} Torr was used [55]. The iron oxide nanoparticles by wet and dry chemical methods (colloid chemical, sol-gel)

Table 8.2. Cobalt oxide thin film deposition by various CVD methods at different temperatures.

Substrate Temp (°C)	Pressure (mbar)	O_2 flow rate (sccm)	Thickness (nm)	Growth rate (nm/min)	Phases
350	10	150	543 10	7	Co_3O_4
350	2	50	20410	3	CoO
400	10	150	41310	9	Co_3O_4
400	2	50	26112	4	CoO^+Co_3Os
450	10	150	62313	10	Co_3O_4
450	2	50	51530	13	Co_3O_4
500	10	150	72633	12	Co_3O_4
500	2	50	127639	21	Co_3O_4

are explained by Hasany et al., as are the applications of iron oxides nanoparticles in various fields, such as water purification, pigments, coating, gas sensors, ion exchangers, catalysts, magnetic data storage devices, resonance imaging, etc. [56]. Oxides of cobalt (Co) are available in a stable form, i.e., CoO, Co_2O_3 (Co^{+2}), Co_3O_4 ($Co^{+2, +3}$). It is a very popular material, especially for Li-Ion batteries (as a cathode material), capacitors, solar thermal energy storage devices (as light absorber layer), and solid-state sensors applications. Maruyama et al. reported cobalt oxide thin film by LT-AP-CVD on borosilicate glass and stainless steel plate substrate [57]. Haniam et al. reported CoO and Co_3O_4 film on Co/Al/Cr/Si-substrate by LCVD. In this case, acetylene and H_2 gas flow rate was kept at 50 and 100 sccm, respectively [58]. The Co_3O_4 has the mixed valency (+2/+3) with normal spinel structure, which is the most stable form of oxides of cobalt. Barreca et al. reported the cobalt oxide thin film by the cold wall, low-pressure CVD method on ITO substrate [59].

The composition and microstructure of cobalt oxide thin films are obtained from a novel cobalt (II) precursor by chemical vapor deposition. Chalhoub et al. reported CoO coating by the MO-CVD method [60] The other cobalt oxide nanomaterials by PE-CVD, AP-CVD, MO-CVD, LCVD, ion-assisted-CVD, and other types have been reported in various research groups [58, 61–64]. Similarly, the other kinds of transition metal oxides, such as Cu_xO_y, Zn_xO_y, Ru_xO_y, etc. can be easily deposited by specific types of CVD methods, which require specific precursors, carrier gases, temperature, and pressure, etc.

2.4.2 Carbon and Its Derivatives

Carbon is the 15th most abundant element on the earth's crust. It has a tetravalent valency and covalent bonds with surrounding atoms. Carbon has its allotropes, such as diamond (sp^3 hybridization in cubic system), graphite (sp^2 hybridization in hexagonal system), buckyball (C-60 or buckminister fullerene), C_{540}, C_{70}, amorphous carbon (having specific sp^2 and sp^3 hybridization ratio at microscopic level), carbon nanotubes (CNTs), lonsdaleite (hexagonal diamond) and its isotopes, such as C_{12} (98.93%), C_{13} (1.07% and C_{14} (radioactive isotope) [65].

2.4.2.1 Carbon Nanotubes (CNTs)

The CNTs are the cylindrical form of graphene (monolayer of graphite) sheet and closed by fullerenoid end-caps. The single, double, and multi-wall CNTs are available, and CNTs have dimensions like inner diameters of 1–3 nm, outer 2–20 nm, length ~ 1 pm, and inter tubular distance of about 340 pm (which is a larger value then inter-planar distance in graphite) [66]. CNTs can be synthesized in various forms, such as CNTs yarns, sheet, sponges, and arrays (well-defined vertically aligned with respect to substrate surface) by using the CVD method. CNTs have a wide range of applications as electrodes in capacitors, supercapacitors, Li-ion batteries (LIBs), diagnostic devices, contrast agents, drug-delivery, microbiology, and antimicrobial therapy of infected diseases. Functionalized CNTs with specific chemical groups that change its physical and biological properties can be used in cancer treatment and drug delivery applications [67–69]. Nowadays, the CVD is the most effective synthesis method for pure CNTs for energy related devices applications. A recent study reported by authors shows the role of MWCNTs coated by the CVD method on the carbon fibers, i.e., carbon felt substrate, for the high performance in binder-free lithium ion battery electrodes application [82]. Saifuddin et al. reported high quality CNTs synthesized by arc-discharge, laser vaporization, and PE-CVD method [70].

2.4.2.2 Graphene

In recent years, graphene is the most eminent material because of its properties and wide range of applications. Graphene shows very good properties as compared to other materials, such as mechanical strength, thermal conductivity, charge mobility, specific surface area, and electrical conductivity. These properties enable possibilities to use this material in the field of nanoelectronics devices and energy. Basically, graphene is one of the derivatives of carbon in a particular arrangement of carbon sp^2 hybridized atoms that changes the properties of carbon material very differently to its original state. In a single layer graphene, six carbons (hexagonal ring) are bonded to each other two-dimensionally. Graphene oxide is the oxidized graphene layer obtained by adding some oxygen functional groups using some particular techniques, such as hummer's method. Reduced graphene oxide (rGO) is the final product one gets after the reduction of graphene oxides (GOs) by using some specific chemical, thermal, and electrical treatments. Still, rGO has some oxygen functional groups in the structure. GO and rGO differ from their functionalized groups. Graphene synthesis using CVD and some deposition parameters are shown below in the Table 8.3.

Table 8.3 shows graphene synthesis by various CVD methods reported by various research groups. Graphene can be synthesized by different CVD methods, such as plasma-enhanced, aerosol pyrolysis CVD with uniform layers with and without substrate [80]. Malesevic et al. reported few layers of graphene synthesized by MW PE-CVD. A wide variety of substrates, such as Si, Ni, Ge, Ti, W, Mo, Ta, quartz, and SS (stainless steel) can be used for depositing graphene by MW PE-CVD [81]. Figure 8.7a shows dimensions and orientation of freestanding FLG in an SEM image. Figure 8.7b with the top view shows the high density of the flakes.

Table 8.3. Showing graphene synthesis by various CVD methods.

Method	Type	Substrate	Precursors	Carrier gas	Temp. range (°C)	Ref.
SWP-CVD	Graphene	Al, Cu foil	CH_4	$Ar + H_2$	320	[71]
Th-CVD	Graphene	Fe-foil	C_2H_2	$Ar + H_2$	700, 750	[72]
CVD	Graphene	Cu-foil	$CH_4 + H_2$	H_2	1000	[73]
CVD	Graphene	Ni-foil	CH_4	$H_2 + Ar$	900	[74]
CVD	Graphene	Ni-foam	CH_4	$H_2 + Ar$	1000	[75]
AP-CVD	Graphene	Polycrystalline Cu foil	CH_4	$H_2 + Ar$	1050	[76]
EA-HFCVD	1D-Graphene	Ni/Si	CH_4	H_2	950	[77]
APRF-CVD	Bilayer Graphene	Cu-foil	Ag, Au, TiO_2-Ethanol	-	1000	[78]
PA-CVD (printing-assisted)	Graphene	Cu-foil	CH_4	Ar, H_2	1045	[79]

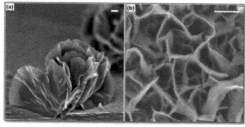

Figure 8.7. (a) Dimensions and orientation of freestanding FLG in SEM image. (b) Top view shows the high density of the flakes. Scale bar in images of both is 1 μm. Reprint with permission [72]. Copyright 2011, Elsevier B.V.

2.5 Advantages of CVD Technique

1. This technique is useful for depositing ultra-thin layers using ALD.
2. Deposition can be done at low temperature even close to ambient temperature. Apart from this, we can control dopant incorporation, substrate temperature, dissociation of precursor, and selective area growth.
3. Precursor is relatively air stable and susceptible to reaction with water.
4. Uses very bulky ligands in the precursor. This improves the vapor pressure and makes it less air/moisture sensitive.
5. Thermal stability of the samples is very high.
6. Particle size for a given metal loading can be controlled by adjusting the precursor vapor pressure, which governs nucleation rate.
7. Low cost, high thermal efficiency, continuous operation, high throughput, and setup is easy to scale up technology.

8. Higher purity can be achieved.

9. High density of nearly 100% of theoretical value can be obtained.

10. Any element and compound can be deposited.

11. Economical in production because many parts can be deposited simultaneously.

Conclusion

In this chapter, we described various CVD techniques for depositing a wide variety of nanomaterials that can be utilized for various applications. It covers detailed information about deposition parameters, i.e., carrier gas, substrate, catalysts, growth rate, temperature, pressure, etc. These parameters play a very important role during the synthesis of nanomaterials using chemical vapor deposition (CVD) technique. Here, we explained various CVD techniques used by various research groups. Each technique has advantages and disadvantages. This study will help the research community to understand various CVD techniques, deposition parameters for synthesizing various carbon-based materials, and other materials for various applications.

Abbreviations

Al	Aluminum
Ar	Argon
AACVD	Aerosol-Assisted Chemical Vapor Deposition
APCVD	Atmospheric Pressure Chemical Vapor Deposition
ATM	Atmospheric Pressure
ALE-CVD	Atomic Layer Epitaxy Chemical Vapor Deposition
ALD	Atomic Layer Deposition
CNT	Carbon Nanotube
CCVD	Catalytic Chemical Vapor Deposition
Cu	Copper
CVD	Chemical Vapor Deposition
CBD	Chemical Bath Deposition
CECVD	Catalysts Enhanced Chemical Vapor Deposition
CVICVD	Chemical Vapor Infiltration Chemical Vapor Deposition
CBE	Chemical Beam Epitaxy
°C	Degree Celsius
DC	Direct Current
ECD	Electrochemical Deposition
ECRCVD	Electron Cyclotron Resonance Chemical Vapor Deposition
Fe	Iron
Ge	Germanium
GHz	Gigahertz
GO	Graphene Oxide
H_2	Hydrogen
kW	Kilowatt

LICVD	Laser-Induced Chemical Vapor Deposition
LPCVD	Low Pressure Chemical Vapor Deposition
MBE	Molecular Beam Epitaxy
mbar	Millibar
MWPECVD	Microwave Plasma-Enhanced Chemical Vapor Deposition
MOCVD	Metal Organic Chemical Vapor Deposition
Mo	Molybdenum
mW	Milliwatts
nm	Nanometer
N_2	Nitrogen
Ni	Nickel
OM-CVD	Organo Metallic Chemical Vapor Deposition
Pa	Pascal
PVD	Physical Vapor Deposition
PACVD	Printing Asserted Chemical Vapor Deposition
PAMBE-CVD	Plasma-Assisted Molecular Beam Epitaxy
PECVD	Plasma-Enhanced Chemical Vapor Deposition
RF	Radio Frequency
RFCVD	Radio Frequency Chemical Vapor Deposition
RLA-CVD	Reactive Layer Assisted Chemical Vapor Deposition
rGO	Reduced Graphene Oxide
RPCVD	Remote Plasma Chemical Vapor Deposition
RT	Room Temperature
Si	Silicon
SCCM	Standard Cubic Centimeters Per Minute
SS	Stainless Steel
Ta	Tantalum
Ti	Titanium
TiC	Carbide of Titanium
TTIP	Titanium Tetra-Isopropoxide
TACVD	Thermally Activated Chemical Vapor Deposition
VLSI	Very Large-Scale Integration
W	Tungsten

References

[1] Carlsson, J.-O. and P.M. Martin. 2010. Chemical vapor deposition. In Handbook of Deposition Technologies for Films and Coatings. Elsevier, 314–363.

[2] Guo, J., J. Li and H. Kou. 2011. Chemical preparation of advanced ceramic materials. In Modern Inorganic Synthetic Chemistry. Elsevier, 429–454.

[3] Wilson, J. 2009. Textile surface functionalisation by chemical vapour deposition (CVD). In Surface Modification of Textiles. Elsevier, 126–138.

[4] Makhlouf, A. 2011. Conventional and advanced coatings for industrial applications: an overview. In Nanocoatings and Ultra-Thin Films. Elsevier, 159–181.

[5] Jones, A.C. and M.L. Hitchman. 2009. Chemical vapour deposition: precursors, processes and applications. Royal Society of Chemistry.

[6] Pierson, H.O. 1999. Handbook of Chemical Vapor Deposition: Principles, Technology and Applications. William Andrew.

[7] Choi, H.S. and S.W. Rhee. 1994. Effect of carrier gases on the chemical vapor deposition of tungsten from WF 6-SiH4. J. Electro. Soc., 141(2): 475–479.
[8] Choi, K.-K. and S.-W. Rhee. 2001. Effect of carrier gas on chemical vapor deposition of copper with (Hexafluoroacetylacetonate) Cu (I) (3, 3 Dimethyl 1 butene). J. Electro. Soc., 148(7): C473–C478.
[9] Lin, Y. and A. Burggraaf. 1992. CVD of solid oxides in porous substrates for ceramic membrane modification. AIChE J., 38(3): 445–454.
[10] Rashid, H.U., K. Yu, M.N. Umar, M.N. Anjum, K. Khan, N. Ahmad and M.T. Jan. 2015. Catalyst role in chemical vapor deposition (CVD) process: a review. Rev. Adv. Mater. Sci., 40(3): 235–248.
[11] Malik, M. and P. O'Brien. 2005. Precursor Chemistry of Advanced Materials. Springer, pp. 173–204.
[12] Curley, R., T. McCormack and M. Phipps. 2018. Low-pressure CVD and plasma-enhanced CVD. Accessed: 09 Nov 2018.
[13] Mankelevich, Y.A., M.N. Ashfold and J. Ma. 2008. Plasma-chemical processes in microwave plasma-enhanced chemical vapor deposition reactors operating with C/H/Ar gas mixtures. J. Appl. Phys., 104(11): 113304.
[14] Palgrave, R.G. and I.P. Parkin. 2006. Aerosol assisted chemical vapor deposition using nanoparticle precursors: a route to nanocomposite thin films. J. Am. Chem. Soc., 128(5): 1587–1597.
[15] Jones, A.C. and M.L. Hitchman. 2009. Overview of chemical vapour deposition. Chemical Vapour Deposition: Precursors, Processes and Applications, 1–36.
[16] Xu, Z., A. Daga and H. Chen. 2001. Microstructure and optical properties of scandium oxide thin films prepared by metalorganic chemical-vapor deposition. Appl. Phys. Lett., 79(23): 3782–3784.
[17] Luo, B., J.W. Johnson, J. Kim, R.M. Mehandru, F. Ren, B.P. Gila and R.D. Briggs. 2002. Influence of MgO and Sc$_2$O$_3$ passivation on AlGaN/GaN high-electron-mobility transistors. Appl. Phys. Lett., 80(9): 1661–1663.
[18] Putkonen, M., M. Nieminen, J. Niinistö, L. Niinistö and T. Sajavaara. 2001. Surface-controlled deposition of Sc$_2$O$_3$ thin films by atomic layer epitaxy using β-diketonate and organometallic precursors. Chem. Mater., 13(12): 4701–4707.
[19] Li, L., X. Wang and X. Zhou. 2003. Advanced Electronic-Ceramic Materials-Proceedings of the 8th IUMRS, Symposium, June 10–14, 2003, Lausanne, Switzerland.
[20] Lee, S.W., A. Daga, Z.K. Xu and H. Chen. 2003. Characterization of MOCVD grown optical coatings of Sc$_2$O$_3$ and Ta-doped SnO$_2$. Mater. Sci. Eng.: B, 99(1-3): 134–137.
[21] Puma, G.L., A. Bono, D. Krishnaiah and J.G. Collin. 2008. Preparation of titanium dioxide photocatalyst loaded onto activated carbon support using chemical vapor deposition: a review paper. J. Hazard. Mater., 157(2-3): 209–219.
[22] Song, Z., J. Hrbek and R. Osgood. 2005. Formation of TiO$_2$ nanoparticles by reactive-layer-assisted deposition and characterization by XPS and STM. Nano Lett., 5(7): 1327–1332.
[23] Pradhan, S.K., P.J. Reucroft, F. Yang and A. Dozier. 2003. Growth of TiO$_2$ nanorods by metalorganic chemical vapor deposition. J. Cryst. Growth, 256(1-2): 83–88.
[24] Shi, J. and X. Wang. 2011. Growth of rutile titanium dioxide nanowires by pulsed chemical vapor deposition. Cryst. Growth Des., 11(4): 949–954.
[25] Xie, H., G. Gao, Z. Tian, N. Bing and L. Wang. 2009. Synthesis of TiO$_2$ nanoparticles by propane/air turbulent flame CVD process. Particuology, 7(3): 204–210.
[26] Nag, J. and R. Haglund Jr. 2008. Synthesis of vanadium dioxide thin films and nanoparticles. J. Phys. Condens. Matter, 20(26): 264016.
[27] Manning, T.D., I.P. Parkin, M.E. Pemble, D. Sheel and D. Vernardou. 2004. Intelligent window coatings: atmospheric pressure chemical vapor deposition of tungsten-doped vanadium dioxide. Chem. Mater., 16(4): 744–749.
[28] Barreca, D., L. Armelao, F. Caccavale, V. Di Noto, A. Gregori, G.A. Rizzi and E. Tondello. 2000. Highly oriented V$_2$O$_5$ nanocrystalline thin films by plasma-enhanced chemical vapor deposition. Chem. Mater., 12(1): 98–103.
[29] Nandakumar, N.K. and E.G. Seebauer. 2011. Low temperature chemical vapor deposition of nanocrystalline V$_2$O$_5$ thin films. Thin Solid Films, 519(11): 3663–3668.
[30] Turel, C.S., I.J. Guilaran, P. Xiong and J.Y.T. Wei. 2011. Andreev nanoprobe of half-metallic CrO$_2$ films using superconducting cuprate tips. Appl. Phys. Lett., 99(19): 192508.

[31] Sousa, P., A. Silvestre and O. Conde. 2011. Cr$_2$O$_3$ thin films grown at room temperature by low pressure laser chemical vapour deposition. Thin Solid Films, 519(11): 3653–3657.

[32] Zhong, Z. and R. Cheng. 2001. Fabrication and characterization of chromium oxide nanoparticles/ thin films. MRS Online Proceedings Library Archive, 635.

[33] Che, G., B.B. Lakshmi, C.R. Martin, E.R. Fisher and R.S. Ruoff. 1998. Chemical vapor deposition based synthesis of carbon nanotubes and nanofibers using a template method. Chem. Mater., 10(1): 260–267.

[34] Li, Y., D. Mann, M. Rolandi, W. Kim, A. Ural, S. Hung and Q. Wang. 2004. Preferential growth of semiconducting single-walled carbon nanotubes by a plasma enhanced CVD method. Nano Lett., 4(2): 317–321.

[35] Chhowalla, M., K.B.K. Teo, C. Ducati, N.L. Rupesinghe, G.A.J. Amaratunga, A.C. Ferrari and W.I. Milne. 2001. Growth process conditions of vertically aligned carbon nanotubes using plasma enhanced chemical vapor deposition. J. Appl. Phys., 90(10): 5308–5317.

[36] Su, M., B. Zheng and J. Liu. 2000. A scalable CVD method for the synthesis of single-walled carbon nanotubes with high catalyst productivity. Chemical Physics Letters, 322(5): 321–326.

[37] Huang, S., M. Woodson, R. Smalley and J. Liu. 2004. Growth mechanism of oriented long single walled carbon nanotubes using "fast-heating" chemical vapor deposition process. Nano Lett., 4(6): 1025–1028.

[38] Baddour, C.E., F. Fadlallah, D. Nasuhoglu, R. Mitra, L. Vandsburger and J.L. Meunier. 2009. A simple thermal CVD method for carbon nanotube synthesis on stainless steel 304 without the addition of an external catalyst. Carbon, 47(1): 313–318.

[39] Colomer, J.-F., C. Stephan, S. Lefrant, G. Van Tendeloo, I. Willems, Z. Konya and J.B. Nagy. 2000. Large-scale synthesis of single-wall carbon nanotubes by catalytic chemical vapor deposition (CCVD) method. Chem. Phys. Lett., 317(1-2): 83–89.

[40] Jung, M., K.Y. Eun, J.K. Lee, Y.J. Baik, K.R. Lee and J.W. Park. 2001. Growth of carbon nanotubes by chemical vapor deposition. Diam. Relat. Mater., 10(3-7): 1235–1240.

[41] Cantoro, M., S. Hofmann, S. Pisana, V. Scardaci, A. Parvez, C. Ducati and J. Robertson. 2006. Catalytic chemical vapor deposition of single-wall carbon nanotubes at low temperatures. Nano Lett., 6(6): 1107–1112.

[42] Huang, L., X. Cui, B. White and S.P. O 'Brien. 2004. Long and oriented single-walled carbon nanotubes grown by ethanol chemical vapor deposition. J. Phys. Chem., 108(42): 16451–16456.

[43] Zhao, N., C. He, Z. Jiang, J. Li and Y. Li. 2006. Fabrication and growth mechanism of carbon nanotubes by catalytic chemical vapor deposition. Mater. Lett., 60(2): 159–163.

[44] Singjai, P., S. Changsarn and S. Thongtem. 2007. Electrical resistivity of bulk multi-walled carbon nanotubes synthesized by an infusion chemical vapor deposition method. Mater. Sci. Eng.: A, 443(1-2): 42–46.

[45] Kato, T., G.H. Jeong, T. Hirata, R. Hatakeyama, K. Tohji and K. Motomiya. 2003. Single-walled carbon nanotubes produced by plasma-enhanced chemical vapor deposition. Chem. Phys. Lett., 381(3-4): 422–426.

[46] Yun, Y., V. Shanov, Y. Tu, S. Subramaniam and M.J. Schulz. 2006. Growth mechanism of long aligned multiwall carbon nanotube arrays by water-assisted chemical vapor deposition. Phys. Chem. B, 110(47): 23920–23925.

[47] Lee, C.J., J. Park, J.M. Kim, Y. Huh, J.Y. Lee and K.S. No. 2000. Low-temperature growth of carbon nanotubes by thermal chemical vapor deposition using Pd, Cr, and Pt as co-catalyst. Chem. Phys. Lett., 327(5-6): 277–283.

[48] Choi, Y.C., D.J. Bae, Y.H. Lee, B.S. Lee, I.T. Han, W.B. Choi and J.M. Kim. 2000. Low temperature synthesis of carbon nanotubes by microwave plasma-enhanced chemical vapor deposition. Synth. Met., 108(2): 159–163.

[49] Hsu, C.M., C.H. Lin, H.L. Chang and C.T. Kuo. 2002. Growth of the large area horizontally-aligned carbon nanotubes by ECR-CVD. Thin Solid Films, 420: 225–229.

[50] Le, H.A., S. Chin, E. Park, L.T. Linh, G.N. Bae and J. Jurng. 2011. Chemical vapor synthesis and characterization of manganese oxides. Chem. Vapor Deposition, 17(7-9): 228–234.

[51] Matsumoto, K., K. Neishi, H. Itoh, H. Sato, S. Hosaka and J. Koike. 2009. Chemical vapor deposition of Mn and Mn oxide and their step coverage and diffusion barrier properties on patterned interconnect structures. Appl. Phys. Express, 2(3): 036503.

[52] Oyedotun, K.O., E. Ajenifuja, B. Olofinjana, B.A. Taleatu, E. Omotoso, M.A. Eleruja and E.O.B. Ajayi. 2015. Metal-organic chemical vapour deposition of lithium manganese oxide thin films via single solid source precursor. Mater. Sci.-Poland, 33(4): 725–731.

[53] Park, S., S. Lim and H. Choi. 2006. Chemical vapor deposition of iron and iron oxide thin films from Fe (II) dihydride complexes. Chemistry of Materials, 18(22): 5150–5152.

[54] Yang, Y., H. Zhang and Y. Yan. 2018. The preparation of Fe_2O_3-ZSM-5 catalysts by metal-organic chemical vapour deposition method for catalytic wet peroxide oxidation of m-cresol. R. Soc. Open Sci., 5(3): 171731.

[55] Yubero, F., M. Ocana, A. Justo, L. Contreras and A.R. González-Elipe. 2000. Iron oxide thin films prepared by ion beam induced chemical vapor deposition: structural characterization by infrared spectroscopy. J. Vac. Sci. Technol. A, 18(5): 2244–2248.

[56] Hasany, S.F., N.H. Abdurahman, A.R. Sunarti and R. Jose. 2013. Magnetic iron oxide nanoparticles: chemical synthesis and applications review. Curr. Nanosci., 9(5): 561–575.

[57] Maruyama, T. and T. Nakai. 1991. Cobalt oxide thin films prepared by chemical vapor deposition from cobalt (II) acetate. Sol. Energy Mater., 23(1): 25–29.

[58] Haniam, P., C. Kunsombat, S. Chiangga and A. Songsasen. 2014. Synthesis of cobalt oxides thin films fractal structures by laser chemical vapor deposition. Sci. World J.

[59] Barreca, D., C. Massignan, S. Daolio, M. Fabrizio, C. Piccirillo, L. Armelao and E. Tondello. 2001. Composition and microstructure of cobalt oxide thin films obtained from a novel cobalt (II) precursor by chemical vapor deposition. Chem. Mater., 13(2): 588–593.

[60] Ehsan, M.A., M.A. Aziz, A. Rehman, A.S. Hakeem, M.A.A. Qasem and O.W. Saadi. 2018. Facile synthesis of gold-supported thin film of cobalt oxide via AACVD for enhanced electrocatalytic activity in oxygen evolution reaction. ECS J. Solid State Sci. Technol., 7(12): P711–P718.

[61] Guyon, C., A. Barkallah, F. Rousseau, K. Giffard, D. Morvan and M. Tatoulian. 2011. Deposition of cobalt oxide thin films by plasma-enhanced chemical vapour deposition (PECVD) for catalytic applications. Surf. Coat. Tech., 206(7): 1673–1679.

[62] Rivera, E.F., B. Atakan and K. Kohse-Höinghaus. 2001. CVD deposition of cobalt oxide (CO_3O_4) from Co(acac)2. Le Journal de Physique IV, 11(PR3): Pr3-629–Pr3-635.

[63] Shirai, H., T. Kinoshita and M. Adachi. 2011. Synthesis of cobalt nanoparticle and fabrication of magnetoresistance devices by ion-assisted aerosol generation method. Aerosol Sci. Tech., 45(10): 1240–1244.

[64] Bahlawane, N., E.F. Rivera, K. Kohse-Höinghaus, A. Brechling and U. Kleineberg. 2004. Characterization and tests of planar Co_3O_4 model catalysts prepared by chemical vapor deposition. Appl. Catal. B: Environ., 53(4): 245–255.

[65] Burchfield, L.A., M. Al Fahim, R.S. Wittman, F. Delodovici and N. Manini. 2017. Novamene: A new class of carbon allotropes. Heliyon, 3(2): e00242.

[66] Burchell, T.D. 1999. Carbon Materials for Advanced Technologies. Elsevier.

[67] Ahmed, W., A. Elhissi, V. Dhanak and K. Subramani. 2018. Carbon nanotubes: Applications in cancer therapy and drug delivery research. Emerg. Nanotech. Dentistry, pp. 371–389.

[68] Subramani, K. and M. Mehta. 2018. Nanodiagnostics in microbiology and dentistry. Emerging Nanotechnologies in Dentistry. William Andrew Publishing, pp. 391–419.

[69] He, Z.-Y., X.-W. Wei and Y.-Q. Wei. 2017. Recent advances of nanostructures in antimicrobial therapy. In Antimicrobial Nanoarchitectonics, Elsevier, pp. 167–194.

[70] Saifuddin, N., A.Z. Raziah and A.R. Junizah. 2013. Carbon nanotubes: a review on structure and their interaction with proteins. J. Chem.

[71] Yamada, T., J. Kim, M. Ishihara and M. Hasegawa. 2013. Low-temperature graphene synthesis using microwave plasma CVD. J. Phys. D: Appl. Phys., 46(6): 063001.

[72] An, H., W.-J. Lee and J. Jung. 2011. Graphene synthesis on Fe foil using thermal CVD. Curr. Appl. Phys., 11(4): S81–S85.

[73] Li, X., W. Cai, J. An, S. Kim, J. Nah, D. Yang and S.K. Banerjee. 2009. Large-area synthesis of high-quality and uniform graphene films on copper foils. Science, 324(5932): 1312–1314.

[74] Juang, Z.Y., C.Y. Wu, A.Y. Lu, C.Y. Su, K.C. Leou, F.R. Chen and C.H. Tsai. 2010. Graphene synthesis by chemical vapor deposition and transfer by a roll-to-roll process. Carbon, 48(11): 3169–3174.

[75] Chen, Z., W. Ren, L. Gao, B. Liu, S. Pei and H.M. Cheng. 2011. Three-dimensional flexible and conductive interconnected graphene networks grown by chemical vapour deposition. Nat. Mater., 10(6): 424.

[76] Yu, Q., L.A. Jauregui, W. Wu, R. Colby, J. Tian, Z. Su and T.F. Chung. 2011. Control and characterization of individual grains and grain boundaries in graphene grown by chemical vapour deposition. Nat. Mater., 10(6): 443.

[77] Wang, X., H. Li, M. Li, C. Li, H. Dai and B. Yang. 2018. Synthesis and characterization of graphene-based nanostructures by electron-assisted hot filament plasma CVD. Diam. Relat. Mater., 86: 179–185.

[78] Ramakrishnan, S., E.J. Jelmy, R. Senthilkumar, M. Rangarajan and N.K. Kothurkar. 2018. One-step RF-CVD method for the synthesis of graphene decorated with metal and metal oxide nanoparticles. J. Nanosci, 18(2): 1089–1096.

[79] Xu, W., W. Wang, Z. Guo and Z. Liu. 2017. Fabrication of submillimeter-sized single-crystalline graphene arrays by a commercial printing-assisted CVD method. RSC Adv., 7(29): 17800–17805.

[80] Park, S. and R.S. Ruoff. 2009. Chemical methods for the production of graphenes. Nat. Nanotechnol., 4(4): 217.

[81] Malesevic, A., R. Vitchev, K. Schouteden, A. Volodin, L. Zhang, G. Van Tendeloo and C. Van Haesendonck. 2008. Synthesis of few-layer graphene via microwave plasma-enhanced chemical vapour deposition. Nanotechnology, 19(30): 305604.

[82] Freitas Neto, D.B., F.F.S. Xavier, E.Y. Matsubara, R. Parmar, R. Gunnella and J.M. Rosolen. 2020. The role of nanoparticle concentration and CNT coating in high-performance polymer-free micro/nanostructured carbon nanotube-nanoparticle composite electrode for Li intercalation. J. Electroanal. Chem., 858: 113826.

CHAPTER 9

CVD Growth of Transition Metal Dichalcogenides MX$_2$ (M: Mo, W, X: Se, S)#

Alejandro Fajardo Peralta,[1] *Jose Valenzuela-Benavides,*[2]
Nestor Perea-Lopez[3,4,*] *and Mauricio Terrones*[3,4]

1. Introduction

Transition metal dichalcogenides (TMDs) in their single-layer (S-L) crystalline form are nowadays an essential piece of the nanotechnology toolset. One single-layer of a TMDs crystal is the thinnest expression that preserves the stoichiometry of the bulk. TMDs are formed in-plane (x,y directions) by covalent bonds between a transition metal (W, Mo, etc.) and a chalcogen atom (S, Se, or Te), while much weaker out-of-plane (z-direction) van der Waals forces dominate the interaction between the atomic layers. Several decades ago, Joensen and coworkers predicted that these ultra-thin crystals could exist isolated from their bulk counterparts. Successful exfoliation of S-L MoS$_2$ was achieved by the intercalation of Li atoms [1]. However, these single-layers restack after removing the solvent, hindering the intrinsic properties of S-L MoS$_2$. It was not until 2010 that two research groups managed to isolate and measure photoluminescence (PL) spectra from microscopic portions of mechanically exfoliated MoS$_2$ crystals. Both groups reported strong PL emission from the thinnest regions of the crystal, a counterintuitive result that revealed a transformation of the band structure of MoS$_2$ from indirect bandgap to direct bandgap semiconductor at

[1] Posgrado en Nanociencias, Centro de Investigación Científica y de Educación Superior de Ensenada, Carretera Ensenada-Tijuana, No. 3918, Zona Playitas, 22860, Ensenada, B.C., Mexico.
[2] Departamento de Física, Centro de Nanociencias y Nanotecnología Universidad Nacional Autonoma de México.
[3] Department of Physics, The Pennsylvania State University, University Park, Pennsylvania 16802, USA.
[4] Center for 2-Dimensional and Layered Materials, The Pennsylvania State University, University Park, Pennsylvania 16802, USA.
* Corresponding author: nup13@psu.edu
# Reviewed by Dr. Humberto R. Gutierrez, Department of Physics, University of South Florida, Tampa, Florida 33620, United States.

the monolayer limit [2, 3]. Similar results to that observed in exfoliated MoS_2 were reported soon after for monolayers of WS_2 synthesized via chemical vapor deposition (CVD) [4], WSe_2 [5], and $MoSe_2$ [6].

The layer dependent transformation of the bandgap observed in S-L TMDs was a breakthrough discovery that could be compared in importance to the discovery of the buckyball (C_{60}) [7], the observation of the chirality induced bandgap variations in single-walled carbon nanotubes, or the isolation of graphene and the subsequent measurement of the behavior of electrons of massless Dirac fermions [8]. All of these discoveries had a profound impact on both technology and basic science.

Mechanical exfoliation (using sticky tape) of single- and few-layers TMDs, although not scalable, can provide high quality samples to enable basic-science studies of the physical properties of TMDs and other 2D materials; both graphene and S-L MoS_2 were first isolated by this method. Mechanical exfoliation yields single-layered crystals with superb quality; however, the resulting single-layer area is quite small and hard to locate. Since large areas are required for more complex applications, such as fabrication of multiple field-effect transistors, optical devices, or sensors, it is necessary to explore other methods of preparation. Here is where CVD plays a crucial role in the development of the science and technology of S-L 2D materials. CVD is a versatile and inexpensive method to grow high-quality crystals, and has been widely adopted for the growth of single- and few-layers TMDs.

In the rest of the chapter, we will explore how CVD is used to grow the four basic semiconducting TMDs—MoS_2, WS_2, $MoSe_2$, and WSe_2. We will review how it is possible to create alloys in single-layer form, as well as heterojunctions and substitutional doping for tuning the bandgap.

2. The CVD Synthesis of 2D Materials

In CVD, a reaction chamber confines the vapors of volatile precursors, and these fumes react chemically by the effect of temperature gradients. The chemical vapors are transported by a controlled flow of gas(es) through the chamber. The variables of the CVD experiment include the choice of chemical precursors, the flow rate(s) of the transport gases, the partial pressure of the reactants, temperature gradients, reaction duration, and the choice of substrate. Under the right conditions, the result is the growth of crystalline solids on the surface of the substrate. The schematic diagram in Figure 9.1 represents perhaps the most basic CVD system that can be used to grow single-layer TMD.

From the plethora of recent reports, the synthesis of TMDs can be obtained using two main routes, the so-called one-step growth and two-step growth [9]. One-step growth means that the gaseous precursors of the chalcogen (X) and the transition metal (M) react within the chamber at the proper formation temperature to obtain MX_2 as the final product [10, 11]. The two-step growth requires a pre-deposited metal or its metal oxide as a thin film (by PVD, sputtering, or other techniques), which is subsequently exposed (within the CVD reactor) to the chalcogen vapors at a high temperature to form the MX_2 compound. Figure 9.2 illustrates the different CVD configurations that can be used.

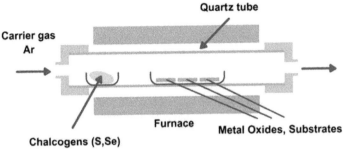

Figure 9.1. Schematic diagram of a generic CVD setup to grow S-L TMD.

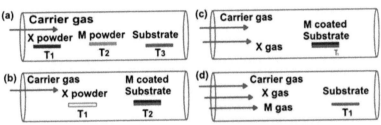

Figure 9.2. Schematic representation of the four primary CVD configurations combining X: the chalcogen powder, M: the transition metal or metal oxide, and the substrates that can have a pre-deposited layer of a metal or metal oxide, adapted from [9].

After many years of research on CVD synthesis of TMDs, the growth parameters are now well established. For example, for the W chalcogenides (WS₂, WSe₂), the growth temperature needs to be higher than Mo chalcogenides when the precursors are their respective oxides. Similarly, the gas flow rates and the initial reactant masses are well known (see Table 9.1 in summary).

2.1 MoS₂

MoS₂ was the first single-layer semiconductor to be synthesized by CVD, and therefore the most studied of the TDMs so far [12, 13, 14, 15]. Usually, CVD growth of MoS₂ monolayers is carried out at atmospheric pressure, a weighed amount of sulfur powder is placed in a crucible at a distance between 10 and 25 cm from the center of the furnace where MoO_3 powder and SiO_2/Si substrates are preloaded. The furnace temperature is slowly raised to 750°C (200°C for sulfur), while the sulfur vapors are carried downstream by the carrier gas (Ar, H_2, H_2S, N_2) [14].

In a variant of the previously described setup, the molybdenum oxide powder precursor is replaced by a molybdenum oxide thin film (1–5 nm) directly deposited on the substrate. The oxide particles in the film promote the nucleation process, where the sulfur atoms react with molybdenum atoms until the seed particles are consumed, liberating oxygen in the process [16]. The crystals obtained by this method vary according to the growth characteristics of the nucleation centers—initially forming island-like areas in the substrate that eventually connect as the crystals grow. This type of growth dynamics only occurs for a Mo oxide film thickness between 1 and 5 nm; films thicker than 5 nm form polycrystalline multilayers, while for films thinner than 1 nm, the emerging islands are unable to connect and form extended monolayers (see Figure 9.3).

Another parameter that has a definitive effect on crystal shape is the concentration ratio of Mo to S atoms available at the edges of the MoS_2 crystal during growth. A hexagonal shaped crystal is obtained when the Mo:S ratio is approximately one. Any disproportion either way leads to triangular shapes [11].

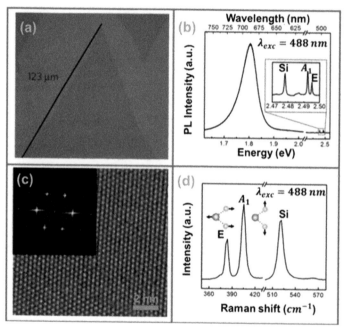

Figure 9.3. (a) Optical micrograph of a CVD grown MoS_2 on a 285 nm SiO_2 on Si substrate. (b) Photoluminescence spectrum of a MoS_2 monolayer. The zoom-in view depicts the Raman peaks. (c) HRTEM image of a triangular island, and its corresponding FFT pattern (insert). (d) Raman spectrum of a triangular-shaped MoS_2 monolayer, with peaks corresponding to vibrational modes (in-plane and out-of-plane).

2.2 WS_2

In some of the early reports on the synthesis of TMDs by CVD, a two-step approach was considered, by exposing a thin film of the transition metal at high temperature to a sulfur vapor, in a process called sulfurization. However, the extremely high temperatures required for evaporating transition metals, such as tungsten and molybdenum, with melting points of 3422°C and 2623°C, respectively, present a serious limitation. On the other hand, the use of the corresponding metal oxide, such as WO_3, offers a viable alternative to thermally deposit these materials as the transition metal source [4]. The melting point of WO_3 is significantly lower than pure W, and in this way, thin oxide films evaporated on SiO_2/Si substrates (~ 10 Å oxide thickness) provide both the precursor and the nucleation centers for the growth of WS_2 monolayer. This process is usually performed at temperatures close to 800°C, and yields 170 μm side WS_2 triangles, as can be seen in Figure 9.4 [17].

In the one-step synthesis of WS_2 monolayers in CVD, the WO_3 powder is placed at the center of the furnace at 950°C. A boat containing sulfur is placed upstream at a lower temperature (~ 200°C), while the substrates (SiO_2/Si, sapphire, etc.) are placed downstream at a lower temperature (600–800°C). The experiment can be carried out either in low-vacuum or at atmospheric pressure using Ar or N_2 as a carrier gas. With this method, 200 μm wide triangular monolayer crystals have been obtained [18].

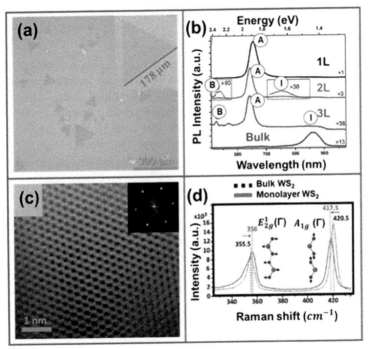

Figure 9.4. (a) Optical image of as-grown WS_2 on 300 nm thick SiO_2/Si. Insert: zoom-in of a large WS_2 triangle. (b) Photoluminescence spectra of WS_2 for 1L, 2L, 3L, and bulk. (c) HRTEM phase-contrast image of S-L WS_2; Insert: diffraction pattern of a WS_2 single-domain. (d) Raman spectra of S-L and bulk WS_2 using the 488 nm laser line excitation, showing a clear shift of the in-plane and out-of-plane vibrational modes, enough to discriminate bulk from S-L.

2.3 *MoSe₂ and WSe₂*

In the CVD synthesis of S-L $MoSe_2$, the selenization of MoO_3 has subtle differences compared to the sulfurization in MoS_2 [19, 20, 21]. The temperature and the carrier gas Ar/H_2 mixture (see Table 9.1) require precise control to obtain large-area single layer selenides. Some selenization processes can proceed rapidly, starting at temperatures as low as 550°C, resulting in vertically aligned crystals (Kong et al.). Other reported methods require higher temperatures and the use of growth promoters ($WO_{2.9}$ MX, M = Na or K; X = Cl, Br or I; Se at 270°C) that result in larger monolayer coverage of the substrate [22]. MoSe$_2$ monolayer crystals, as large as few millimeters of good quality were synthesized at atmospheric pressure using glass substrates [20].

Similarly, CVD processes have been reported for WSe$_2$ monolayers grown on graphene [5], sapphire [23], and SiO$_2$/Si [24], with crystal sizes in the 200 μm range and large area coverage [25]. Here, the carrier gas mixture played an essential role in the removal of residual oxides. Of particular technological interest, is the attainment of a preferential growth orientation of triangular-shaped WSe$_2$ with the help of native defects present in hexagonal boron nitride (h-BN) [26]. On the downside, TMDs selenides are fragile in an oxygen atmosphere. An important note about using selenium-compounds containing selenium can produce SeO$_2$ in CVD that has toxic effects, and thus require special care, such as proper handling of vapor residues.

2.4 *Choice of Substrates*

TMD growth by CVD by its nature is a process that requires temperatures between 400°C and 900°C (see Table 9.1). Due to this, the substrates must be thermally stable at such temperatures. These substrates should have a high melting point and low roughness (ultra-flat crystals). The most commonly used substrates are SiO$_2$/Si, bare Si, sapphire, and other 2D crystals, such as graphene and h-BN.

Silicon is the most widely used substrate for TMDs and many other crystals. Silicon is ubiquitous in the crystal growth processes because of the well-defined process that makes super high-quality wafers affordable. Silicon substrates with less than 300 nm SiO$_2$ layer create a superb optical contrast with monolayer 2D crystals, such as graphene and TMDs [27, 28]. Sapphire is also used for CVD growth of TMDs. Like silicon, it has a flat surface and a hexagonal symmetry that promotes highly oriented TMDs films [29]. Van der Waals epitaxy in sapphire is preferred because when TMD islands coalesce, lower angle boundaries are formed. Another desirable characteristic of sapphire is its high optical transmittance, which makes it an ideal substrate for optoelectronic applications. A pre-annealed treatment of sapphire substrates before the CVD process itself has some advantages. It has been shown that the c-plane of sapphire pre-annealed at 1000°C works well as a template for WSe$_2$ growth [30].

It is technologically relevant to create devices where every material is atomically thin. In such a system, graphene could act as the electrical conductor where electrons can move freely in the 2D plane, and through the interfaces with a van der Waals material. TMDs can play the part of the semiconductor with h-BN as the insulator. Growing TMDs on graphene by CVD has already been demonstrated [15, 31]. Graphene is also an excellent substrate material for the preparation of High-resolution transmission electron microscopy (HRTEM) samples, and as a photovoltaic collector [32, 33]. Hexagonal boron nitride can also be used as a substrate, and in this case its insulating properties promote valley spin electronics phenomena in TMDs [34].

2.5 *Characterization Basics*

A key aspect boosting research on S-L TMDs is the availability of ad hoc characterization techniques. Starting with optical methods, we can mention photoluminescence (PL) spectroscopy—it uses an excitation beam with photon energy higher than the bandgap of the material of interest. Monolayer TMDs are direct bandgap semiconductors; hence, the photon emission is intense regardless of the infinitesimal thickness of the film [35, 3, 2]. Raman spectroscopy is another optical technique—it probes the vibrational modes of the crystals using laser excitation, a monochromator, and a detector to monitor the energy losses of the excitation beam. The Raman shift peaks appear at energies commensurate with the vibrational modes of the lattice, and serve as a fingerprint of the material [36, 37]. For TMDs, PL and Raman represent excellent non-destructive techniques with high sensitivity to the material type and the number of layers [38, 39].

Another characterization technique that complements PL and Raman spectroscopies is atomic force microscopy (AFM), suitable for measuring layer thickness of TMDs and other 2D materials. It has angstrom resolution in the z-direction and can easily resolve a single layer TMD [40]. On the downside, since the height information is extracted from a topography image, the TMD must be supported on an ultra-flat substrate for AFM to work accurately [41].

High-resolution transmission electron microscopy (HRTEM) might be the most powerful technique available so far. Using AC-HRTEM, it is possible to see the lattice, edges, vacancies [42], and grain boundaries of S-L TMD [43, 44]. However, it requires either the growth of the TMD directly on the TEM grid, or a complicated and usually dirty transfer procedure [45].

3. Heterostructures and Doping

As presented in the previous sections, CVD is an excellent technique to grow high purity single-layer TMDs. However, it is of great interest to be able to synthesize heterostructures and introduce impurities in the lattice of TMDs. TMDs monolayers have many properties that may be very useful for optoelectronic devices. By analogy with conventional semiconductor materials, the TMDs bandgap can be easily modified by staking heterostructures, surface straining, or by chemically changing the lattice with vacancies and doping.

3.1 Out–of-plane Heterostructures

The creation of heterostructures is of prime importance for the fabrication of electronic and optoelectronic devices based on semiconductors. The first approach for generating heterostructures made of TMDs is by vertical stacking. Artificial stacking of different S-L TMDs was achieved using dry or wet transfer methodologies [46, 45]. Heterostacking of semiconductor TMDs generates a type II band alignment where the conduction band resides in one of the materials and the valence band in the other one. Bilayer systems of combinations of sulfides and selenides of tungsten and molybdenum offer another option to tune the bandgap of TMDs [47, 48]. CVD offers a clean alternative to transfer methods, and CVD grown TMD heterostructures exhibit a perfectly clean interface between the two single layers [49, 50]. Clean interfaces are essential for the observation of new physical phenomena in these heterostructures [51].

3.2 In-plane Heterostructures

A valuable strategy in the synthesis of CVD is the implementation of a methodology that permits control of the lateral growth of heterostructures of alternately different MX_2 compositions. The growth can follow the crystalline boundaries in-plane as a guide for the manufacture of concentric triangles of WS_2-WSe_2 structures, as shown in Figure 9.5a [52], or different lateral structures, such as WS_2-MoS_2 [53]. Even intentionally implanted defects in the substrate can guide the in-plane orientation of triangular crystal growth, such as the alignment of WSe_2 on h–BN [26].

In-plane junctions can be synthesized by sequential stages of CVD growth, each with different precursors and gas flows to promote the growth of the desired TMD. This has a direct impact on the technological trend of manufacturing devices that make use of such properties as valley electronics, spin-flip changes, and 2D transport in semiconductor-junctions that can be obtained by a bandgap alignment between the different compositions merged at the boundaries of triangular MX_2 crystals.

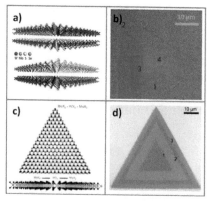

Figure 9.5. (a) Vertically grown WS_2/MoS_2 heterostructures [47]. (b) Optical microscopy image of the WS_2/MoS_2 heterostructure [50]. (c) Schematic illustration of an in-plane MoX_2-WX_2 junction. (d) Optical image of a triple junction heterostructure consisting of S-L $MoSe_2$-WSe_2 [52].

3.3 Doping and Defects

Defects and doping are the most straightforward way to tune the electronic band structure of a semiconductor [54]. For silicon, boron, and phosphorus, doping produces the p-type and n-type Si used to form the first solid-state electronic devices [55]. CVD makes possible the incorporation of heteroatoms in the lattice of semiconducting TMDs. The simplest example of doping is by forming alloys of TMDs, such as WS_2 and MoS_2. This can be achieved by substituting the chalcogen in the form $MoS_{2(1-x)}Se_{(x)}$ [56, 57] or the transition metal in the form $Mo_{(1-x)}W_xS_2$ [58, 59]. In both of these alloys, the bandgap can be tuned within the limits of the bandgaps of the TMDs being alloyed.

As a further step in the substitutional inclusion of heteroatoms outside the traditional transition metal or chalcogen groups, CVD has enabled the integration of manganese [60], rhenium [61], and carbon [62].

4. Summary

After almost a decade of research on single-layer TMDs, CVD has played a crucial role, enabling the production of good quality single-layer TMD crystals. Continuous improvement of the CVD method has enabled a substantial increase in the area coverage and quality of the crystals. Table 9.1 summarizes the growth conditions for the semiconducting TMDs discussed in this chapter.

Table 9.1. CVD synthesis conditions and variables to grow single-layer MX_2 (M: W, Mo; X: S, Se) TMDs systems and alloys.

TMD	Precursors	Substrate	Temperature	Pressure/gas	Morphology/size	Reference
MoS_2	$(NH_4)_2MoS_4$, S	SiO_2/Si, $SrTiO_3$	400°C	200–2 mTorr, precursor gas	2–5 nm Hexagonal flakes	[12]
MoS_2	$(NH_4)_2MoS_4$, S	SiO_2/Si	500–1000°C	1–500 Torr/ precursor gas	Bilayer and trilayer thin films	[63]
MoS_2	$(NH_4)_2 MoS_4$	Graphene	400°C	10 mTorr to 1 atm/70 sccm Ar	Few layer TMDs on graphene	[64]
MoS_2	MoO_3, S 180°C	SiO_2/Si (300 nm)	650°C	Atmosphere	Monolayer	[16]
MoS_2	Nanoribbons MoO_3, S 180°C	SiO_2/Si (300 nm)	850°C	Atmosphere	Monolayer and few layers	[65]
MoS_2	$MoCl_3$, S 300°C	SiO_2/Si (300 nm)	700°C,	Atmosphere	Triangular monolayers and Stars, 10–20 μm	[43]
MoS_2	MoO_3, S 100°C	SiO_2/Si (300 nm)	850°C	2 Torr/Ar	Monolayer and few layers	[66]
MoS_2	MoO_3, S 130°C	Mica	530°C	225 mTorr/ Ar	Monolayers	[67]
MoS_2	MoO_3, S 130°C	Graphene/ SiC, SiO_2/ Si	670°C	5 Torr/Ar	Monolayers	[31]
MoS_2	MoS_2 powders, S (200°C)	SiO_2/Si (300 nm)	900°C	20 Torr/ar	Monolayer and few layer TMDs	[28]
WS_2	~ 1 to 2.8 nm WO_X thin film, S (200°C)	SiO_2/Si	400–600°C	Atmosphere/ Ar	Mono, bi, and trilayer	[68]
WS_2	WO_3 film by ALD, H_2S gas	SiO_2/Si	750–900°C	450 mTorr	Mono, bi, and tetra-layers	[69]
WS_2	WO_3 thin films (5–20Å), S (200°C)	SiO_2/Si	400–1000°C	Atmosphere/ Ar	Monolayers and nanotubes	[70]

Table 9.1 Contd. ...

...Table 9.1 Contd.

TMD	Precursors	Substrate	Temperature	Pressure/gas	Morphology/size	Reference
WS$_2$	WO$_3$, SiO$_2$/Si, S (200°C)	SiO$_2$/Si (300 nm)	700–950°C	Atmosphere/Ar	Triangular monolayers 200 μm	[18]
WS$_2$	WO$_3$ film (5 to10 Å) on SiO$_2$/Si	SiO$_2$/Si (300 nm)	800°C	Atmosphere/Ar	Triangular Monolayers	[4]
MoSe$_2$	225°C MoCl$_3$	SiO$_2$/Si (300 nm)	450–600°C	Atmosphere	Thin films	[71]
MoSe$_2$	MoO$_3$, Se, Ar/H$_2$ (15% H$_2$)	SiO$_2$/Si (275 nm)	750°C	1 Atm/50 sccm Ar	Triangular Monolayers 135 μm	[6]
MoSe$_2$	MoO$_3$, Se (270°C), Ar/H$_2$ (15% H$_2$)	Sapphire	700–900°C	40–70 Ar-10 H$_2$ sccm, 25–350 Torr	Triangular Monolayers 5–10 μm	[19]
MoSe$_2$	MoO$_3$, Se (270°C), Ar	SiO$_2$/Si (300 nm), mica	820°C	1 Atm/Ar 10 sccm	Triangular Monolayers 20–40 μm	[22]
MoSe$_2$	MoO$_3$, Se (270°C), H$_2$	SiO$_2$/Si (100 nm)	800°C	20 Torr	Mono, bi and tri layer full 1 cm^2 area	[20]
MoSe$_2$	5 nm Mo film, Se (220–300°C)	SiO$_2$/Si	820°C	1 Atm/Ar 10 sccm	Films with vertically aligned layers	[72]
MoSe$_2$	Se 20 cm from center, MoO$_3$ powder	Molten glass	450–600°C	Atmosphere/Ar	Triangular monolayers 10–50 μm to 1 mm	[73]
WSe$_2$	5 nm WO$_3$, Se (500°C)	graphene/SiC	800°C	Atmosphere/Ar	Monolayers, and multilayers	[5]
WSe$_2$	WO$_3$, Se (270°C), Ar, H$_2$	Sapphire	750–850°C in Sapphire, 925°C center	Ar 80 sccm, H$_2$ 20 sccm, 1 Torr	Monolayer 1 mm	[23]
WSe$_2$	WO$_{2.9}$ MX (M = Na or K; X = Cl, Br or I), Se (270°C)	SiO$_2$/Si	750–850°C	1 Atm, 80/20 sccm Ar/H$_2$	20–200 μm depend on growth promoters	[25]
WSe$_2$	WO$_3$, Se (400°C), Ar	SiO$_2$/Si (300 nm)	900°C	Ar 14 sccm 1 Atm	Triangle–hexagon 1–5 μm	[24]
WS$_{2(1–x)}$ Se$_{2x}$	S, WO$_3$, Se	SiO$_2$/Si (300 nm)	450–600°C	Atmosphere/Ar	Triangle-hexagon 5–40 μm, Mo:W variable ratio	[74]

Table 9.1 Contd. ...

...Table 9.1 Contd.

TMD	Precursors	Substrate	Temperature	Pressure/gas	Morphology/ size	Reference
$WS_{2(1-x)}$ Se_{2x}	W, Se, S	SiO_2/Si (300 nm)	650–700°C	Atmosphere/ Ar	Triangle-hexagon 5–40 µm, Mo:W depends ratio	[75]
$Mo_{1-x}W_xS_2$	S, WO_3, MoO_3	SiO_2/Si (300 nm)	450–600°C	Atmosphere/ Ar	Triangle-Shaped 40 µm, Mo:W depends ratio	[68]
$Mo_{1-x}W_xS_2$	S, WO_3, MoO_3	SiO_2/Si (300 nm)	850°C	Atmosphere/ Ar	Triangle Monolayers	[76]
$Mo_{1-x}W_xS_2$	MoS_2 or MoS_2/ WO_X	Sapphire	650–700°C	Atmosphere/ Ar	Triangle Monolayers	[58]
$Mo_{1-x}W_xS_2$	SiO_2/Si,S, WO_3, MoO_3	SiO_2/Si (300 nm)	300°C	Atmosphere/ Ar	Triangle Monolayers	[59]
$Mo_{1-x}W_xSe_2$	Se, WO_3, MoO_3	SiO_2/Si (300 nm)	450–800°C	Atmosphere/ Ar	Triangle-hexagon 5–40 µm, Mo:W depends ratio	[77]
$Mo_{1-x}W_xSe_2$	W, Mo, Se	SiO_2/Si (300 nm)	650–850°C	Atmosphere/ Ar	Triangle-hexagon 5–40 µm, Mo:W depends ratio	[78]
$MoS_{2(1-x)}$ Se_{2x}	Se, S, MoO_3	SiO_2/Si (300 nm)	700–800°C	Atmosphere/ Ar	Triangle-hexagon 5–40 µm, S:Se depends ratio	[57]
$MoS_{2(1-x)}$ Se_{2x}	MoS_2, $MoSe_2$ powders 940–945°C	SiO_2/Si (300 nm)	600–700°C	60 mTorr/Ar	Monolayers alloys	[75]
$MoS_{2(1-x)}$ Se_{2x}	MoS_2 layers (by CVD), Se (270°C)	SiO_2/Si (300 nm)	600–900°C	Atmosphere/ Ar	Triangle-hexagon 5–40 µm, S:Se depends ratio	[79]
$MoS_{2(1-x)}$ Se_{2x}	thiophenol dissolved in tetrahydrofuran, diphenyl-diselenide, MoO_3	SiO_2/Si (300 nm)	650–700°C	H_2/ Atmosphere	Single-layer triangular or hexagonal 1–10 µm	[56]

References

[1] Joensen, P., R.F. Frindt and S.R. Morrison. 1986. Single-Layer MoS_2. Mater. Res. Bull., 21(4): 457–461.

[2] Mak, Kin Fai, Changgu Lee, James Hone, Jie Shan and Tony F. Heinz. 2010. Atomically thin MoS_2: A new direct-gap semiconductor. Phys. Rev. Lett., 105(13): 2–5.

[3] Andrea Splendiani, Liang Sun, Yuanbo Zhang, Tianshu Li, Jonghwan Kim, Chi-Yung Chim, Giulia Galli and Feng Wang. 2010. Emerging photoluminescence in monolayer MoS_2. Nano Lett., 10(4): 1271–1275.

[4] Humberto R. Gutiérrez, Nestor Perea-López, Ana Laura Elías, Ayse Berkdemir, Bei Wang, Ruitao Lv, Florentino López-Urías, Vincent H. Crespi, Humberto Terrones and Mauricio Terrones. 2013. Extraordinary room-temperature photoluminescence in triangular WS_2 monolayers. Nano Lett., 13(8): 3447–3454.

[5] Jing-Kai Huang, Jiang Pu, Chang-Lung Hsu, Ming-Hui Chiu, Zhen-Yu Juang, Yung-Huang Chang, Wen-Hao Chang, Yoshihiro Iwasa, Taishi Takenobu and Lain-Jong Li. 2014. Large-area synthesis of highly crystalline WSe_2 mono layers and device applications. Acs Nano, 8(1): 923–930.

[6] Wang, X., Y. Gong, Gang Shi, W.L. Chow, K. Keyshar, G. Ye, R. Vajtai, J. Lou, Z. Liu, E. Ringe, B.K. Tay and P.M. Ajayan. 2014. Chemical vapor deposition growth of crystalline mono layer $MoSe_2$. Acs Nano, 8(5): 5125–5131.

[7] Liu, Y., S.C. Obrien, Q. Zhang, J.R. Heath, F.K. Tittel, R.F. Curl, H.W. Kroto and R.E. Smalley. 1986. Negative carbon cluster ion-beams—new evidence for the special nature of C60. Chem. Phys. Lett., 126(2): 215–217.

[8] Raimond, J.M. and M. Brune. 2004. Quantum Computation, F. De Martini, and C. Monroe. 2004_ Science_Novoselov.pdf. 306(October): 666–669.

[9] Bosi, M. 2015. Growth and synthesis of mono and few-layers transition metal dichalcogenides by vapour techniques: a review. RSC Adv., 5(92): 75500–75518.

[10] Keng-Ku Liu, Wenjing Zhang, Yi-Hsien Lee, Yu-Chuan Lin, Mu-Tung Chang, Ching-Yuan Su, Chia-Seng Chang, Hai Li, Yumeng Shi, Hua Zhang, Chao-Sung Lai and Lain-Jong Li. 2012. Growth of large-area and highly crystalline MoS_2 thin layers on insulating substrates. Nano Lett., 12(3): 1538–1544.

[11] Wang, X.S., H.B. Feng, Y.M. Wu and L.Y. Jiao. 2013. Controlled synthesis of highly crystalline MoS_2 flakes by chemical vapor deposition. J. Am. Chem. Soc., 135(14): 5304–5307.

[12] Zhan, Y.J., Z. Liu, S. Najmaei, P.M. Ajayan and J. Lou. 2012. Large-area vapor-phase growth and characterization of MoS_2 atomic layers on a SiO_2 substrate. Small, 8(7): 966–971.

[13] Yu-Chuan Lin, Wenjing Zhang, Jing-Kai Huang, Keng-Ku Liu, Yi-Hsien Lee, Chi-Te Liang, Chih-Wei Chu and Lain-Jong Li. 2012. Wafer-scale MoS_2 thin layers prepared by MoO_3 sulfurization. Nanoscale, 4(20): 6637–6641.

[14] Han Wang, Lili Yu, Yi-Hsien Lee, Wenjing Fang, Allen Hsu, Patrick Herring, Matthew Chin, Madan Dubey, Lain-Jong Li, Jing Kong and Tomas Palacios. 2012. Large-scale 2D electronics based on single-layer MoS_2 grown by chemical vapor deposition. 2012 IEEE International Electron Devices Meeting (Iedm).

[15] Yumeng Shi, Wu Zhou, Ang-Yu Lu, Wenjing Fang, Yi-Hsien Lee, Allen Long Hsu, Soo Min Kim, Ki Kang Kim, Hui Ying Yang, Lain-Jong Li, Juan-Carlos Idrobo and Jing Kong. 2012. van der Waals epitaxy of MoS_2 layers using graphene as growth templates. Nano Lett., 12(6): 2784–2791.

[16] Yi-Hsien Lee, Xin-Quan Zhang, Wenjing Zhang, Mu-Tung Chang, Cheng-Te Lin, Kai-Di Chang, Ya-Chu Yu, Jacob Tse-Wei Wang, Chia-Seng Chang, Lain-Jong Li and Tsung-Wu Lin. 2012. Synthesis of large-area MoS_2 atomic layers with chemical vapor deposition. Adv. Mater., 24(17): 2320–2325.

[17] Chunxiao Cong, Jingzhi Shang, Xing Wu, Bingchen Cao, Namphung Peimyoo, Caiyu Qiu, Litao Sun and Ting Yu. 2014. Synthesis and optical properties of large-area single-crystalline 2D semiconductor WS_2 monolayer from chemical vapor deposition. Adv. Opt. Mater., 2(2): 131–136.

[18] Alharbi, A. and D. Shahrjerdi. 2016. Electronic properties of monolayer tungsten disulfide grown by chemical vapor deposition. Appl. Phys. Lett., 109 (19).

[19] Yung-Huang Chang, Wenjing Zhang, Yihan Zhu, Yu Han, Jiang Pu, Jan-Kai Chang, Wei-Ting Hsu, Jing-Kai Huang, Chang-Lung Hsu, Ming-Hui Chiu, Taishi Takenobu, Henan Li, Chih-I Wu, Wen-Hao Chang, Andrew Thye Shen Wee and Lain-Jong Li. 2014. Monolayer $MoSe_2$ grown by chemical vapor deposition for fast photodetection. Acs Nano, 8(8): 8582–8590.

[20] Xin Lu, M. Iqbal Bakti Utama, Junhao Lin, Xue Gong, Jun Zhang, Yanyuan Zhao, Sokrates T. Pantelides, Jingxian Wang, Zhili Dong, Zheng Liu, Wu Zhou and Qihua Xiong. 2014. Large-area synthesis of monolayer and few-layer $MoSe_2$ films on SiO_2 substrates. Nano Lett., 14(5): 2419–2425.

[21] Jonathan C. Shaw, Hailong Zhou, Yu Chen, Nathan O. Weiss, Yuan Liu, Yu Huang and Xiangfeng Duan. 2014. Chemical vapor deposition growth of monolayer $MoSe_2$ nanosheets. Nano Res., 7(4): 511–517.

[22] Jing Xia, Xing Huang, Ling-Zhi Liu, Meng Wang, Lei Wang, Ben Huang, Dan-Dan Zhu, Jun-Jie Li, Chang-Zhi Gu and Xiang-Min Meng. 2014. CVD synthesis of large-area, highly crystalline MoSe$_2$ atomic layers on diverse substrates and application to photodetectors. Nanoscale, 6(15): 8949–8955.

[23] Zhang, W., M.H. Chiu, C.H. Chen, W. Chen, L.J. Li and A.T.S. Wee. 2014. Role of metal contacts in high-performance phototransistors based on WSe$_2$ monolayers. Acs Nano, 8(8): 8653–8661.

[24] Liang Chen, Bilu Liu, Ahmad N. Abbas, Yuqiang Ma, Xin Fang, Yihang Liu and Chongwu Zhou. 2014. Screw-dislocation-driven growth of two-dimensional few-layer and pyramid-like WSe$_2$ by sulfur-assisted chemical vapor deposition. Acs Nano, 8(11): 11543–11551.

[25] Shisheng Li, Shunfeng Wang, Dai-Ming Tang, Weijie Zhao, Huilong Xu, Leiqiang Chu, Yoshio Bando, Dmitri Golberg and Goki Eda. 2015. Halide-assisted atmospheric pressure growth of large WSe$_2$ and WS$_2$ monolayer crystals. Appl. Mater. Today, 1(1): 60–66.

[26] Xiaotian Zhang, Fu Zhang, Yuanxi Wang, Daniel S. Schulman, Tianyi Zhang, Anushka Bansal, Nasim Alem, Saptarshi Das, Vincent H. Crespi, Mauricio Terrones and Joan M. Redwing. 2019. Defect-controlled nucleation and orientation of WSe$_2$ on hBN: A route to single-crystal epitaxial monolayers. ACS Nano, 13(3): 3341–3352.

[27] Hai Li, Jumiati Wu, Xiao Huang, Gang Lu, Jian Yang, Xin Lu, Qihua Xiong and Hua Zhang. 2013. Rapid and reliable thickness identification of two-dimensional nanosheets using optical microscopy. Acs Nano, 7(11): 10344–10353.

[28] Sanfeng Wu, Chunming Huang, Grant Aivazian, Jason S. Ross, David H. Cobden and Xiaodong Xu. 2013. Vapor–solid growth of high optical quality MoS$_2$ monolayers with near-unity valley polarization. ACS Nano, 7(3): 2768–2772.

[29] Xiaoli Li, Yafang Shi, Shuai Li, Wei Shi, Wenpeng Han, Chuan Zhou, Xiaohui Zhao and Baolai Liang. 2018. Layer-number dependent reflection spectra of MoS$_2$ flakes on SiO$_2$/Si substrate. Opt. Mater. Express, 8(10): 3082–3082.

[30] Areej Aljarb, Zhen Cao, Hao-Ling Tang, Jing-Kai Huang, Mengliu Li, Weijin Hu, Luigi Cavallo and Lain-Jong Li. 2017. Substrate lattice-guided seed formation controls the orientation of 2D transition-metal dichalcogenides. ACS Nano, 11(9): 9215–9222.

[31] Yu-Chuan Lin, Ning Lu, Nestor Perea-Lopez, Jie Li, Zhong Lin, Xin Peng, Chia Hui Lee, Ce Sun, Lazaro Calderin, Paul N. Browning, Michael S. Bresnehan, Moon J. Kim, Theresa S. Mayer, Mauricio Terrone and Joshua A. Robinson. 2014. Direct synthesis of van der Waals solids. Acs Nano, 8(4): 3715–3723.

[32] Joeson Wong, Deep Jariwala, Giulia Tagliabue, Kevin Tat, Artur R. Davoyan, Michelle C. Sherrott and Harry A. Atwater. 2017. High photovoltaic quantum efficiency in ultrathin van der Waals heterostructures. Acs Nano, 11(7): 7230–7240.

[33] Jariwala, D., T.J. Marks and M.C. Hersam. 2017. Mixed-dimensional van der Waals heterostructures. Nat. Mater., 16(2): 170–181.

[34] Mitsuhiro Okada, Takumi Sawazaki, Kenji Watanabe, Takashi Taniguch, Hiroki Hibino, Hisanori Shinohara and Ryo Kitaura. 2014. Direct chemical vapor deposition growth of WS$_2$ atomic layers on hexagonal boron nitride. Acs Nano, 8(8): 8273–8277.

[35] Eda, G., H. Yamaguchi, D. Voiry, T. Fujita, M.W. Chen and M. Chhowalla. 2011. Photoluminescence from chemically exfoliated MoS$_2$. Nano Lett., 11(12): 5111–5116.

[36] Lee, C., H. Yan, L.E. Brus, T.F. Heinz, J. Hone and S. Ryu. 2010. Anomalous lattice vibrations of single- and few-layer MoS$_2$. Acs Nano, 4(5): 2695–2700.

[37] Ayse Berkdemir, Humberto R. Gutiérrez, Andrés R. Botello-Méndez, Néstor Perea-López, Ana Laura Elías, Chen-Ing Chia, Bei Wang, Vincent H. Crespi, Florentino López-Urías, Jean-Christophe Charlier, Humberto Terrones and Mauricio Terrones. 2013. Identification of individual and few layers of WS$_2$ using Raman spectroscopy. Sci. Rep., 3.

[38] Elena del Corro, Humberto Terrones, Ana Elias, Cristiano Fantini, Simin Feng, Minh An Nguyen, Thomas E. Mallouk, Mauricio Terrones and Marcos A. Pimenta. 2014. Excited excitonic states in 1L, 2L, 3L, and bulk WSe$_2$ observed by resonant Raman spectroscopy. Acs Nano, 8(9): 9629–9635.

[39] Terrones, H., E. Del Corro, S. Feng, J.M. Poumirol, D. Rhodes, D. Smirnov, N.R. Pradhan, Z. Lin, M.A.T. Nguyen, A.L. Elías, T.E. Mallouk, L. Balicas, M.A. Pimenta and M. Terrones. 2014. New first order raman-active modes in few layered transition metal dichalcogenides. Sci. Rep., 4.

[40] Ly, Thuc Hue, Seok Joon Yun, Quoc Huy Thi and Jiong Zhao. 2017. Edge delamination of monolayer transition metal dichalcogenides. ACS Nano, 11(7): 7534–7541.

[41] Deyi Fu, Jian Zhou, Sefaattin Tongay, Kai Liu, Wen Fan, Tsu Jae King Liu and Junqiao Wu. 2013. Mechanically modulated tunneling resistance in monolayer MoS_2. Appl. Phys. Lett., 103(18).

[42] Jothi Priyanka Thiruraman, Kazunori Fujisawa, Gopinath Danda, Paul Masih Das, Tianyi Zhang, Adam Bolotsky, Néstor Perea-López, Adrien Nicolaï, Patrick Senet, Mauricio Terrones and Marija Drndić. 2019. Angstrom-size defect creation and ionic transport through pores in single-layer MoS_2. Nano Lett., 18(3): 1651–1659.

[43] Arend M. Van Der Zande, Pinshane Y. Huang, Daniel A. Chenet, Timothy C. Berkelbach, Yumeng You, Gwan Hyoung Lee, Tony F. Heinz, David R. Reichman, David A. Muller and James C. Hone. 2013. Grains and grain boundaries in highly crystalline monolayer molybdenum disulphide. Nat. Mater., 12(6): 554–561.

[44] Jinhuan Wang, Xiaozhi Xu, Ruixi Qiao, Jing Liang, Can Liu, Bohao Zheng, Lei Liu, Peng Gao, Qingze Jiao, Dapeng Yu, Yun Zhao and Kaihui Liu. 2014. Dislocation motion and grain boundary migration in two-dimensional tungsten disulphide. Nat. Commun., 5.

[45] Tianyi Zhang, Kazunori Fujisawa, Tomotaroh Granzier-Nakajima, Fu Zhang, Zhong Lin, Ethan Kahn, Néstor Perea-López, Ana Laura Elías, Yin-Ting Yeh and Mauricio Terrones. 2019. Clean transfer of 2D transition metal dichalcogenides using cellulose acetate for atomic resolution characterizations. Acs Appl. Nano Mater., 2(8): 5320–5328.

[46] Andres Castellanos-Gomez, Michele Buscema, Rianda Molenaar, Vibhor Singh, Laurens Janssen, Herre S.J. van der Zant and Gary A. Steele. 2014. Deterministic transfer of two-dimensional materials by all-dry viscoelastic stamping. 2D Mater., 1(1).

[47] Terrones, Humberto, Florentino López-Urías and Mauricio Terrones. 2013. Novel hetero-layered materials with tunable direct band gaps by sandwiching different metal disulfides and diselenides. Sci. Rep., 3: 1–7.

[48] Xiaoping Hong, Jonghwan Kim, Su Fei Shi, Yu Zhang, Chenhao Jin, Yinghui Sun, Sefaattin Tongay, Junqiao Wu, Yanfeng Zhang and Feng Wang. 2014. Ultrafast charge transfer in atomically thin MoS_2/WS_2 heterostructures. Nat. Nanotechnol., 9(9): 682–686.

[49] Yongji Gong, Junhao Lin, Xingli Wang, Gang Shi, Sidong Lei, Zhong Lin, Xiaolong Zou, Gonglan Ye, Robert Vajtai, Boris I. Yakobson, Humberto Terrones, Mauricio Terrones, Beng Kang Tay, Jun Lou, Sokrates T. Pantelides, Zheng Liu, Wu Zhou and Pulickel M. Ajayan. 2014. Vertical and in-plane heterostructures from WS_2/MoS_2 monolayers. Nat. Mater., 13(12): 1135–1142.

[50] Yongji Gong, Sidong Lei, Gonglan Ye, Bo Li, Yongmin He, Kunttal Keyshar, Xiang Zhang, Qizhong Wang, Jun Lou, Zheng Liu, Robert Vajtai, Wu Zhou and Pulickel M Ajayan. 2015. Two-step growth of two-dimensional $WSe_2/MoSe_2$ heterostructures. Nano Lett., 15(9): 6135–6141.

[51] Yu, Y., Z. Wang, J. Wei, W. Zhao, X. Lin, Z. Jin, W. Liu and G. Ma. 2018. Ultrafast formation and dynamics of interlayer exciton in a large-area CVD-grown WS_2/WSe_2 heterostructure. J. Condens. Matter Phys., 30(49).

[52] Sahoo, Prasana K., Shahriar Memaran, Yan Xin, Luis Balicas and Humberto R. Gutiérrez. 2018. One-pot growth of two-dimensional lateral heterostructures via sequential edge-epitaxy. Nature, 553(7686): 63–67.

[53] Kun Chen, Xi Wan, Jinxiu Wen, Weiguang Xie, Zhiwen Kang, Xiaoliang Zeng, Huanjun Chen, and Jian-Bin Xu. 2015. Electronic properties of MoS_2-WS_2 heterostructures synthesized with two-step lateral epitaxial strategy. Acs Nano, 9(10): 9868–9876.

[54] Kittel, Charles. 1953. Introduction to Solid State Physics. New York, Wiley.

[55] Streetman, Ben G. 1972. Solid State Electronic Devices, Prentice-Hall Series in Solid State Physical Electronics. Englewood Cliffs, N.J., Prentice-Hall.

[56] John Mann, Quan Ma, Patrick M. Odenthal, Miguel Isarraraz, Duy Le, Edwin Preciado, David Barroso, Koichi Yamaguchi, Gretel von Son Palacio, Andrew Nguyen, Tai Tran, Michelle Wurch, Ariana Nguyen, Velveth Klee, Sarah Bobek, Dezheng Sun, Tony F. Heinz, Talat S. Rahman, Roland Kawakami and Ludwig Bartels. 2014. 2-Dimensional transition metal dichalcogenides with tunable direct band gaps: $MoS_{2(1-x)}Se_{2x}$ monolayers. Adv. Mater., 26(9): 1399–1404.

[57] Yongji Gong, Zheng Liu, Andrew R. Lupini, Gang Shi, Junhao Lin, Sina Najmaei, Zhong Lin, Ana Laura Elías, Ayse Berkdemir, Ge You, Humberto Terrones, Mauricio Terrones, Robert Vajtai,

Sokrates T. Pantelides, Stephen J. Pennycook, Jun Lou, Wu Zhou and Pulickel M. Ajayan. 2014. Band gap engineering and layer-by-layer mapping of selenium-doped molybdenum disulfide. Nano Lett. 14(2): 442–449.

[58] Liu, H.F., K.K.A. Antwi, S. Chua and D.Z. Chi. 2014. Vapor-phase growth and characterization of $Mo_{1-x}W_xS_2$ ($0 <= x <= 1$) atomic layers on 2-inch sapphire substrates. Nanoscale, 6(1): 624–629.

[59] Zhong Lin, Michael T. Thee, Ana Laura Elías, Simin Feng, Chanjing Zhou, Kazunori Fujisawa, Néstor Perea-López, Victor Carozo, Humberto Terrones and Mauricio Terrones. 2014. Facile synthesis of MoS_2 and $Mo_xW_{1-x}S_2$ triangular monolayers. Apl. Materials, 2(9).

[60] Kehao Zhang, Simin Feng, Junjie Wang, Angelica Azcatl, Ning Lu, Rafik Addou, Nan Wang, Chanjing Zhou, Jordan Lerach, Vincent Bojan, Moon J. Kim, Long-Qing Chen, Robert M. Wallace, Mauricio Terrones, Jun Zhu and Joshua A Robinson. 2015. Manganese doping of monolayer MoS_2: The substrate is critical. Nano Lett., 15(10): 6586–6591.

[61] Kehao Zhang, Brian M. Bersch, Jaydeep Joshi, Rafik Addou, Christopher R. Cormier, Chenxi Zhang, Ke Xu, Natalie C. Briggs, Ke Wang, Shruti Subramanian, Kyeongjae Cho, Susan Fullerton-Shirey, Robert M. Wallace, Patrick M. Vora and Joshua A. Robinson. 2018. Tuning the electronic and photonic properties of monolayer MoS_2 via *in situ* rhenium substitutional doping. Adv. Funct. Mater., 28(16).

[62] Zhang, F., Y. Lu, D.S. Schulman, T. Zhang, K. Fujisawa, Z. Lin, Y. Lei, A.L. Elias, S. Das, S.B. Sinnott and M. Terrones. 2019. Carbon doping of WS_2 monolayers: Bandgap reduction and p-type doping transport. Sci. Adv., 5(5).

[63] Liu, K.K., Y.H. Lee, Y.C. Lin, J.K. Huang, X.-Q. Zhang, T.-W. Lin and L.J. Li. 2012. Chemical vapor deposited MoS_2 thin layers and their applications. ECS Trans., 50(6): 61–63.

[64] Loan, P.T.K., W.J. Zhang, C.T. Lin, K.H. Wei, L.J. Li and C.H. Chen. 2014. Graphene/MoS_2 heterostructures for ultrasensitive detection of DNA hybridisation. Adv. Mater., 26(28): 4838–+.

[65] Sina Najmaei, Zheng Liu, Wu Zhou, Xiaolong Zou, Gang Shi, Sidong Lei, Boris I. Yakobson, Juan-Carlos Idrobo, Pulickel M. Ajayan and Jun Lou. 2013. Vapour phase growth and grain boundary structure of molybdenum disulphide atomic layers. Nat. Mater., 12(8): 754–759.

[66] Yu, Y.F., C. Li, Y. Liu, L.Q. Su, Y. Zhang and L.Y. Cao. 2013. Controlled scalable synthesis of uniform, high-quality monolayer and few-layer MoS_2 films. Sci, Rep., 3.

[67] Qingqing Ji, Yanfeng Zhang, Teng Gao, Yu Zhang, Donglin Ma, Mengxi Liu, Yubin Chen, Xiaofen Qiao, Ping-Heng Tan, Min Kan, Ji Feng, Qiang Sun and Zhongfan Liu. 2013. Epitaxial monolayer MoS_2 on mica with novel photoluminescence. Nano Lett., 13(8): 3870–3877.

[68] Ana Laura Elías, Néstor Perea-López, Andrés Castro-Beltrán, Ayse Berkdemir, Ruitao Lv, Simin Feng, Aaron D. Long, Takuya Hayashi, Yoong Ahm Kim, Morinobu Endo, Humberto R. Gutiérrez, Nihar R. Pradhan, Luis Balicas, Thomas E. Mallouk, Florentino López-Urías, Humberto Terrones and Mauricio Terrones. 2013. Controlled synthesis and transfer of large-area WS2 sheets: From single layer to few layers. Acs Nano, 7(6): 5235–5242.

[69] Jeong-Gyu Song, Jusang Park, Wonseon Lee, Taejin Choi, Hanearl Jung, Chang Wan Lee, Sung-Hwan Hwang, Jae Min Myoung, Jae-Hoon Jung, Soo-Hyun Kim, Clement Lansalot-Matras and Hyungjun Kim. 2013. Layer-controlled, wafer-scale, and conformal synthesis of tungsten disulfide nanosheets using atomic layer deposition. ACS Nano, 7(12): 11333–11340.

[70] Yu Zhang, Yanfeng Zhang, Qingqing Ji, Jing Ju, Hongtao Yuan, Jianping Shi, Teng Gao, Donglin Ma, Mengxi Liu, Yubin Chen, Xiuju Song, Harold Y. Hwang, Yi Cui and Zhongfan Liu. 2013. Controlled growth of high-quality monolayer WS_2 layers on sapphire and imaging its grain boundary. Acs Nano, 7(10): 8963–8971.

[71] Nicolas D. Boscher, Claire J. Carmalt, Robert G. Palgrave, Jesus J. Gil-Tomas and Ivan P. Parkin. 2006. Atmospheric pressure CVD of molybdenum diselenide films on glass. Chem. Vap. Deposition 12: 692–698.

[72] Desheng Kong, Haotian Wang, Judy J. Cha, Mauro Pasta, Kristie J. Koski, Jie Yao and Yi Cui. 2013. Synthesis of MoS_2 and $MoSe_2$ films with vertically aligned layers. Nano Lett., 13(3): 1341–1347.

[73] Jianyi Chen, Xiaoxu Zhao, Sherman J.R. Tan, Hai Xu, Bo Wu, Bo Liu, Deyi Fu, Wei Fu, Dechao Geng, Yanpeng Liu, Wei Liu, Wei Tang, Linjun Li, Wu Zhou, Tze Chien Sum and Kian Ping Loh. 2017. Chemical vapor deposition of large-size monolayer $MoSe_2$ crystals on molten glass. J. Am. Chem. Soc., 139(3): 1073–1076.

[74] Qi Fu, Lei Yang, Wenhui Wang, Ali Han, Jian Huang, Pingwu Du, Zhiyong Fan, Jingyu Zhang and Bin Xiang. 2015. Synthesis and enhanced electrochemical catalytic performance of monolayer $WS_{2(1-x)}Se_{2x}$ with a tunable band gap. Adv. Mater., 27(32): 4732–4738.

[75] Qingliang Feng, Yiming Zhu, Jinhua Hong, Mei Zhang, Wenjie Duan, Nannan Mao, Juanxia Wu, Hua Xu, Fengliang Dong, Fang Lin, Chuanhong Jin, Chunming Wang, Jin Zhang and Liming Xie. 2014. Growth of large-area 2D $MoS_{2(1-x)}Se_2$, semiconductor. Adv. Mater., 26(17): 2648–2653.

[76] Shoujun Zheng, Linfeng Sun, Tingting Yin, Alexander M. Dubrovkin, Fucai Liu, Zheng Liu, Ze Xiang Shen and Hong Jin Fan. 2015. Monolayers of $W_xMo_{1-x}S_2$ alloy heterostructure with in-plane composition variations. Appl. Phys. Lett., 106(6).

[77] Sefaattin Tongay, Wen Fan, Jun Kang, Joonsuk Park, Unsal Koldemir, Joonki Suh, Deepa S. Narang, Kai Liu, Jie Ji, Jingbo Li, Robert Sinclair and Junqiao Wu. 2014. Tuning interlayer coupling in large-area heterostructures with CVD-grown MoS_2 and WS_2 monolayers. Nano Lett., 14(6): 3185–3190.

[78] Mei Zhang, Juanxia Wu, Yiming Zhu, Dumitru O. Dumcenco, Jinhua Hong, Nannan Mao, Shibin Deng, Yanfeng Chen, Yanlian Yang, Chuanhong Jin, Sunil H. Chaki, Ying-Sheng Huang, Jin Zhang and Liming Xie. 2014. Two-dimensional molybdenum tungsten diselenide alloys: photoluminescence, Raman scattering, and electrical transport. Acs Nano, 8(7): 7130–7137.

[79] Sheng-Han Su, Yu-Te Hsu, Yung-Huang Chang, Ming-Hui Chiu, Chang-Lung Hsu, Wei-Ting Hsu, Wen-Hao Chang, Jr-Hau He and Lain-Jong Li. 2014. Band gap-tunable molybdenum sulfide selenide monolayer alloy. Small, 10(13): 2589–2594.

CHAPTER 10

Metal Oxide/CNT/Graphene Nanostructures for Chemiresistive Gas Sensors

Sanju Rani,[1,2] *Manoj Kumar,*[1,2] *Yogesh Singh,*[1,2] *Rahul Kumar*[1,2]
and V.N. Singh[1,2,*]

1. Introduction

Gas sensors are one of the important aspects of modern day lifestyle. People are becoming more and more health and environment conscious. The presence of dangerous gases, such as NO_2, CO, ammonia, H_2S, and SO_2 in the atmosphere has become a matter of concern for people in general. The presence of even a trace amount of these elements can be harmful to the health of humans and other living entities, and therefore, recently, a lot of efforts are being put into making efficient gas sensors (such as MOS, catalytic metal, conducting polymers, optical based gas sensors) [1], which can detect ppm and ppb levels of toxic gases. It is interesting to note that toxic gases, such as ammonia and H_2S, etc., not only harm humans and other living entities, but also electrical appliances, such as the condenser unit of chillier, and so on, resulting in a huge loss of resources. There are several reviews on gas sensing based on metal oxides. CNT and graphene, being semiconducting in nature, transport the produced electron to the nearest electrode very easily, which helps in the collection of electrons. Thus, sensors based on metal oxide combined with CNT and graphene show enhanced sensitivity and better response time. There are very few reviews on metal oxide/CNT/graphene-based gas sensors [2–7]. In this review, we are trying to summarize the most important results of gas sensors based on metal oxides/CNT/graphene (their synthesis and gas sensing performance with different gases). Apart from this, gas sensing mechanism involving CNT/graphene

[1] CSIR-National Physical Laboratory, Dr. K.S. Krishnan Marg, New Delhi-110012, India.
[2] AcSIR-Academy of Scientific and Innovative Research (AcSIR), India.
* Corresponding author: singhvn@nplindia.org

with metal oxide has also been explained in this review. This chapter will serve the researchers in the field of gas sensors with all aspects of such gas sensors.

2. Synthesis of Metal Oxides/CNT Based Gas Sensor

A brief summary of synthesis methods of metal oxide and CNTs are given below.

2.1 Metal Oxides

Metal oxides are one of the most significant and generally characterized solid catalysts. In recent years, research on nanoparticles has provided a wide scope for the development of novel solutions in the field of electronics, cosmetics, optics, and healthcare, as it has unique properties, such as optical, electrical, mechanical, magnetic, and catalytic. At present, MOS nanoparticles are used in gas sensors because of their small size (1 nm–100 nm) and having high surface area per unit mass [1], which enhances the gas sensing properties. Metal oxides typically contain an anion of oxygen in the oxidation state of –2. MO can easily lose electrons, and are good conductors of heat and electricity, because they have low ionization potential. Metal oxides are of two types: n-type and p-type. Table 10.1 shows how different target gases affect the resistance of metal oxide. Researchers have shown great interest in metal oxide-based gas sensors for the detection of toxic gases due to their excellent sensitivity, high chemical stability, and low cost [8].

MOs also have large band gaps (> 1); for example, ZnO has a band gap of 3.37 eV and detects various gases (such as LPG, ethanol, NH_3, toluene, acetone, formaldehyde, O_2, CO, NO_2) [9–17]. Table 10.2 summarizes metal oxides, such as SnO_2 [18–27], TiO_2 [28–32], WO_3 [33–40], Fe_2O_3 [41–47], In_2O_3 [48–53], La_2O_3 [54], $CuCrO_2$ [55], NiO [56, 57], Co_3O_4 [58–60], their band gap, and gases it can detect.

Table 10.1. Behavior of various MOs in the presence of reducing and oxidizing gases.

Type of sensitive materials	Metal oxide	Type of target gas		Response	
		Reducing gas	Oxidizing gas		
n-type	MgO, ZnO, CaO, WO_3, TiO_2, In_2O_3, SnO_2, Fe_2O_3	Resistance decrease	Resistance increase	R_a/R_g	R_g/R_a
p-type	La_2O_3, CeO_2, $CuCrO_2$, Mn_2O_3, NiO, Y_2O_3, Co_3O_4	Resistance increase	Resistance decrease	R_g/R_a	R_a/R_g

2.2 Synthesis of Metal Oxides

Some of the techniques used for making metal oxide nanoparticles/thin films are summarized below. Details are provided in the references: (i) Sputtering [61], (ii) Wet process [62], (iii) Hydrothermal techniques [63], (iv) Flame Transport Synthesis [64], (v) Reflux method [12], (vi) Thermal evaporation [65], (vii) Cross linking deposition [66], (viii) Sonication method [67], etc.

Table 10.2. Metal oxides, their band gap, and gases that they can detect.

Oxide type	Band gap (eV)	Detectable gas	Reference
ZnO	3.37	LPG, Ethanol, NH_3, Toluene, Acetone, Formaldehyde, O2, CO, NO_2	[9–17]
SnO_2	3.6	H_2, CO, Ethanol, Formaldehyde, H_2S, CH_4, O_3, SO_2, NH_3, Acetone	[18–27]
TiO_2	3.2 ~ 3.35	Ethanol, O_2, H_2, NO_2, NH_3	[28–32]
WO_3	2.7	NO, Acetone, SO_2, O_3, NH_3, CO, H_2S, NO_2	[33–40]
Fe_2O_3	2.2	CO, Ethanol, Acetone, LPG, NH_3, Formaldehyde, NO_2	[41–47]
In_2O_3	3	NO_2, O_3, Ethanol, Acetone, NH_3, H_2S	[48–53]
La_2O_3	5.8	CO_2	[54]
$CuCrO_2$	3.2	O_3	[55]
NiO	3.6 ~ 4	Ethanol, Formaldehyde	[56–57]
Co_3O_4	2	Acetone, Formaldehyde, CO	[58–60]

2.2.1 Process of Growth of Nanomaterial

In the process of crystal formation, there are mainly three steps:

i) Nucleation: For a specific solvent, a substance has a certain solubility, and when it's supersaturated, the solute is precipitated, which in turn forms crystal nuclei.

ii) Growth: After the nuclei is formed, they adhere to each other and the rate of super-saturation becomes less so that nuclei is formed again and an equilibrium is obtained. If the nucleation time is short, it will lead to a more uniform grain size.

iii) Ripening: In this process, which is typically called Ostwald ripening, larger particles are grown over a period of time, and on the other hand, small particles become more small and finally dissolve.

2.3 Carbon Nanotube (CNT)

Recently, carbon nanostructure has made a big impact in material science. Ijima et al. found long tubular structures, called carbon nanotube, which have unique electronic properties that depend on the growth of carbon lattice into the tube. If the carbon lattice grows in a cylindrical form, then the zig-zag, armchair, and chiral shape of the carbon atom along the top of the cylinder is observed. Carbon nanotubes are of two types (i) single-walled carbon nanotubes (SWCNT), (ii) multi-walled carbon nanotubes (MWCNTs). The properties (such as conductivity, larger surface area, porosity, strong interaction among adsorbed molecules), make MWCNTs ideal candidates for designing gas sensors in different configurations. Elastic modulus (strength) of an individual perfect nanotube is around ~ 1 TPa or 150 GPa [68].

2.4 Synthesis of CNT

There are three main techniques to make CNTs [69]:

i) CVD (chemical vapor deposition): In this technique, a hydrocarbon source gas flows into an enclosed chamber in which the molecules of hydrocarbon are broken into reactive species and react in the presence of catalysts (usually Ni, Fe, or Co) at 500–1000°C temperature range. CNTs get coated on the substrate. In comparison to the other two techniques, CNT is formed at a relatively lower temperature in CVD.

This technique is more efficient in producing high quantities of MWCNT as well as SWCNT. High defect density of MWCNT is the biggest disadvantage seen in this deposition method.

ii) Arc discharge technique: This method is done in a vacuum chamber which consists of two carbon electrodes acting as a carbon source. An inert gas, in general helium, is applied to hasten the rate of carbon deposition. When DC voltage is applied between the carbon anode and cathode, plasma of the inert gas is generated to evaporate the carbon atoms. The ejected carbon atoms are deposited on the negative electrode to form CNTs. Both SWCNTs and MWCNTs can be grown by this method. The growth of SWCNTs requires a catalyst. It is the principle method to produce high quality CNTs, with nearly perfect structures.

iii) Laser ablation technique: In this method, a target made of carbon is etched by a laser pulse beam of high intensity in a closed furnace with flow of inert gas and a catalyst. After this, the CNT are formed on a substrate.

Both the ablation method and arch discharge require high temperature, which is one of the disadvantages, as a high amount of energy is used.

2.5 Synthesis of Graphene

Graphene is a 2D material consisting of sp^2 hybridized carbon atoms in a hexagonal structure. It is a highly conductive material. There are many techniques for synthesizing graphene. Here in Table 10.3, a few general methods are given and compared on the scale of cost, complexity, scalability, etc. [70].

Table 10.4 enlists recent advancements in the synthesis of metal oxide nanostructures/CNT/GO composites, the gases that are sensed, etc. An et al. had grown WO_3 nanotube by using thermally evaporated TeO_2 nanowire as a sacrificial core, which was removed by thermally evaporating TeO_2 in the presence of air. The thicknesses of the nanowalls were 20–30 nm, and diameters were 100–120 nm. These were used to detect NO_2 gas upto 1–5 ppm at 300°C [61]. Hashishin and Tamaki synthesized CNT-WO_3 composite. They used the wet process technique, which yielded composite with grain size of 50 ~ 200 nm, and were used for detecting NO_2 gas up to 5 ppm at 200°C [62]. Zhang et al. synthesized α-Fe_2O_3 nanosheet by solvothermal method, with an average thickness of 50 nm and diameter of 200 nm. It

Table 10.3. The comparison among the different techniques of producing graphene [70].

	Mechanical exfoliation (Scotch-tape method)	Chemical exfoliation/GO chemical reduction	Chemical vapor deposition
Quality of synthesized graphene	Very high	Poor	High
Maximum electron mobility (cm²/Vs)	200,000	10	350,000
Feasibility of SLG production	Yes, with difficulties	No	Yes
Scalability	Not scalable	Scalable	Scalable
Maximum sample size	Approximately 100 µm	Arbitrary	7.5 m²
Production price (per 1 cm²)	–	< 0.1 USD	< 1 USD
Production method complexity	Low	Medium	Very high

was used to detect H_2S gas up to 5 ppm at 135°C [71]. In 2014, Muthukumaran et al. produced α-Fe_2O_3/CNT nanocomposite by hydrothermal method. The thickness of the composite was 1–10 µm, and was used for detecting NH_3 gas up to 2000 ppm at 100°C [63]. In 2017, Schütt et al. synthesized a network of ZnO-CNT hybid tetrapods by using flame transport technique. The length of the tetrapod arms varied in the range of 5–60 µm, and the diameter of the tetrapod arms were in the range of 0.8–8 µm. This was used to detect NH_3 up to 100 ppm at room temperature. SEM image of this paper is also shown in Figure 10.3a, which shows the tetrapodal structure of ZnO-CNT [64]. Septiani et al. synthesized MWCNT-ZnO nanocomposite by reflux method. The particle size of ZnO lies from 30 nm to 46 nm. These were used to detect toluene up to 200 ppm at 150°C [12]. Seo et al. prepared TiO_2 nanotubes with diameter and length of 70 nm and 600 nm, respectively, by hydrothermal treatment for the detection of toluene up to 50 ppm at 500°C [72]. Zheng et al. synthesized In_2O_3 nanotower with an octahedral cap size of 600 nm by using of thermal evaporation technique and this sensor used for detecting H_2 gas up to 1000 ppm at 240°C [65]. Co_3O_4/carbon hybrid hollow nanospheres with diameter of 40 ~ 60 nm were synthesized by Liu et al. in 2019 by using cross-linking deposition. It was used to detect H_2S gas up to 50 ppm at 92°C [66]. Choi et al. synthesized SnO_2 nanowires with a diameter of ~ 50 nm by using thermal evaporation technique. It was used to detect NO_2 gas up to 5 ppm at 200°C [73]. Jia et al. prepared SnO_2/CNT by using sonication method, where he achieved the crystallite size of 4 nm, and it was used for detecting ethanol up to 100 ppm at 200°C. TEM and XRD images of SnO_2/CNT nanocomposites and porous SnO_2 nanotubes are shown in Figures 10.1b and 10.1c [67].

Table 10.4. Synthesis of some metal oxides and CNT/metal oxides.

Metal oxide nano structure	Technique & size	Details	Gas detected with concentration	Ref.
WO_3 Nanotube	sputtering 20 ~ 30 nm (wall thickness) 100 ~ 120 nm (diameter)	Te was evaporated on Si substrate, and then they are put in horizontal tube furnace to obtain TeO_2 nanowires. These were used a sacrificial core and WO_3 was deposited over it using sputtering with constant rotation. TeO_2 core was removed by thermally annealing these structures in air and WO_3 nanotube was obtained.	NO_2 gas 5 ppm	[61]
CNT-WO_3 Composite	By Wet process 50 ~ 200 nm (grain size)	$(NH4)_{10}W_{12}O_{14}.5H_2O$ neutralized by dilute Nitric acid sol^n. Obtained precipitate (CH_2WO_4) washed with deionized water (4 ~ 5 times) and dried. Micro drop of H_2SO_4 (0.1 to 7 wt%) dropped on surface of CNT_s between Au electrodes with micro gap (5μ), by using micro injection. Dried and calcined at 400°C for 3 h under inert gas of Ar. Obtained CNT-WO_3 composite.	NO_2 gas 5 ppm	[62]
α-Fe_2O_3 Nanosheet	By Solvothermal method 50 nm (thickness) 200 nm (diameter)	Mixed (2.5 mM) $Fe_2(C_5H_7O_2)$. H_2O and (35 ml) ethanol to form a clear sol^n. Stirred vigorously for 1 h and transferred into 50 ml Teflon lined stainless steel autoclave. Heated at 150°C for 10 h and cooled down to RT. Red precipitates obtained. Centrifuged 3 times with deionized water and absolute ethanol water. Dried in a vacuum at 60°C for 7 h. Obtained α-Fe_2O_3 nanosheet.	H_2S gas 5 ppm	[71]
α-Fe_2O_3/CNT Nanocomposite	By Hydrothermal method 1 ~ 10 μm (thickness)	Mixed (95%) H_2SO_4, (69%) HNO_3 and CNT with ratio 1:1. Magnetically stirred at 60°C for 24 h. Washed several times with deionized water and centrifuged 3 times. Dried at 80°C for 24 h. Mixture mixed with red hued colored α-Fe_2O_3 nanostructures by 2:1. Sonication for a few min at ambient temp. Obtained α-Fe_2O_3/CNT nanocomposite.	NH_3 gas 2000 ppm	[63]

Table 10.4 Contd. ...

...Table 10.4 Contd.

Metal oxide nano structure	Technique & size	Details	Gas detected with concentration	Ref.
ZnO Tetrapodal Network	By Flame Transport synthesis $3 \sim 5$ μm (diameter) $15 \sim 20$ μm (length)	First step: Mixed Zinc powder (grain size $\sim$ 1–5 μm) and Poly vinyl butyral (1:2 by mass). Heated in a muffle furnace at 900°C for 30 min (heating rate $\sim$ 60°C/min). Powder pressed into cylindrical pellets (h = 3 mm, d = 6 mm) with density 0.3 g/ cm^3. Reheated at 1150°C for 4 h. Second step: 0.1 wt% CNT_S dispersed in distilled water. Use ultra-sonication bath for 20 min. Sol^n filled in computer controlled syringe and dropped slowly (130 μL) on the templates. Dried template at least 1 h.	NH_3 gas 100 ppm	[64]
MWCNT-ZnO Nanocomposite	By Reflux method $30 \sim 46$ nm	Mixed Zn $(NO_3).4H_2O$, $MWCNT_S$ and Ethylene glycol. Refluxed at 197°C for 12 h. Yellow precipitate obtained. Cooled at RT and washed several times. ZnO in mixture formed in Zinc glycolate. Calcined at 400°C for 3 h. MWCNT-ZnO nanocomposite was obtained.	Toluene gas 200 ppm	[12]
TiO_2 Nanotube	By Hydrothermal method 600 nm (length) 70 nm (diameter)	Mixed TiO_2 powder and NaOH sol^n was hydrothermally treated at 230°C. It was washed with HCl sol^n, filtered and dried. Obtained TiO_2 nanotube was calcined at $600 \sim 700$°C and ball milled.	Toluene gas 50 ppm	[72]
In_2O_3 Nano Tower	By Thermal Evaporation 600 nm (octahedral cap size)	Active carbon and In_2O_3 powder in 1:1 was heated rapidly in a horizontal tube furnace up to 1050°C (heating rate $\sim$ 40°C/ min) to obtain In_2O_3 nanotube.	H_2 gas 1000 ppm	[65]
CO_3O_4/C Hollow Nano sphere	By Cross-linking Deposition $40 \sim 60$ nm (diameter)	A gel using cobalt acetate was made. It was washed and dried and carbonized at 600°C under N_2 atmosphere to obtain CO_3O_4/C hollow nanosphere.	H_2S gas 50 ppm	[66]
SnO_2 Nanowire	By Thermal Evaporation $\sim$ 50 nm (diameter)	Sn metal powder was heated in a horizontal type furnace at 750°C. The flow of Ar was controlled to obtain SnO_2 nanowire.	NO_2 gas 5 ppm	[73]

Table 10.4 Contd. ...

...Table 10.4 Contd.

Metal oxide nano structure	Technique & size	Details	Gas detected with concentration	Ref.
SnO₂/CNT Composite	By Sonication method ~ 4 nm (crystallite size)	Dissolved SnCl₂.2H₂O in distilled water was mixed with HCl and as treated CNT. These were sonicated and stirred. After washing they were calcined at 350°C to achieve SnO₂/CNT nanocomposite.	Ethanol 100 ppm	[67]

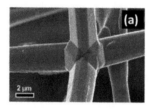
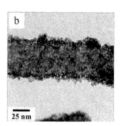

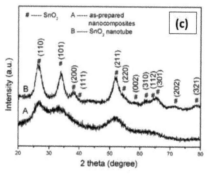

Figure 10.1. (a) SEM images of ZnO-tetrapod−CNT networks at 2.0 wt% [64], (b) TEM images of the SnO₂/CNT nanocomposites [67], (c) XRD patterns of the SnO₂/CNT nanocomposites and the porous SnO₂ nanotubes [67] (permission granted).

3. Metals Oxide/CNT/Graphene Based Gas Sensors

The interesting part of this chapter is the comparison between metal oxides and CNT/metal oxide or graphene/metal oxide for different gas sensing parameters (such as concentration of gas, operating temperature, sensor response, response time, recovery time, etc.).

3.1 SnO₂-Based Gas Sensor

SnO₂ is widely used in gas sensors. It has a wide band gap of ~ 3.6 eV, and has a tetragonal structure. Some SnO₂, SnO₂/noble metals, SnO₂/CNT, SnO₂/graphene-based gas sensors are presented in Table 10.5. Liu et al., using a combustion chemical vapor deposition process, fabricated nanostructured SnO₂ thin-film gas sensors.

Table 10.5. Comparison between SnO$_2$-based gas sensor.

Material	Structure	Sensing gas	Gas conc. (ppm)	Operating temperature (°C)	Response (%)	Response time (sec)	Recovery time (sec)	Ref.
SnO$_2$/CNT	Composite	Ethanol	100	200	130	?	?	[74]
GO@SnO$_2$	Nanofiber/Nanosheet	Formaldehyde	100	60	280	?	?	[75]
SnO$_2$/CuO	Nanowire	NH$_3$	1%	RT	300%^{c}	~125	~500	[76]
Au@SnO$_2$	Core-shell	Formaldehyde	50	RT	2.9	80	69	[77]
La$_2$O$_3$-SnO$_2$	Nanowire	Ethanol	100	400	57.3	1	110	[78]
SnO$_2$-Pt/MWCNTs	Nanocomposites	Methane	100	RT	28.25	176	763	[79]
SnO$_2$-NiO	Nanofiber	Formaldehyde	10	200	6^c	50	80	[80]
SnO$_2$/Ag	Nanowire	Ethanol	100	450	228.1	5	100	[81]
Pt–SnO$_2$	Porous nano solid	CO	100	RT	64.5	144	?	[19]
SnO$_2$/Rgo	Composite	H$_2$S	50	RT	33	2	292	[82]
Pd/SnO$_2$	Thin film	Formaldehyde	10	250	1.55	50	50	[83]
SnO$_2$/MWCNTs	Film	NH$_3$	60–800	RT	2–32	<300	<300	[84]

They used Pt interdigitated electrodes. The thickness of the film was < 1 µm, and it consisted of nanocrystalline particles with size < 30 nm. At 300°C, it showed a sensor response of 1075 to 500 ppm ethanol vapor, with response time and recovery time of 31 and 8 s, respectively. The detection limit was below 1 ppm. The sensor responses of SnO_2 thin-film-based gas sensor for different concentration of ethanol are shown in Figure 10.2a with permission [74]. In 2020, Wan et al. fabricated graphene oxide (GO)@SnO_2 nanofiber/nanosheets (NF/NSs) nanocomposites, which sensed 100 ppm formaldehyde gas with response of 280% at 60°C temperature [75]. In 2012, Mashock et al. reported significant enhancement in the gas sensing properties of CuO nanowire by using SnO_2 nanocrystals (NCs). The sensing performance was observed at room temperature (RT), having sensitivity of ~ 300% for 1% NH_3 in air. The response and recovery times were ~ 125 s and ~ 500 s, respectively [76]. Chunga et al. in 2014, fabricated Au@SnO_2 core–shell structure using sol-gel method, and enhanced the sensor response to 2.9 for 50 ppm of formaldehyde gas. The response and recovery times were 80 s and 62 s, respectively [77]. Hieu et al. in 2008, fabricated SnO_2 nanowires (NWs) and coated it with a functional La_2O_3 layer by solution deposition, and found enhanced sensor response of ~ 57.3 at 400°C for ethanol with response and recovery time of ~ 1 and ~ 110 s, respectively [78]. Navazani et al. in 2020, fabricated SnO_2-Pt/MWCNTs-based gas sensor with response of 28.25% for 100 ppm methane at room temperature. The response and recovery times were 176 s and 763 s, respectively [79]. Zheng et al. in 2011, fabricated sensors from NiO/SnO_2 nanofibers which exhibited good sensing properties to 10 ppm formaldehyde gas at an operating temperature of 200°C, and the minimum detection limit was 0.08 ppm with the response of ~ 6, response time and recovery time of the sensors were about 50 s and 80 s, respectively [80]. Hwang et al. in 2011 decorated Ag layers of thicknesses ~ 550 nm on the surface of SnO_2 NWs via e-beam evaporation and enhanced the response to ~ 228.1 for 100 ppm ethanol (operating at 450°C) with response and recovery times ~ 5 sec and ~ 100 s, respectively [81]. Wang et al. in 2013 prepared Pt-loaded SnO_2 porous nanosolid (SnO_2:Pt PNS), which had the response of 64.5 to 100 ppm CO at room temperature with a response time of

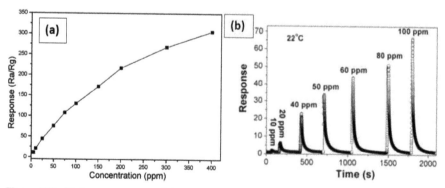

Figure 10.2. (a) Sensor responses of the sensor device upon exposure to different concentrations of ethanol at a working temperature of 200°C [74] (permission granted), (b) Response curves towards different concentrations of H_2S [82] (permission granted).

~ 144 s [19]. Song et al. in 2016, demonstrated gas sensors based on SnO_2 quantum wires, which were anchored on reduced graphene oxide (rGO) nanosheets, and had response of 33 for 50 ppm for H_2S gas at room temperature with response and recovery times of 2 s and 292 s, respectively. The response curve of SnO_2/Rgo-based gas sensors for different gas concentrations of H_2S gas is shown in Figure 10.2b [82]. Wang et al. in 2009, fabricated 1 mol% Pd-doped SnO_2 thin films by sol–gel method, and detected 10 ppm formaldehyde at 250°C with a response of 1.55, and response time and recovery time of 50 s each [83]. Hieu et al. in 2007 fabricated SnO_2/MWCNTs composite by thin film microelectronic technique, which detected 60–800 ppm NH_3 at room temperature with a response of 2–30, and response time and recovery time around less than 300 sec [84].

3.2 ZnO-Based Gas Sensors

ZnO has a great impact in gas sensors because of its wide band gap of ~ 3.37 eV. Some ZnO, ZnO/Noble metals, ZnO/CNT, ZnO/Graphene-based gas sensors are presented in Table 10.6. Tian et al. used a facile solution-processing technique to grow ZnO nanorods arrays. At 300°C, it showed a sensor response of 70% to 100 ppm ethanol vapor, with response time and recovery time of 10 and 10 s, respectively. The detection limit was around 1 ppm [85]. In 2014, Chung et al. fabricated Au@ZnO core-shell structure, which sensed 5 ppm formaldehyde gas with a response of 10.57% at room temperature. The response and recovery times were 138 s and 104 s, respectively [86]. In 2011, Yu et al. reported sensing performances of ZnO–CuO and Pt/ZnO–CuO. At room temperature (RT), it showed a response of ~ 2.64% for 1000 ppm CO in air. The response and recovery times for both were ~ 80 s, respectively [87]. Alfano et al. in 2019, fabricated graphene functionalized with ZnO nanoparticles using microwave irradiation method, and found enhanced sensor response of ~ 26 at room temperature for 1 ppm NO_2 gas [88]. Schutt et al. in 2017, fabricated sensors from tetrapodal-ZnO (ZnO-T) networks functionalized with CNTs to form (ZnO-T-CNT), which exhibited good sensing properties to 100 ppm NH_3 gas at room temperature with a response of ~ 330, response time and recovery time of the sensors were about 18 s and 35 s, respectively. The sensor response of ZnO-T with different gas concentration are shown in Figure 10.3a [64]. Wang et al. in 2019, fabricated a composite prepared from zinc oxide and graphene oxide nanoribbons (ZnO/GONR)-based gas sensor with a response of ~ 18% for 50 ppm NO_2 gas at room temperature. The response and recovery times were 31 s and 27 s, respectively [89]. Das et al. in 2010, fabricated sensors using Pt/ZnO single nanowire Schottky diodes by using e-beam lithography, which exhibited good sensing properties to 2500 ppm hydrogen gas at room temperature, and with a response of ~ 80. The sensing properties of Pt/ZnO single nanowire with different temperature are shown in Figure 10.3b [90]. Iftekhar Uddin et al. in 2014, synthesized a silver (Ag)-loaded zinc oxide (ZnO)-reduced graphene oxide-based gas sensor with a response of ~ 21.2 for 100 ppm acetylene gas at 150°C. The response and recovery times were 25 s and 80 s, respectively [91].

Table 10.6. Comparison among ZnO-based gas sensors.

Material	Structure	Sensing gas	Gas conc. (ppm)	Operating temperature (°C)	Response (%)	Response time (sec)	Recovery time (sec)	Ref.
ZnO	Nanorods array	Ethanol	100	370	70	10	10	[85]
Au/ZnO	Core shell	Formaldehyde	5	RT	10.57	138	104	[86]
ZnO-CuO/Pt	Particles	CO	1000	RT	2.64	80	80	[87]
ZnO-G	Nanoparticles	NO_2	1	RT	26	~	~	[88]
ZnO-CNT	Tetrapods Network	NH_3	100	RT	330	18	35	[64]
ZnO/GO	Nanorods	NO_2	50	RT	~18	31	27	[89]
ZnO/Pt	Nanowire	H_2	2500	RT	~80	~	~	[90]
Ag-ZnO-Rgo	Composite	Acetylene	100	150	21.2	25	80	[91]

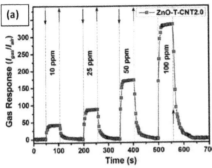
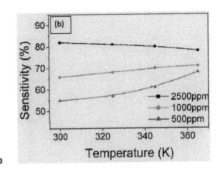

Figure 10.3. (a) Dynamic NH_3 response (100 ppm) for ZnO-T samples with content of CNTs: 2.0 wt% [64] (permission granted), (b) Temperature-dependent sensitivity for three representative H_2 concentrations [90] (permission granted).

3.3 TiO₂ Based Gas Sensor

TiO_2 is also widely used in gas sensors. It has a wide band gap of 3.2–3.35 eV. Some TiO_2, TiO_2/noble metals, TiO_2/CNT, TiO_2/graphene-based gas sensors are presented in Table 10.7. Seo et al., using a hydrothermal treatment process, fabricated porous gas sensing films composed of TiO_2 nanotubes gas sensors. At 500°C, it showed a sensor response of 50 to 50 ppm toluene gas with some response time and recovery time [72]. In 2008, Liang et al. fabricated a thick-film of NASICON (sodium super ionic conductor), which was formed on the outer side hollow Al_2O_3 tube, and after that, a thin layer of V_2O_5-doped TiO_2 was adhered on the NASICON as an electrode for sensing, which sensed 50 ppm SO_2 gas with a response of 90% at 400°C temperature. The response time was 10 s and the recovery time was 35 s [92]. In 2010, Moon et al. reported the gas sensing properties of pure and Pd-doped TiO_2 nanofiber mats. The sensing performance was observed at an operating temperature of 180°C, having sensitivity ~ 38% for 2.1 ppm NO_2 [31]. Acharyya et al. in 2016, fabricated sensors from TiO_2 nanotubes and reduced graphene oxide (RGO), which exhibited good sensing properties, to 800 ppm methanol room temperature, with the response of 96.93%, response and recovery times of the sensors were about 18 s and 61 s, respectively [93]. Hu et al. in 2010, decorated Ag-TiO_2 heterostuctures on TiO_2 nanobelts surface by photoreduction, and enhanced the response to ~ 46.153% for 500 ppm ethanol operating at 200°C, with response and recovery times of ~ 2 s and ~ 2 s, respectively. The sensitivity of TiO_2 nanobelts, Ag-TiO_2 nanobelts, surface-coarsened TiO_2 nanobelts, and surface-coarsened Ag-TiO_2 nanobelts with different concentrations of ethanol are shown in Figure 10.4 [94]. Karthik et al. in 2020 prepared reduced graphene oxide/titanium dioxide (rGO/TiO_2) composite thin films by using spray pyrolysis technique, which had the high sensitivity of 92% to 1500 ppm CO_2 at 673°C, with response and recovery times of ~ 30 s and 25 s, respectively [95]. Fort et al. in 2019, demonstrated gas sensors based on gold and TiO_2 nanoparticles decorated with SWCNT, which have sensitivity of 10% for 1 ppm of NO_2 gas at 240°C [96].

Table 10.7. Comparison among TiO_2-based gas sensors.

Material	Structure	Sensing gas	Gas conc. (ppm)	Operating temperature (°C)	Response (%)	Response time (sec)	Recovery time (sec)	Ref.
TiO_2	Nanotube	Toluene	50	500	50	~	~	[72]
V_2O_5-TiO_2	Film	SO_2	50	300	90	10	35	[92]
Pd-doped TiO_2	Nanofiber mats	NO_2	2.1	180	38	~	~	[31]
TiO_2 NT-rGO	Hybrid	Methanol	800	RT	96.93	18	61	[93]
Ag-TiO_2	Nano belt	Ethanol	500	200	46.153	~2	~2	[94]
rGO/TiO_2	Nanocomposite	CO_2	1500	673	92	~30	~25	[95]
Au/TiO_2/SWCNT	Network	NO_2	1	240	10	~	~	[96]

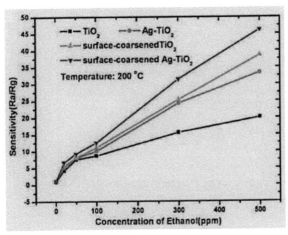

Figure 10.4. The sensitivity profiles of ethanol vapor sensors based on TiO_2 nanobelts, Ag-TiO_2 nanobelts, surface-coarsened TiO_2 nanobelts, and surface-coarsened Ag-TiO_2 nanobelts upon exposure to different concentrations of ethanol vapor at 200°C [96] (permission granted).

3.4 α-Fe₂O₃-Based Gas Sensor

$α$-Fe_2O_3 have great impact on gas sensors because of its wide band gap of ~ 2.2 eV. Some $α$-Fe_2O_3, $α$-Fe_2O_3/noble metals, $α$-Fe_2O_3/CNT, $α$-Fe_2O_3/graphene-based gas sensors are presented in Table 10.8. Liang et al. fabricated ultrafine $α$-Fe_2O_3 nanoparticles by a facile reverse micro-emulsion method. At 340°C, it showed a sensor response of 36% to 3000 ppm acetone gas [97]. In 2015, Mirzaei et al. fabricated Ag@$α$-Fe_2O_3 nanocomposite, which sensed 100 ppm ethanol gas with a response of 6% at 250°C. The response and recovery times were 5.5 s and 16 s, respectively. The change in resistance of Ag@$α$-Fe_2O_3 nanocomposite with different concentrations of acetone are shown in Figure 10.5a [98]. In 2018, Kim et al. reported Fe_2O_3 particles uniformly decorated on carbon nanotubes (Fe_2O_3@ CNT). The sensing performance was observed at room temperature (RT), having response of ~ 1.5% for 1 ppm H_2S gas [99]. Sen et al. in 2014, fabricated Polyaniline/ $γ$-ferric oxide (PANi/$γ$-Fe_2O_3) nanocomposite films, and found sensor response of ~ 0.5 at 28°C for 50 ppm LPG gas with a response time of 60 s [100]. Muthukumaran et al. in 2014, fabricated sensors from iron oxide nanostructures with a dendritic-like pine tree morphology, which showed good sensing properties to 4000 ppm NH_3 gas at an operating temperature of 100°C, with the response of ~ 2.65%, and the response and recovery times of the sensors were about 300 s and 900 s, respectively [63]. Wang et al. in 2014, fabricated a hierarchical $α$-Fe_2O_3/NiO composite with a hollow nanostructure-based gas sensor with response of ~ 18.68% for 100 ppm toluene gas at 300°C. The response of pure NiO and $α$-Fe_2O_3/NiO composites with different concentrations of toluene gas are shown in Figure 10.5b [101]. Zhang et al. in 2017, prepared reduced graphene oxide/$α$-Fe_2O_3 (rGO/$α$-Fe_2O_3) hybrid nanocomposites, which exhibited good sensing properties to 5 ppm NO_2 gas at room temperature with a response of ~ 3.68. The response and recovery times were ~ 32 s and 1435 s respectively [102]. Basu et al. in 2019, demonstrated gas sensors based on

Table 10.8. Comparison among α-Fe₂O₃-based gas sensors.

Material	Structure	Sensing gas	Gas conc. (ppm)	Operating temperature (°C)	Response	Response time (sec)	Recovery time (sec)	Ref.
α-Fe₂O₃	Nanoparticals	Acetone	3000	340	36	~	~	[97]
Ag@α-Fe₂O₃	Core shell	Ethanol	100	250	6	5.5	16	[98]
Fe₂O₃@CNT	Decorated nanotube	H₂S	1	RT	~1.5	~	~	[99]
Polyaniline/α-Fe₂O₃	Composite	LPG	50	28	0.5	60	~	[100]
α-Fe₂O₃/CNT	Composite	NH₃	4000	100	2.65	300	900	[63]
α-Fe₂O₃/NiO	Composite	Toluene	100	300	18.68	~	~	[101]
rGO/ α-Fe₂O₃	Hybrid Nano composite	NO₂	5	RT	3.68	32	1435	[102]
rGO-α-Fe₂O₃	Nano composite	CO	10	RT	~400	21	8	[41]

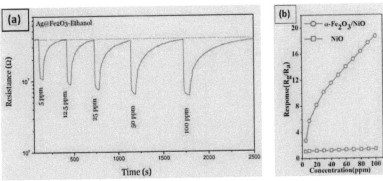

Figure 10.5. (a) Sensing response of Ag@α-Fe$_2$O$_3$ towards different concentrations of ethanol vapor at 250°C [98] (permission granted), (b) Response of pure NiO and α-Fe$_2$O$_3$/NiO composites at 300°C vs toluene concentration [101] (permission granted).

nanocomposites of rGO-α-Fe$_2$O$_3$, which showed a response of ~ 400% for 10 ppm of CO gas at room temperature. The response and recovery times were 21 s and 8 s, respectively [41].

3.5 In$_2$O$_3$-Based Gas Sensors

In$_2$O$_3$ has a great impact on gas sensors because of its wide band gap of ~ 3 eV. Some In$_2$O$_3$, In$_2$O$_3$/noble metals, In$_2$O$_3$/CNT, In$_2$O$_3$/graphene-based gas sensors are presented in Table 10.9. Nie et al. synthesized indium nitrate/polyvinyl pyrrolidone (In(NO$_3$)$_3$/PVP) composite nanofibers. At room temperature, it showed a sensor response of 53.20% to 1000 ppm NH$_3$ gas. The response of In$_2$O$_3$/PANI with different concentrations of NH$_3$ is shown in Figure 10.6 [103]. In 2016, Karmaoui et al. fabricated sub-spherical In$_2$O$_3$-Pt nanoparticles (NPs), which sensed 1.56 ppm acetone gas with a response of ~ 20% at 150°C [27]. Shruthi et al. in 2019, prepared Y$_2$O$_3$-In$_2$O$_3$ nanocomposites (NCs), which exhibited good sensing properties to 200 ppm methanol gas at room temperature, with a response of ~ 923. The response and recovery times were ~ 13 s and 18 s respectively [104]. In 2012, Pashchank et al. reported fabrication of In$_2$O$_3$/multi-walled carbon nanotubes (CNTs) composite microstructures. The sensing performance was observed at 200–400°C, having response of ~ 100% for 500–5000 ppm H$_2$ gas [105]. Liu et al. in 2013, fabricated a Pd-decorated In$_2$O$_3$ film, and found enhanced sensor response of ~ 4.6% at room temperature for1 vol% H$_2$ gas. The response and recovery times were 28 s and 32 s, respectively [106]. Liu et al. in 2017, fabricated sensors based on the materials composed of hierarchical flower-like In$_2$O$_3$ and reduced graphene oxide (rGO), which exhibited good sensing properties to 1 ppm NO$_2$ gas at an operating temperature of 74°C, with a response of ~ 1337%, response time and recovery time of the sensors were about 208 s and 39 s, respectively [107]. Inyawilert et al. in 2019 fabricated 0–1.0 wt% PdO$_x$-doped In$_2$O$_3$ nanoparticles-based gas sensors with a response of ~ 3526% for 10000 ppm H$_2$ gas at 250°C [108]. Ma et al. in 2010, fabricated sensors from Pr-doped In$_2$O$_3$ nanoparticles, which exhibited good sensing properties to

Table 10.9. Comparison among In_2O_3-based gas sensors.

Material	Structure	Sensing gas	Gas conc. (ppm)	Operating temperature (°C)	Response (%)	Response time (sec)	Recovery time (sec)	Ref.
In_2O_3/PANI	Composite Nanofiber	NH_3	1000	RT	53.20	?	?	[103]
In_2O_3/Pt	Nanoparticles	Acetone	1.56	150	~20	?	?	[27]
Y_2O_3-In_2O_3	Nanocomposites	Methanol	200	RT	~923	13	18	[104]
In_2O_3/CNT	Composite	H_2	500–5000	200–400	100	?	?	[105]
Pd-In_2O_3	Particles	H_2	1 vol.%	RT	4.6	28	32	[106]
rGO-In_2O_3	Composite	NO_2	1	74	1337	208	39	[107]
PdO_x-doped In_2O_3	Nanoparticles	H_2	10000	250	3526	?	?	[108]
Pr-doped In_2O_3	Nanoparticles	Ethanol	50	240	106	16.2	10	[109]

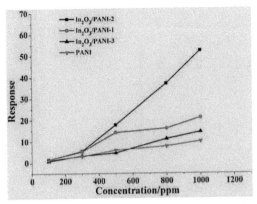

Figure 10.6. The response values of pure PANI and three In₂O₃/PANI nanofibers sensors to different concentrations of NH₃ at room temperature [103] (permission granted).

50 ppm ethanol gas at an operating temperature of 240°C, and with a response of ~ 106. The response and recovery times were 16.2 s and 10 s, respectively [109].

3.6 WO₃-based Gas Sensor

WO_3 also has a great impact in gas sensors because of its wide band gap of ~ 2.7 eV. Some WO_3, WO_3/noble metals, WO_3/CNT, WO_3/graphene-based gas sensors are presented in Table 10.10. Ke et al. fabricated a WO_3 thin film-based gas sensor. At 200°C, it showed a sensor response of 12.5% to 20 ppm benzene gas. The response and recovery times were 70 s and 120 s, respectively. Figure 10.7 shows the change in resistance with different concentrations of benzene [110]. In 2015, Kim et al. fabricated WO_3 nanorods decorated with both Pd and Au nanoparticles, which sensed 200 ppm acetone gas with a response of 152.4% at 300°C. The response and recovery times were 101 s and 96 s, respectively [111]. In 2019, Zhao et al., reported Au@WO_3 core–shell nanospheres (CSNSs)-based gas sensors. The sensing performance was observed at 100°C, having a response of ~ 136% for 5 ppm NO_2 gas, and the response and recovery times were 218 s and 2649 s, respectively [112]. Wang et al. in 2017, fabricated Pt-loaded mesoporous WO_3-based gas sensors, and found sensor response of ~ 13.61 at 125°C for 200 ppm NH_3 gas, with a response time of 43 s and recovery time of 272 s [113]. Hashishin et al. in 2008, fabricated sensors from CNTs-WO_3 composite, which exhibited good sensing properties to 5 ppm NO_2 gas at an operating temperature of 300°C, with a response of ~ 677%, and the response time and recovery time of the sensors were about 30 s and 75 s, respectively [62]. Shi et al. in 2016, fabricated a reduced graphene oxide/hexagonal WO_3 (rGO/h-WO_3) nanosheets composites-based gas sensor with a response of ~ 168.58% for 10 ppm H_2S gas at 330°C, and also having a response time of ~ 7 s, and recovery time of ~ 55 s [114]. Le et al. in 2019, fabricated a sensor from 0.5 wt% CNT/WO_3 nanoplate, which exhibited good sensing properties to 30 ppm NH_3 gas at an operating temperature of 50°C, and the minimum-detection-limit was 3 ppb, with a response of ~ 45 [115]. Zhao et al. in 2019, prepared mesoporous WO_3@ graphene aerogel nanocomposites, which exhibited good sensing properties to

Table 10.10. Comparison among WO_3-based gas sensors.

Material	Structure	Sensing gas	Gas conc. (ppm)	Operating temperature (°C)	Response	Response time (sec)	Recovery time (sec)	Ref.
WO_3	Thin film	Benzene	20	200	12.5	70	120	[110]
$AuPd/WO_3$	Nanorods	Acetone	200	300	152.4	101	96	[111]
$Au@WO_3$	core–shell	NO_2	5	100	136	218	2649	[112]
WO_3/Pt	Mesoporous	NH_3	200	125	13.61	43	272	[113]
CNT/WO_3	Composite	NO_2	5	300	677	30	75	[62]
rGO/WO_3	Composite	H_2S	10	330	168.58	7	55	[114]
WO_3/CNT	Nanoplate	NH_3	30	50	45	~	~	[115]
$WO_3@$ graphene	aerogel nanocomposites	Acetone	50	150	15	13	65	[116]

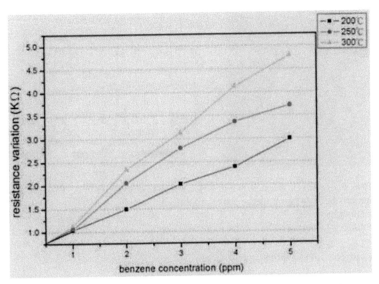

Figure 10.7. Effect of the sensor temperature on benzene sensitivity [110] (permission granted).

50 ppm acetone gas at an operating temperature of 150°C, and with a response of ~ 15%. The response and recovery times were ~ 13 s and 65 s, respectively [116].

4. Summary of Recent Advances

Innovations in Gas Sensing

i) **Using the CNT-based paper, it is possible to make flexible gas sensors**

CNT paper mixed with metal oxide can be pasted on the substrate, and electrode would be deposited over it. Thus, the idea of flexible gas sensor can be realized.

ii) **CNT mixed with metal oxide**

When CNT is used in combination with metal oxide, it improves its gas sensing properties. CNT is decorated with metal oxide, CNT is kept in less quantity just to help in enhancing the conduction, sensing performance, and also help in reducing temperature. There are several types of CNTs, and in this review, we are mostly focusing on the MWCNTs. These can be functionalized and CNTs can work as a sensing material, but in such cases the responses are very poor, and therefore we have focused on composites of metal oxides with CNT. For example, Schütt et al. synthesized ZnO-T-CNT network by using flame transport technique. This was used to detect NH_3 up to 100 ppm at room temperature, and have a response of 330, with response and recovery times of 18 s and 35 s, respectively. Figures 10.8a and b show the sensing mechanism and response of ZnO-T samples with different contents of CNTs: 0.0, 0.8, 2.0 wt%, and (j–l) 4.0 wt% for NH_3 gas [64] (permission granted).

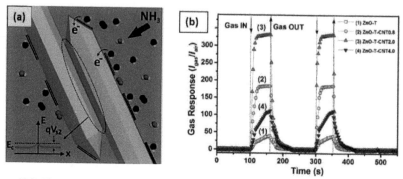

Figure 10.8. Shows (a) On the exposure to NH_3 vapor, a higher change in the height of the potential barrier between the ZnO-T-arms can be obtained due to adsorbed NH_3 molecules on the CNT, leading to an electron transfer to ZnO through the CNT, (b) Dynamic NH_3 response (100 ppm) for ZnO-T samples with different contents of CNTs: 0.0, 0.8, 2.0 wt%, and (j–l) 4.0 wt% [64] (permission granted).

iii) Graphene mixed with metal oxide

It has been observed that presence of graphene helps in improving the gas sensing properties. For example, Song et al. demonstrated gas sensors based on SnO_2 quantum wires, which were anchored on reduced graphene oxide (rGO) nanosheets, and had a response of 33 for 50 ppm of H_2S gas at room temperature, with response and recovery times of 2 s and 292 s, respectively. The gas sensing mechanism is also shown in Figures 10.9a and b [82] (permission granted).

iv) Presence of noble metals

Noble metals affect the gas sensing properties by influencing the electronics of the chemical environment at the surface of the metal oxide. For example, Chung et al. fabricated Au@ZnO core–shell structure-based gas sensors, which exhibited a good response of around 10.57 for formaldehyde gas up to 5 ppm at room temperature, with response and recovery times of 138 s and 104 s, respectively [86]. Figures 10.10a,b, and c show the gas sensing mechanism and sensor response of pure SnO_2 NWs and Ag-decorated SnO_2 NWs [81] (permission granted).

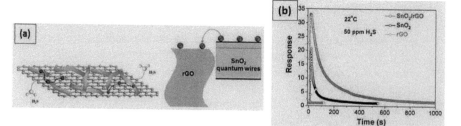

Figure 10.9. (a) Schematic illustration of H_2S sensing mechanism of gas sensors employing SnO_2 quantum wire/rGO nanocomposites, (b) Response curves of gas sensors based on pristine rGO, SnO_2 quantum wires (8 h) and SnO_2/rGO nanocomposites (8 h) [82] (permission granted).

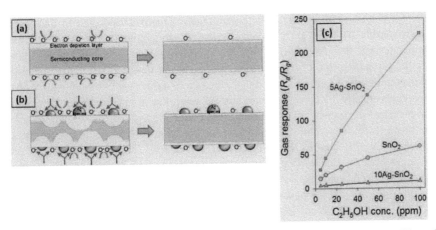

Figure 10.10. Schematic diagrams on the gas sensing mechanism of (a) pure SnO_2 NWs and (b) Ag-decorated SnO_2 NWs, and (c) C_2H_5OH responses (Ra/Rg) [81] (permission granted).

v) 1D, 2D, thin film

One-dimensional nanostructures, such as nanowires, nanotubes, nanorods, nanopillars, nanoneedles, nanocylinders, and nanowhiskers, have great attention because of their potential applications in catalysts as carrier materials. For example, in 2011, Hwang et al. fabricated a sensor via e-beam evaporation techniques. Ag layers were uniformly coated on the surface of SnO_2 NWs with thicknesses ranging from 5 to 50 nm, which were converted into islands of Ag by heat treatment at 450°C for 2 h. They have a good response of $\sim$ 228.1 for 100 ppm ethanol (operating at 450°C), with response and recovery times of $\sim$ 5 sec and $\sim$ 100 s, respectively. The TEM image of Ag-coated SnO_2 NWs is shown in Figure 10.11 [81] (permission granted).

Figure 10.11. Transmission electron micrographs of 5Ag-SnO_2 NW networks [81].

5. (i) Gas Sensing Mechanism

In order to explain the gas sensing mechanism, n-type semiconductor oxide materials with reducing gas are considered, as these are easier to explain. Similar will be the case for p-type oxide semiconductor materials with oxidizing gas. In the case of nanomaterials, first they are annealed at a certain temperature to form a neck between nanomaterials. It is well known that metal oxides are oxygen deficient by nature, resulting in adsorption of oxygen at its service in the form of O_2, O^-, O^{2-} depending on the temperature of the surrounding. When these oxides are exposed to air, oxygen molecules engage the electrons, and thus the resistance of nanomaterials increases in the presence of oxygen. In the presence of reducing gases, these oxygen molecules are taken away by it, and thus the electrons are released back, which increases the conductivity or decreases the resistance of the material. This can also be explained in terms of formation of depletion layer. In the presence of oxygen, electrons are bound, therefore it can be said that the width of the depletion layer increased, and in the presence of reducing gases, the depletion region width is reduced. The adsorption of gases at the surface of the metal oxide can be understood with the help of the following equations [103]

$$O_2(gas) \rightarrow O_2(ads) \tag{1a}$$

At $(T < 100°C)$, $$O_2(ads) + e^- \rightarrow O_2^- \ (ads) \tag{1b}$$

At $(100°C \leq T < 300°C)$, $$O_2^- \ (ads) + e^- \rightarrow 2O^- \ (ads) \tag{1c}$$

At $(T > 300°C)$, $$O^- \ (ads) \ e^- \rightarrow O^{2-} \ (ads) \tag{1d}$$

In the case of nanoparticles, when the size of the nanoparticles is in the range of Debye length, the sensor response increases drastically. The Debye length L_D is defined as—

$$L_D = (\varepsilon\varepsilon_o kT/e^2 n_b)^{1/2}$$

where ε, K,T, e, and n_b are the dielectric constant, Boltzmann constant, temperature, electronic charge, and density of electrons, respectively. This is due to the creation of a complete depletion region at the neck of the particles (become oxygen deficient), resulting in increase in the resistance to a very large amount. In the presence of reducing gases, almost all of the oxygen is taken away, resulting in the opening of the complete channel at the neck region, resulting in very low resistance. Thus, when nanomaterials with smaller size are used, the response of the device increases drastically.

(ii) Role of Metal Catalyst

When noble metals are used for enhancing the gas sensing response, it acts in two ways (i) Chemical sensitization: When noble metals such as platinum or gold are used, they adsorb oxygen molecules on their surface, thus more oxygen is adsorbed compared to bare metal oxide. Thus, the resistance of the material increases by using noble metals. In the presence of reducing gases, they take away almost all the oxygen, thus decreasing the resistance. In such cases, the presence of noble metal provides electrons for conduction, thus decreasing the resistance further. (ii) Electronic sensitization: On the other hand, some metals such as palladium, silver, etc. form metal oxide in the presence of oxygen. Thus, p-n type junction is formed between p-type oxide catalyst and n-type sensing metal oxide, which increases the width of the depletion layer. Thus, the overall resistance of the material in the presence of oxygen is increased. In the presence of reducing gases, these noble metal oxides turn to metal, and provide electrons to decrease the resistance further. Thus, it increases the resistance in the presence of oxygen and decreases the resistance in the presence of reducing gases. Thus, they enhance the overall sensor response. The gas sensing mechanism of Pt/ZnO single nanowire for H_2 gas is shown in Figures 10.12a,b, and c [90] (permission granted).

(iii) Role of CNT

Since CNTs are longer and conducting, they help in transporting the electron to the nearest electrode. Thus, when CNT is present, it can increase the response time. Similarly, CNTs are generally p-type materials, and therefore they take part in increasing or decreasing the concentration of electrons, and thus changing the sensing properties, as discussed in the case of noble metals.

(iv) Role of Graphene

Graphene being a two-dimensional material and also having good conductivity, helps in a similar way as carbon nanotube in affecting the sensing properties, but with better efficiency.

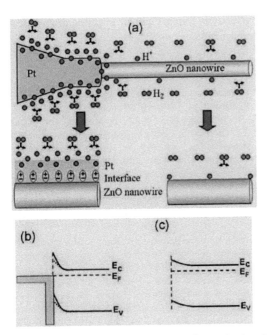

Figure 10.12. (a) Diagram of the Pt/ZnO single nanowire H_2-sensing mechanism. Energy-level diagram of (b) Pt/ZnO nanowire interface and (c) ZnO nanowire surface [90].

6. Conclusion and Further Remark

The main advantage of this kind of metal oxide CNT/graphene composite gas sensor is that the sensing can be done at a relatively low temperature. By optimizing the conditions, sensing can also be done at room temperature in certain cases for certain gases. The disadvantage of this method is that the amount of CNT should be very less, otherwise they will short the electrodes. Thus, the amount of CNT needs to be optimized.

Need of standardization of gas sensor

One of the main concerns of gas sensor research is lack of standardization, e.g., definition of limit of detection. The limit of detection is determined by further extrapolating the R_g/R_a versus concentration curve to $3\sigma/R_a$. Where σ is defined as the standard deviation of R_a.

There are other problems in the standardization of gas sensors: The researchers working in this field do not adhere to a uniform limit of detection while claiming the detection limit of their gas sensor. Also, there is a lot of non-uniformity in the working temperature selection at different places and environment. It ranges from 30°C to 400°C. Another point is that it is well known that water vapor produces resistance change in metal oxides, and therefore it is also mandatory to provide humidity data. It is also to be noted that very few researchers have worked away from the linear range or selectivity to meet industrial application. Due to these reasons, gas sensors are not standardized.

References

[1] Dey, A. 2017. Semiconductor metal oxide gas sensors: A review. Mater. Sci. Eng. B, 229: 206–217. December 2017, doi: 10.1016/j.mseb.2017.12.036.
[2] Bezzon, V.D.N., T.L.A. Montanheiro, B.R.C. de Menezes, R.G. Ribas, V.A.N. Righetti, K.F. Rodrigues and G.P. Thim. 2019. Carbon nanostructure-based sensors: a brief review on recent advances. Adv. Mater. Sci. Eng., vol. 2019, doi: 10.1155/2019/4293073.
[3] Vikrant, K., V. Kumar and K.H. Kim. 2018. Graphene materials as a superior platform for advanced sensing strategies against gaseous ammonia. J. Mater. Chem. A, 6(45): 22391–22410. doi: 10.1039/c8ta07669c.
[4] Sun, D., Y. Luo, M. Debliquy and C. Zhang. 2018. Graphene-enhanced metal oxide gas sensors at room temperature: A review. Beilstein J. Nanotechnol., 9(1): 2832–2844. doi: 10.3762/bjnano.9.264.
[5] Abideen, Z.U., Jae-H. Kim, Jae-H. Lee, Jin-Y. Kim, A. Mirzaei, H.W. Kim and S.S. Kim. 2017. Electrospun metal oxide composite nanofibers gas sensors: A review. J. Korean Ceram. Soc., 54(5): 366–379. doi: 10.4191/kcers.2017.54.5.12.
[6] Donarelli, M. and L. Ottaviano. 2018. 2d materials for gas sensing applications: A review on graphene oxide, MoS_2, WS_2 and phosphorene. Sensors, 18(11). doi: 10.3390/s18113638.
[7] Xu, K., C. Fu, Z. Gao, F. Wei, Y. Ying, C. Xu and G. Fu. 2018. Nanomaterial-based gas sensors: A review. Instrum. Sci. Technol., 46(2): 115–145. doi: 10.1080/10739149.2017.1340896.
[8] Song, Z., S. Xu, J. Liu, Z. Hu, N. Gao, J. Zhang, F. Yi, G. Zhang, S. Jiang and H. Liu. 2018. Enhanced catalytic activity of SnO_2 quantum dot films employing atomic ligand-exchange strategy for fast response H2S gas sensors. Sensors Actuators, B Chem., 271: 147–156. February, doi: 10.1016/j.snb.2018.05.122.

[9] Patil, L.A., A.R. Bari, M.D. Shinde and V. Deo. 2010. Ultrasonically prepared nanocrystalline ZnO thin films for highly sensitive LPG sensing. Sensors Actuators B. Chem., 149(1): 79–86. doi: 10.1016/j.snb.2010.06.027.

[10] Zhou, X., Y. Zhu, W. Luo, Y. Ren, P. Xu, A.A. Elzatahry, X. Cheng, A. Alghamdi, Y. Deng and D. Zhao. 2016. Chelating-assisted soft-templating synthesis of ordered mesoporous zinc oxides for low concentration gas sensing. Mater. Chem. A, (4): 15064–15071. doi: 10.1039/C6T A05687C.

[11] Anh, V., L. Anh, T. Quang, V. Ngoc and N. Van Quy. 2013. Applied surface science enhanced NH_3 gas sensing properties of a QCM sensor by increasing the length of vertically orientated ZnO nanorods. Appl. Surf. Sci., 265: 458–464. doi: 10.1016/j.apsusc.2012.11.028.

[12] Septiani, N.L.W., B. Yuliarto, Nugraha and H.K. Dipojono. 2017. Multiwalled carbon nanotubes–zinc oxide nanocomposites as low temperature toluene gas sensor. Appl. Phys. A Mater. Sci. Process., 123(3): 1–9. doi: 10.1007/s00339-017-0803-y.

[13] Zeng, Y., T. Zhang, M. Yuan, M. Kang, G. Lu and R. Wang. 2009. Chemical growth and selective acetone detection based on ZnO nanorod arrays. Sensors and Actuators B, 143: 93–98. doi: 10.1016/j.snb.2009.08.053.

[14] Peng, L., J. Zhai, D. Wang, Y. Zhang, P. Wang, Q. Zhao and T. Xie. 2010. Size and photoelectric characteristics-dependent formaldehyde sensitivity of ZnO irradiated with UV light. Sensors and Actuators B: Chem., 148: 66–73. doi: 10.1016/j.snb.2010.04.045.

[15] Minaee, H., S.H. Mousavi, H. Haratizadeh and P.W. De Oliveira. 2013. Oxygen sensing properties of zinc oxide nanowires, nanorods, and nanoflowers: The effect of morphology and temperature. Thin Solid Films, 545: 8–12. doi: 10.1016/j.tsf.2013.05.155.

[16] Duc, N., H. Si, D. Dang, N. Van Duy and N. Duc. 2013. Chemical on-chip growth of wafer-scale planar-type ZnO nanorod sensors for effective detection of CO gas. Sensors and Actuators B, 181: 529–536. doi: 10.1016/j.snb.2013.02.047.

[17] Öztürk, S., N. Kılınç and Z. Ziya. 2013. Fabrication of ZnO nanorods for NO_2 sensor applications: Effect of dimensions and electrode position. 581(2): 196–201. doi: 10.1016/j.jallcom.2013.07.063.

[18] Wang, Z., Z. Li, T. Jiang, X. Xu and C. Wang. 2021. Ultrasensitive hydrogen sensor based on PdO-loaded SnO_2 electrospun nanofibers at room temperature. ACS Appl. Mater. Interfaces, 5(6): 2013–2021, doi: 10.1021/am3028553.

[19] Wang, K., T. Zhao, G. Lian, Q. Yu, C. Luan, Q. Wang and D. Cui. 2013. Room temperature CO sensor fabricated from Pt-loaded SnO_2 porous nanosolid. Sensors Actuators B. Chem., 184: 33–39. doi: 10.1016/j.snb.2013.04.054.

[20] Liu, S., M. Xie, Y. Li, X. Guo and W. Ji. 2010. Novel sea urchin-like hollow core–shell SnO_2 superstructures: Facile synthesis and excellent ethanol sensing performance. Sensors Actuators B. Chem., 151(1): 229–235. doi: 10.1016/j.snb.2010.09.015.

[21] Li, Z., Q. Zhao, W. Fan and J. Zhan. 2011. Porous SnO_2 nanospheres as sensitive gas sensors for volatile organic compounds detection. Nanoscale, 1646–1652. doi: 10.1039/c0nr00728e.

[22] Xue, X., L. Xing, Y. Chen, S. Shi, Y. Wang and T. Wang. 2008. Synthesis and H_2S sensing properties of CuO-SnO_2 core/shell PN-junction nanorods. J. Phys. Chem. C, 4: 12157–12160.

[23] Chang, J., H. Kwon, J. Huh and D.D. Lee. 1997. Tin oxide-based methane gas sensor promoted by alumina-supported Pd catalyst. Sensors and Actuators B, 45: 271–277.

[24] Korotcenkov, G., I. Blinov, M. Ivanov and J.R. Stetter. 2007. Ozone sensors on the base of SnO_2 films deposited by spray pyrolysis. Sensors and Actuators B, 120(2): 679-686 doi: 10.1016/j.snb.2006.03.029.

[25] Nio, S., P. Hidalgo, R.H.R. Castro, C.V Coelho and D. Gouve. 2005. Surface segregation and consequent SO_2 sensor response in SnO_2-NiO. Chem. Mater., 17(16): 4149–4153. Doi: 10.1021/cm049020g.

[26] Yang, W., P. Wan, X. Zhou, J. Hu, Y. Guan and L. Feng. 2014. Self-assembled In_2O_3 truncated octahedron string 2 and its sensing properties for formaldehyde. Sensors Actuators B. Chem., 20(1): 228–233 doi: 10.1016/j.snb.2014.05.003.

[27] Mohamed, K., G.L. Salvatore, L. Mariangela, M.T. David, D. Nicola, C.P. Robert, P.S. Maria, A.L. Joao and N. Giovanni. 2016. Pt-decorated In_2O_3 nanoparticles and their ability as a highly sensitive (< 10 ppb) acetone sensor for biomedical applications. Sensors Actuators B. Chem., 230: 697–705. doi: 10.1016/j.snb.2016.02.100.

[28] Pandeeswari, R., R.K. Karn and B.G. Jeyaprakash. 2014. Ethanol sensing behaviour of sol-gel dip-coated TiO_2 thin films. Sensors Actuators, B Chem., 194: 470–477. doi: 10.1016/j.snb.2013.12.122.

[29] Wang, H., L. Chen, J. Wang, Q. Sun and Y. Zhao. 2014. A micro oxygen sensor based on a nano sol-gel TiO_2 thin film. Sensors, 14(9): 16423–16433. doi: 10.3390/s140916423.

[30] Azhar, A.H., X. Lijuan, W. Zhe, Z. Ali, L. Zhong, P. Tomas, G. Meros, R. Tomas and P. Andrej. 2019. Remarkable improvement in hydrogen sensing characteristics with Pt/TiO_2 interface control. ACS Sens., 4(11): 2997–3006. doi: 10.1021/acssensors.9b01537.

[31] Moon, J., J.A. Park, S.J. Lee, T. Zyung and I.D. Kim. 2010. Pd-doped TiO_2 nanofiber networks for gas sensor applications. Sensors Actuators, B Chem., 149(1): 301–305. doi: 10.1016/j.snb.2010.06.033.

[32] Tai, H., Y. Jiang, G. Xie, J. Yu, X. Chen and Z. Ying. 2008. Influence of polymerization temperature on NH_3 response of $PANI/TiO_2$ thin film gas sensor. Sensors Actuators, B Chem., 129(1): 319–326. doi: 10.1016/j.snb.2007.08.013.

[33] Cai, Z.X., H.Y. Li, J.C. Ding and X. Guo. 2017. Hierarchical flowerlike WO_3 nanostructures assembled by porous nanoflakes for enhanced NO gas sensing. Sensors Actuators, B Chem., 246: 225–234. doi: 10.1016/j.snb.2017.02.075.

[34] Mehta, S., D. Nadargi, M. Tamboli, V. Patil, I. Mulla and S. Suryavanshi. 2019. Macroporous WO_3: Tunable morphology as a function of glycine concentration and its excellent acetone sensing performance. Ceram. Int., 45(1): 409–414. doi: 10.1016/j.ceramint.2018.09.181.

[35] Aijun, Y., W. Dawei, L. Tiansong, C. Jifeng, L. Weijuan, P. Jianbin, L. Zhu, W. Xiaohua and R. Mingzhe. 2020. Single ultrathin WO_3 nanowire as a superior gas sensor for SO_2 and H_2S: Selective adsorption and distinct I-V response. Mater. Chem. Phys., 240: 122165. doi: 10.1016/j.matchemphys.2019.122165.

[36] Kaur, N., D. Zappa, N. Poli and E. Comini. 2019. Integration of VLS-Grown WO_3 nanowires into sensing devices for the detection of H_2S and O_3. ACS Omega, 4(15): 16336–16343. doi: 10.1021/acsomega.9b01792.

[37] Kulkarni, S.B., Y.H. Navale, S.T. Navale, F.J. Stadler, N.S. Ramgir and V.B. Patil. 2019. Hybrid polyaniline-WO_3 flexible sensor: A room temperature competence towards NH_3 gas. Sensors Actuators, B Chem., 288: 279–288. March 2018, doi: 10.1016/j.snb.2019.02.094.

[38] Junhao, M., R. Yuan, Z. Xinran, L. Liangliang, Z. Yoangheng, C. Xiaowei, X. Pengcheng, L. Xinxin, D. Yonghui and Z. Dongyuan. 2018. Pt nanoparticles sensitized ordered mesoporous WO_3 semiconductor: gas sensing performance and mechanism study. Adv. Funct. Mater., 28(6): 1–12. doi: 10.1002/adfm.201705268.

[39] Yuhui, L., L. Wei, Q. Nan, L. Dr. Junping, W. Jing, L. Wei, F. Shanshan, C. Junchen, X. Jiaqiang, A.E. Ahmed, H.E. Mahir, D. Yonghui and Z. Dongyuan. 2014. Highly ordered mesoporous tungsten oxides with a large pore size and crystalline framework for H_2S sensing. Angew. Chemie–Int. Ed., 53(34): 9035–9040. doi: 10.1002/anie.201403817.

[40] Yinglin, W., C. Xiaobiao, Y. Qiuyue, L. Jie, G. Yuan, S. Peng and L. Geyu. 2016. Preparation of Ag-loaded mesoporous WO_3 and its enhanced NO_2 sensing performance. Sensors Actuators, B Chem., 225(2): 544–552. doi: 10.1016/j.snb.2015.11.065.

[41] Basu, A.K., P.S. Chauhan, M. Awasthi and S. Bhattacharya. 2019. α-Fe_2O_3 loaded rGO nanosheets based fast response recovery CO gas sensor at room temperature. Appl. Surf. Sci., 465: 56–66. doi: 10.1016/j.apsusc.2018.09.123.

[42] Nguyen, T.A.T., D.C. Nguyen, C.N. Le, Q.K. Dinh, N. Pham, V.T. Nguyen, M.H. Chu and V.H. Nguyen. 2018. Fe_2O_3 nanoporous network fabricated from Fe_3O_4/reduced graphene oxide for high-performance ethanol gas sensor. Sensors Actuators, B Chem., 255: 3275–3283. doi: 10.1016/j.snb.2017.09.154.

[43] Yusuf, V.K., M. Julien, L. Minsu, Y. Yuan, Z. Quadir, J. Xuchuan and Y. Aibing. 2015. Hydrothermal Synthesis of Ternary α-Fe_2O_3-ZnO-Au nanocomposites with high gas-sensing performance. Sensors and Actuators B, 209: 889–897. doi: 10.1016/j.snb.2014.12.065.

[44] Chaitongrat, B. and S. Chaisitsak. 2018. Fast-LPG sensors at room temperature by α-Fe_2O_3/CNT nanocomposite thin films. J. Nanomater., pp. 1–11. doi: 10.1155/2018/9236450.

[45] Haiwei, F., W. Qiqi, D. Jijun and Z. Yi. 2020. Fe_2O_3 nanotube coating micro-fiber interferometer for ammonia detection. Sensors Actuators, B Chem., 303: 127186. doi: 10.1016/j.snb.2019.127186.

[46] He, Y., D. Wang, F. Ge and L. Liu. 2015. SnO_2-doped α-Fe_2O_3 patulous microtubes for high performance formaldehyde sensing. J. Semicond., 36(8). doi: 10.1088/1674-4926/36/8/083005.

[47] Hjiri, M., M.S. Aida and G. Neri. 2019. NO_2 selective sensor based on α-Fe_2O_3 nanoparticles synthesized via hydrothermal technique. Sensors, 19(1). doi: 10.3390/s19010167.

[48] Haining, M., Y. Lingmin, Y. Xiong, L. Yuan, L. Chun, Y. Mingli and F. Xinhui. 2019. Room temperature photoelectric NO_2 gas sensor based on direct growth of walnut-like In_2O_3 nanostructures. J. Alloys Compd., 782: 1121–1126. doi: 10.1016/j.jallcom.2018.12.180.

[49] Daniele, Z., P. Paola, T. Jean-Marc, S. Anna, O. Alexandru, W. Udo and B. Nicolae. 2019. Investigation of the film thickness influence on the sensor response of In_2O_3-based sensors for O_3 detection at low temperature and Operando DRIFT study. Proceedings, 14(1): 45. doi: 10.3390/proceedings2019014045.

[50] Xuewen, W., G. Haibo, Z. Chunxue, D. Yuqi, M. Jiaqi, L. Zhenjie, L. Wenwen, Z. Fan, Z. Wu, Y. Junfeng, Z. Zhiyong and W. Zhaoke. 2019. Preparation of In_2O_3 thin film and the study of gas sensitivity to ethanol. Integr. Ferroelectr., 199(1): 58–68. doi: 10.1080/10584587.2019.1592598.

[51] Feng, C., Y. Man, W. Xi, S. Yang, G. Lanlan, X. Ning, K. Xueying, X. Xiumei, S. Yanfeng and L. Geyu. 2019. Template-free synthesis of cubic-rhombohedral-In_2O_3 flower for ppb level acetone detection. Sensors Actuators, B Chem., 290: 459–466. doi: 10.1016/j.snb.2019.04.013.

[52] Andre, R.S., L.A. Mercante, M.H.M. Facure, L.H.C. Mattoso and D.S. Correa. 2019. Enhanced and selective ammonia detection using In_2O_3/reduced graphene oxide hybrid nanofibers. Appl. Surf. Sci., 473: 133–140. doi: 10.1016/j.apsusc.2018.12.101.

[53] Zhijie, L., Y. Shengnan, S. Mengxuan, L. Hao, W. Zhonglin, W. Junqiang, S. Wenzhong and Q.F. Yong. 2019. Significantly enhanced temperature-dependent selectivity for NO_2 and H_2S detection based on In_2O_3 nano-cubes prepared by CTAB assisted solvothermal process. J. Alloys Compd., p. 152518. doi: 10.1016/j.jallcom.2019.152518.

[54] Jinesh, K.B. V.A.T. Dam, J. Swerts, C. de Nooijer, S. van Elshocht, S.H. Brongersma and M. Crego-Calama. 2011. Room-temperature CO_2 sensing using metal-insulator-semiconductor capacitors comprising atomic-layer-deposited La_2O_3 thin films. Sensors Actuators, B Chem., 156(1): 276–282. doi: 10.1016/j.snb.2011.04.033.

[55] Zanhong, D., F. Xiaodong, L. Da, Z. Shu, T. Ruhua, D. Weiwei, W. Tao, M. Gang and Z. Xuebin. 2009. Room temperature ozone sensing properties of p-type transparent oxide $CuCrO_2$. J. Alloys Compd., 484(1-2): 619–621. doi: 10.1016/j.jallcom.2009.05.001.

[56] Cho, N.G., I.S. Hwang, H.G. Kim, J.H. Lee and I.D. Kim. 2011. Gas sensing properties of p-type hollow NiO hemispheres prepared by polymeric colloidal templating method. Sensors Actuators, B Chem., 155(1): 366–371. doi: 10.1016/j.snb.2010.12.031.

[57] Castro-Hurtado, I., C. Malagù, S. Morandi, N. Pérez, G.G. Mandayo and E. Castaño. 2013. Properties of NiO sputtered thin films and modeling of their sensing mechanism under formaldehyde atmospheres. Acta Mater., 61(4): 1146–1153. doi: 10.1016/j.actamat.2012.10.024.

[58] Zhang, C., L. Li, L. Hou and W. Chen. 2019. Fabrication of Co_3O_4 nanowires assembled on the surface of hollow carbon spheres for acetone gas sensing. Sensors Actuators, B Chem., 291: 130–140. April, doi: 10.1016/j.snb.2019.04.064.

[59] Liu, Y., G. Zhu, B. Ge, H. Zhou, A. Yuan and X. Shen. 2012. Concave Co_3O_4 octahedral mesocrystal: Polymer-mediated synthesis and sensing properties. CrystEngComm., 14(19): 6264–6270. doi: 10.1039/c2ce25788b.

[60] Busacca, C., A. Donato, M. Lo Faro, A. Malara, G. Neri and S. Trocino. 2020. CO gas sensing performance of electrospun Co_3O_4 nanostructures at low operating temperature. Sensors Actuators, B Chem., 303: 127193. doi: 10.1016/j.snb.2019.127193.

[61] An, S., S. Park, H. Ko and C. Lee. 2014. Fabrication of WO_3 nanotube sensors and their gas sensing properties. Ceram. Int., 40(1): 1423–1429. Part B, doi: 10.1016/j.ceramint.2013.07.025.

[62] Hashishin, T. and J. Tamaki. 2008. Conductivity-type sensor based on CNT-WO_3 composite for NO_2 detection. J. Nanomater., (1): 0–4. doi: 10.1155/2008/352854.

[63] Muthukumaran, P., C. Sumathi, J. Wilson, C. Sekar, S.G. Leonardi and G. Neri. 2014. Fe_2O_3/carbon nanotube-based resistive sensors for the selective ammonia gas sensing. Sens. Lett., 12(1): 17–23. doi: 10.1166/sl.2014.3220.

[64] Fabian, S., P. Vasile, A. Rainer and L. Oleg. 2017. Single and networked ZnO-CNT hybrid tetrapods for selective room temperature high-performance ammonia sensors. ACS Appl. Mater. Interfaces, 9(27): 23107–23118. doi: 10.1021/acsami.7b03702.

[65] Zheng, Z.Q., L.F. Zhu and B. Wang. 2015. In$_2$O$_3$ nanotower hydrogen gas sensors based on both schottky junction and thermoelectronic emission. Nanoscale Res. Lett., 10(1): 1–14. doi: 10.1186/s11671-015-1002-4.

[66] Lihong, L., Y. Ming, Z. Hui, X. Yingming, C. Xiaoli, Z. Xianfa, G. Shan, S. Haiyan and H. Lihua. 2019. Co$_3$O$_4$/carbon hollow nanospheres for resistive monitoring of gaseous hydrogen sulfide and for nonenzymatic amperometric sensing of dissolved hydrogen peroxide. Microchimica Acta, 186: 184. doi: 10.1007/s00604-019-3253-8.

[67] Yong, J., H. Lifang, G. Zheng, C. Xing, M. Fanli, L. Tao, L. Minqiang and L. Jinhuai. 2010. Preparation of porous tin oxide nanotubes using carbon nanotubes as templates and their gas-sensing properties. J. Phys. Chem. C, 113(22): 9581–9587. doi: 10.1021/jp9001719.

[68] Ajayan, B.P.M., L.S. Schadler, C. Giannaris and A. Rubio. 2000. Single-walled carbon nanotube ± polymer composites: strength and weakness. Adv. Mater. 12(10): 750–753. doi: 10.1002/(SICI)1521-4095(200005)12.

[69] Yeow, J.T.W. and Y. Wang. 2009. A review of carbon nanotubes-based gas sensors. J. Sensors, 2009: 493904. doi: 10.1155/2009/493904.

[70] Control, J., Z. Ereš and S. Hrabar. 2018. Low-cost synthesis of high-quality graphene in do-it-yourself CVD reactor. Automatika, 59(3): 255–261, doi: 10.1080/00051144.2018.1528691.

[71] Zhang, H.J., F.N. Meng, L.Z. Liu and Y.J. Chen. 2019. Convenient route for synthesis of alpha-Fe$_2$O$_3$ and sensors for H$_2$S gas. J. Alloys Compd., 774: 1181–1188. doi: 10.1016/j.jallcom.2018.09.384.

[72] Seo, M.H., M. Yuasa, T. Kida, J.S. Huh, N. Yamazoe and K. Shimanoe. 2009. Detection of organic gases using TiO$_2$ nanotube-based gas sensors. Procedia Chem., 1(1): 192–195. doi: 10.1016/j.proche.2009.07.048.

[73] Choi, Y.J., I.S. Hwang, J.G. Park, K.J. Choi, J.H. Park and J.H. Lee. 2008. Novel fabrication of an SnO$_2$ nanowire gas sensor with high sensitivity. Nanotechnology, 19(9). doi: 10.1088/0957-4484/19/9/095508.

[74] Liu, Y., E. Koep and M. Liu. 2005. A highly sensitive and fast-responding SnO$_2$ sensor fabricated by combustion chemical vapor deposition. Chem. Mater. 17(15): 3997–4000. doi: 10.1021/cm050451o.

[75] Kechuang, W., Y. Jialin, W. Ding and W. Xianying. 2019. Graphene oxide@3D hierarchical SnO$_2$ nanofiber/nanosheets nanocomposites for highly sensitive and low-temperature formaldehyde detection. Molecules, 25(1): 35. doi:10.3390/molecules25010035.

[76] Mashock, M., K. Yu, S. Cui, S. Mao, G. Lu and J. Chen. 2012. Modulating gas sensing properties of CuO nanowires through creation of discrete nanosized p-n junctions on their surfaces. ACS Appl. Mater. Interfaces, 4: 4192–4199. doi: 10.1021/am300911z.

[77] Chung, F., R. Wu and F. Cheng. 2014. Fabrication of a Au @SnO$_2$ core–shell structure for gaseous formaldehyde sensing at room temperature. Sensors Actuators B. Chem., 190: 1–7. doi: 10.1016/j.snb.2013.08.037.

[78] Van Hieu, N., H. Kim, B. Ju and J. Lee. 2008. Enhanced performance of SnO$_2$ nanowires ethanol sensor by functionalizing with La$_2$O$_3$. Sensors Actuators B. Chem., 133: 228–234. doi: 10.1016/j.snb.2008.02.018.

[79] Navazani, S., M. Hassanisadi, M.M. Eskandari and Z. Talaei. 2020. Design and evaluation of SnO$_2$-Pt/MWCNTs hybrid system as room temperature-methane sensors. Synth. Met., 260: 116267. doi: 10.1016/j.synthmet.2019.116267.

[80] Wang, J., Y. Zheng, J. Wang and P. Yao. 2011. Formaldehyde sensing properties of electrospun NiO-doped SnO$_2$ nanofibers. Sensors Actuators B. Chem., 156(2): 723–730. doi: 10.1016/j.snb.2011.02.026.

[81] In-Sung, H., C. Joong-Ki, W. Hyung-Sik, K. Sun-Jung, J. Se-Yeon, S. Tae-Yeon, K. Il-Doo and L. Jong-Heun. 2011. Facile control of C$_2$H$_5$OH sensing characteristics by decorating discrete Ag nanoclusters on SnO$_2$ nanowire networks. ACS Appl. Mater. Interfaces, 3: 3140–3145. doi: 10.1021/am200647f.

[82] Zhilong, S., W. Zeru, W. Baocun, L. Zhen, X. Songman, Z. Wenkai, Y. Haoxiong, L. Min, H. Zhao, Z. Jianfeng, Y. Fei and L. Huan. 2016. Sensitive room-temperature H$_2$S gas sensors employing SnO$_2$ quantum wire/reduced graphene oxide nanocomposites. Chem. Mater., 28(4): 1205–1212. doi: 10.1021/acs.chemmater.5b04850.

[83] Wang, J., P. Zhang, J.Q. Qi and P.J. Yao. 2009. Silicon-based micro-gas sensors for detecting formaldehyde. Sensors Actuators, B Chem., 136(2): 399–404. doi: 10.1016/j.snb.2008.12.056.

[84] Van Hieu, N., L.T.B. Thuy and N.D. Chien. 2008. Highly sensitive thin film NH_3 gas sensor operating at room temperature based on SnO_2/MWCNTs composite. Sensors Actuators, B Chem., 129(2): 888–895. doi: 10.1016/j.snb.2007.09.088.

[85] Tian, S., F. Yang, D. Zeng and C. Xie. 2012. Solution-processed gas sensors based on ZnO nanorods array with an exposed (0001) facet for enhanced gas-sensing properties. J. Phys. Chem. C, 116(19): 10586–10591. doi: 10.1021/jp2123778.

[86] Chung, F.C., Z. Zhu, P.Y. Luo, R.J. Wu and W. Li. 2014. Au@ZnO core-shell structure for gaseous formaldehyde sensing at room temperature. Sensors Actuators, B Chem., 199: 314–319. doi: 10.1016/j.snb.2014.04.004.

[87] Yu, M.R., R.J. Wu and M. Chavali. 2011. Effect of 'Pt' loading in ZnO-CuO hetero-junction material sensing carbon monoxide at room temperature. Sensors Actuators, B Chem., 153(2): 321–328. doi: 10.1016/j.snb.2010.09.071.

[88] Brigida, A., L.M. Maria, P. Tiziana, M. Ettore, B. Alfonsina, D.F. Girolamo and D.V. Paola. 2019. Improvement of NO_2 detection: Graphene decorated with ZnO nanoparticles. IEEE Sens. J., 19(19): 8751–8757. doi: 10.1109/JSEN.2019.2922412.

[89] Chao, W., Z. Long, H. Hui, X. Rui, J. Da-Peng, Z. Shao-Hui, W. Lu-Jia, C. Zi-Yang and P. Ge-Bo. 2019. A nanocomposite consisting of ZnO decorated graphene oxide nanoribbons for resistive sensing of NO_2 gas at room temperature. Microchim. Acta, 186(8). doi: 10.1007/s00604-019-3628-x.

[90] Das, S.N., J.P. Kar, J.H. Choi, T. Lee, K.J. Moon and J.M. Myoung. 2010. Fabrication and characterization of ZnO single nanowire-based hydrogen sensor. J. Phys. Chem. C, 114(3): 1689–1693. doi: 10.1021/jp910515b.

[91] Iftekhar Uddin, A.S.M., D.T. Phan and G.S. Chung. 2015. Low temperature acetylene gas sensor based on Ag nanoparticles-loaded ZnO-reduced graphene oxide hybrid. Sensors Actuators, B Chem., 207: 362–369. Part A, doi: 10.1016/j.snb.2014.10.091.

[92] Liang, X., T. Zhong, B. Quan, B. Wang and H. Guan. 2008. Solid-state potentiometric SO_2 sensor combining NASICON with V_2O_5-doped TiO_2 electrode. Sensors Actuators, B Chem., 134(1): 25–30. doi: 10.1016/j.snb.2008.04.003.

[93] Acharyya, D. and P. Bhattacharyya. 2016. Highly efficient room-temperature gas sensor based on TiO_2 nanotube-reduced graphene-oxide hybrid device. IEEE Electron Device Lett., 37(5): 656–659. doi: 10.1109/LED.2016.2544954.

[94] Peiguang, H., D. Guojun, Z. Weijia, C. Jingjie, L. Jianjian, L. Hong, L. Duo, W. Jiyang and C. Shaowei. 2010. Enhancement of ethanol vapor sensing of TiO_2 nanobelts by surface engineering. ACS Appl. Mater. Interfaces, 2(11): 3263–3269. doi: 10.1021/am100707h.

[95] Karthik, P., P. Gowthaman, M. Venkatachalam and A.T. Rajamanickam. 2020. Propose of high performance resistive type H_2S and CO_2 gas sensing response of reduced graphene oxide/titanium oxide (rGO/TiO_2) hybrid sensors. J. Mater. Sci. Mater. Electron., 31(4): 3695–3705. doi: 10.1007/s10854-020-02928-4.

[96] Fort, A., E. Panzardi, A. Al-hamry, V. Vignoli and M. Mugnaini. 2019. Highly sensitive detection of NO_2 by Au and TiO_2. Sensors, (2): 1–17. doi:10.3390/s20010012.

[97] Liang, S., J. Li, F. Wang, J. Qin, X. Lai and X. Jiang. 2017. Highly sensitive acetone gas sensor based on ultrafine α-Fe_2O_3 nanoparticles. Sensors Actuators, B Chem., 238: 923–927. doi: 10.1016/j.snb.2016.06.144.

[98] Ali, M., J. Kamal, H. Babak, B. Anna, B. Maryam, G.L. Salvatore and N. Giovanni. 2015. Synthesis, characterization and gas sensing properties of ag@α-Fe_2O_3 core–shell nanocomposites. Nanomaterials, 5(2): 737–749. doi: 10.3390/nano5020737.

[99] Kim, W., J.S. Lee and J. Jang. 2018. Facile synthesis of size-controlled Fe_2O_3 nanoparticle-decorated carbon nanotubes for highly sensitive H_2S detection. RSC Adv., 8(56): 31874–31880. doi: 10.1039/c8ra06464d.

[100] Sen, T., N.G. Shimpi, S. Mishra and R. Sharma. 2014. Polyaniline/γ-Fe_2O_3 nanocomposite for room temperature LPG sensing. Sensors Actuators, B Chem., 190: 120–126. doi: 10.1016/j.snb.2013.07.091.

[101] Chen, W., C. Xiaoyang, Z. Xin, S. Peng, H. Xiaolong, S. Kengo, L. Geyu and Y. Noboru. 2014. Hierarchical α-Fe₂O₃/NiO composites with a hollow structure for a gas sensor. ACS Appl. Mater. Interfaces, 6(15): 12031–12037. doi: 10.1021/am501063z.

[102] Zhang, H., L. Yu, Q. Li, Y. Du and S. Ruan. 2017. Reduced graphene oxide/α-Fe₂O₃ hybrid nanocomposites for room temperature NO₂ sensing. Sensors Actuators, B Chem., 241(2): 109–115. doi: 10.1016/j.snb.2016.10.059.

[103] Nie, Q., Z. Pang, H. Lu, Y. Cai and Q. Wei. 2016. Ammonia gas sensors based on In₂O₃/PANI hetero-nanofibers operating at room temperature. Beilstein J. Nanotechnol., 7(1): 1312–1321. doi: 10.3762/bjnano.7.122.

[104] Shruthi, J., N. Jayababu, P. Ghosal and M.V. Ramana Reddy. 2019. Ultrasensitive sensor based on Y₂O₃-In₂O₃ nanocomposites for the detection of methanol at room temperature. Ceram. Int., 45(17): 21497–21504. doi: 10.1016/j.ceramint.2019.07.141.

[105] Pashchanka, M., J. Schneider and A. Gurlo. 2012. Inkjet printed In₂O₃ and In₂O₃/CNT hybrid microstructures for future gas sensing application. 14th Int. Meet. Chem. Sensors, pp. 791–794. doi: 10.5162/IMCS2012/P1.0.14.

[106] Liu, B., Y. Liu, H. Li and Q. Li. 2013. High-performance room-temperature hydrogen sensors based on combined effects of Pd decoration and Schottky barriers. Nanoscale, 5: 2505. doi: 10.1039/c3nr33872j.

[107] Jie, L., L. Shan, Z. Bo, W. Yinglin, G. Yuan, L. Xishuang, W. Yue and L. Geyu. 2017. Flower-like In₂O₃ modified by reduced graphene oxide sheets serving as a highly sensitive gas sensor for trace NO₂ detection. J. Colloid Interface Sci., 504(2): 206–213. doi: 10.1016/j.jcis.2017.05.053.

[108] Inyawilert, K., A. Wisitsoraat, C. Liewhiran, A. Tuantranont and S. Phanichphant. 2019. H₂ gas sensor based on PdOx-doped In₂O₃ nanoparticles synthesized by flame spray pyrolysis. Appl. Surf. Sci., 475: 191–203. December 2018, doi: 10.1016/j.apsusc.2018.12.274.

[109] Ma, Z.H., R.T. Yu and J.M. Song. 2020. Facile synthesis of Pr-doped In₂O₃ nanoparticles and their high gas sensing performance for ethanol. Sensors Actuators, B Chem., 305: 127377. doi: 10.1016/j.snb.2019.127377.

[110] Ke, M.-T., M.-T. Lee, C.-Y. Lee and L.-M. Fu. 2009. A MEMS-based benzene gas sensor with a self-heating WO₃ sensing layer. Sensors, 9(4): 2895–2906. doi: 10.3390/s90402895.

[111] Kim, S., S. Park, S. Park and C. Lee. 2015. Acetone sensing of Au and Pd-decorated WO₃ nanorod sensors. Sensors Actuators, B Chem., 209: 180–185. doi: 10.1016/j.snb.2014.11.106.

[112] Sikai, Z., S. Yanbai, Z. Pengfei, Z. Xiangxi, H. Cong, Z. Qiang and W. Dezhou. 2019. Design of Au@WO₃ core–shell structured nanospheres for ppb-level NO2 sensing. Sensors and Actuators, B: Chem., 282: 917–926. doi: 10.1016/j.snb.2018.11.142.

[113] Yinglin, W., L. Jie, C. Xiaobiao, G. Yuan, M. Jian, S. Yanfeng, S. Peng, L. Fengmin, L. Xishuang, Z. Tong and L. Geyu. 2017. NH₃ gas sensing performance enhanced by Pt-loaded on mesoporous WO₃. Sensors Actuators, B Chem., 238: 473–481. doi: 10.1016/j.snb.2016.07.085.

[114] Shi, J., Z. Cheng, L. Gao, Y. Zhang, J. Xu and H. Zhao. 2016. Facile synthesis of reduced graphene oxide/hexagonal WO₃ nanosheets composites with enhanced H₂S sensing properties. Sensors Actuators, B Chem., 230: 736–745. doi: 10.1016/j.snb.2016.02.134.

[115] Le, X.V., V.T. Duong, L. Anh, L. Thi, V.T. Pham and H. Lam. 2019. Composition of CNT and WO₃ nanoplate: Synthesis and NH₃ gas sensing characteristics at low temperature. 29(4): 61–68. doi: 10.14456/jmmm.2019.48.

[116] Tao, Z., R. Yuan, J. Guangyou, Z. Yuye, F. Yuchi, Y. Jianping, Z. Xin, J. Wan, W. Lianjun and L. Wei. 2019. Facile synthesis of mesoporous WO₃@graphene aerogel nanocomposites for low-temperature acetone sensing. Chinese Chem. Lett., 30(12): 2032–2038. doi: 10.1016/j.cclet.2019.05.006.

CHAPTER 11

Chemical Route Synthesis and Properties of CZTS Nanocrystals for Sustainable Photovoltaics

Shefali Jain, Pooja Semalti, Vidya Nand Singh and
Shailesh Narain Sharma*

1. Introduction

1.1 Foundation of Photovoltaic Solar Cell

Huge amount of energy demand and environmental concerns for our earth (such as large demand of fossil fuels for energy production, climate change, carbon emission, etc.) forces us to use renewable and clean sources of energy [1]. Sun provides us a clean and inexhaustible source of energy [2]. Photovoltaic (PV) cells made up of semiconductor materials have the capability to capture and convert this solar energy to electricity through an electronic process, as shown in Figure 11.1. Energy received from the sun frees the electrons in the semiconductor materials. These charges are then forced to travel through an electrical circuit for driving different electrical devices. The PV effect was first observed by Alexandre Edmund Becquerel in 1839. Bell Labs situated in the U.S. were the first who introduced the solar cells for some scientific and commercial applications [3].

1.2 Efficiency of Solar Cells

The maximum efficiency of any solar cell is the maximum power that can be collected from the cell. The maximum theoretical limit of an efficiency was

CSIR-National Physical Laboratory (NPL), Dr. K.S. Krishnan Road, New Delhi 110012, India, and Academy of Scientific and Innovative Research (AcSIR), Council of Scientific & Industrial Research (CSIR)-Human Resource Development Centre, (CSIR-HRDC) Campus, Ghaziabad, Uttar Pradesh 201002, India.
Emails: jshefali85@yahoo.com, phynobel9@gmail.com, singhvn@nplindia.org
* Corresponding author: shailesh@nplindia.org; snsharma99@gmail.com

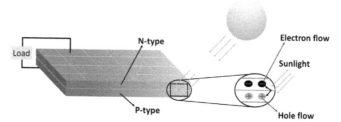

Figure 11.1. Working of photovoltaic device.

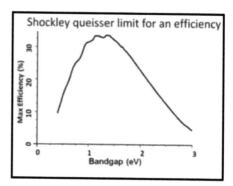

Figure 11.2. Shockley-Queisser limit for the efficiency of a solar cell.

determined in 1961 by William Shockley and Hans Queisser. It was found that for a single p-n junction having a band gap of 1.34 eV, the maximum limit for solar conversion efficiency (under 1000 W/m², AM 1.5, Figure 11.2) is around 33.7% [4]. The presently commercialized silicon-based solar cell, having a band gap of 1.1 eV, has a theoretical maximum efficiency of 29%. However, the practically achieved efficiency for these mono-crystalline silicon solar cells is about 25% due to different losses, mainly by reflection off the front surface and light blockage. According to the Shockley Queisser (SQ) limit, the maximum efficiency of CZTS-based photovoltaic materials would be ~ 32.2% [5]. This concept is valid only for a single p-n junction cell, however, the cell having multiple p-n junctions can outperform this limit with a maximum efficiency limit of 86% [6].

1.3 Future Photovoltaic Cell

The commercialized solar cells available in the market are based on crystalline silicon and thin-film semiconductor materials. Silicon solar cells are highly efficient, having higher manufacturing costs, however, thin-film solar cells have a low manufacturing cost but lower efficiencies. Other types of solar cells, such as multi-junction tandem cells, have specialized applications in satellites and military applications due to their light weight and very high efficiencies. The present and future scenarios of the efficiencies for different type of solar cells are shown in Figure 11.3. On the basis of the different materials used, solar cells are divided into three different generations.

Best Research-Cell Efficiencies

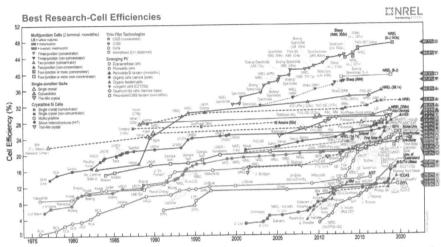

Figure 11.3. Cell efficiency levels for different generation photovoltaic cells (Single junction wafer Si approaching practical performance limits and challenge facing thin films: closing the gap between laboratory and production devices). Source: National Renewable Energy Laboratory (NREL), July 2018.

1.3.1 First Generation Photovoltaic Solar Cell

This is the oldest form of solar cells using single/multi crystalline silicon wafers as an absorber material. They have a payback time of 1–2 years. Presently, these types of solar cells are used for both domestic and commercial purposes. They are the most popular as they have a good performance (25%) with high stability. However, the production of these types of solar cells requires high cost and lot of energy.

1.3.2 Second Generation Photovoltaic Solar Cell (Thin-Film Technology)

These types of solar cells have lower efficiency (15–22%) as compared to the first generation. However, the production cost for this thin film technology is lower. These thin film technology-based solar cells use amorphous silicon, CdTe, and CIGS/CZTS inks as an absorber material. They have many advantages, such as low materials consumption, low production cost, and production on flexible substrates (Figure 11.4). However, the production of these type of solar cells includes mainly

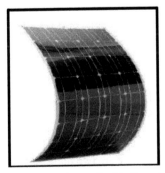

Figure 11.4. Flexible substrate for thin film solar cell [8].

vacuum synthesis at high temperature. Researchers are working on non-vacuum processes for their production.

1.3.3 Third Generation Photovoltaic Solar Cell

These types of solar cells are most efficient solar cells which use organic materials, such as polymers, as active materials. The multijunction tandem cells having high efficiency also belong to this category, however, they are very expensive. The thin film photovoltaic cells based on low bandgap perovskite materials are another category of solar cells gaining the attention of the scientific community. This type of solar cell has many advantages, such as being simple, quick, low cost, flexible, and having a large-scale roll-to-roll production for a commercial purpose. However, they suffer from low efficiency and high instability.

2. Introduction to CZTS Nanoparticles

2.1 Brief Summary of Properties of CZTS

Photovoltaic electricity generation based on silicon technology with a record efficiency of 25% does not fulfill the energy demands of the earth [9]. To overcome the high cost of silicon wafers, the materials such as CdSe, CdTe, $CuInSe_2$, $CuInGa(SSe)_2$, CIGS offering stable and efficient ($\sim$ 22%) [9] photovoltaic modules (second generation photovoltaic materials) have been studied. However, the availability and expensiveness of indium is also one of the concerns. To meet the increased demands of energy, the major issues, such as increasing the use of the traditionally available materials, using cost effective methods, and searching for other alternatives for both the materials and cost-effective methods maintaining the eco-friendly environments need to be addressed. Hence, it is necessary to develop the solar cell with high efficiency, long term stability, less environmental damaging, and lowest possible cost. The quaternary compounds, such as Cu_2ZnSnS_4 (CZTS), $Cu_2ZnSnSe_4$ (CZTSe), and $Cu_2ZnSn(Se_xS_{1-x})_4$ (CZTSSe) have attracted escalating consideration of researchers for photovoltaic devices [10]. They have very good properties, such as P-type conductivity, ideal direct bandgap, and large absorption coefficient for photoactive applications. Among these, the CZTS absorber layer is of immense importance because of its direct and low band gap of 1.4–1.5 eV with a higher absorption coefficient of 10^4 cm^{-1} [11]. The availability of the constituent elements of these particles is $\sim$ 50, 75, 2.2, and 260 ppm, respectively, as compared to indium, which is $\sim$ 0.049 ppm only [12]. Due to the toxic constituents, such as Cd and low abundance of In and Te, these quaternary chalcopyrite materials act as an absorber layer by replacing half with tin and half with zinc. Therefore, CZTS compound has been studied as a new developing absorber material in inorganic, low-cost solar cell devices.

2.2 Structural Analysis of CZTS

CZTS is a I_2–II–IV–V_4 quaternary chalcopyrite semiconductor compound. On the basis of crystal modifications, there are two main tetragonal structures: kesterite type

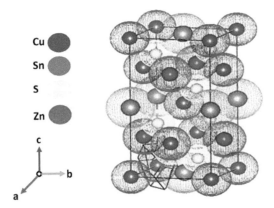

Figure 11.5. Schematic representations of the kesterite CZTS structure.

and stannite type [13]. Both the kesterite and stannite structures constitute a cubic closed packing arrangement of the anion (S), and cations (Cu^+, Zn^{2+}, Sn^{4+}), such that cations occupy 50% tetrahedral voids. Thus, due to different positions of the cations present in the structure, they have different space groups.

Kesterite are the minerals with $AI_2BIICIVXVI_4$ structure (with A = Cu, B = Zn, Fe, C = Sn, and X = S, Se) which are found in nature with iron ($Cu_2(Zn, Fe)$ SnS_4) content in them [14]. The kesterite phased CZTS is mostly found because of its higher thermodynamic stability, i.e., lower energy as depicted by first principle calculations by Chen et al. as compared to stannite phased CZTS [15, 16].

The kesterite structure is constituted of alternate layers of cations Sn and Zn coordinated with Cu along z axis at z = 0, 1/4, ½, and ¾ respectively (Figure 11.5). Thus, in kesterite phase CZTS with space group $I4^-$ the coordinates of Cu with Zn are 2a (0,0,0), while another Cu was arranged at 2c (0,1/2,1/4) and 2d (0,1/2,3/4), while Sn occupies the 2b site (0, 0, 1/2). The S anions occupies 8g (x,y,z) in the (110) mirror plane [17, 18].

There exists one more type of structure for CZTS, i.e., wurtzite structure. In this structure, Zn is replaced by Zn, Cu, and Sn in wurtzite ZnS, such that each sulfur atom is coordinated by one Zn, two Cu, and one Sn atom. It is just a cationic disorder in kesterite CZTS structure.

2.3 Pure Phase Engineering and Defect States of CZTS

The valance band consists of a linear combination of S-3p and Cu-3d (antibonding). However, the conduction band is constituted of an isolated energy band made by S-3p and Sn-5s states [17, 19]. They explain the most stable symmetry in case of kesterite phase CZTS having a high absorption coefficient in visible spectrum, which is dependent on dielectric function. Neutron powder diffraction by Schorr et al. determines that Cu_2ZnSnS_4 crystallizes in the most stable kesterite structure. They have low formation energy with point defects Zn_{Cu} and Cu_{Zn} anti-sites in the Cu-Zn (001) plane at Z = 1/4 and ¾ which makes CZTS p-type [20].

In literature, the CZTS synthesized in Zn-rich and Cu-poor configurations was observed to have a high efficiency solar cell, as it makes CZTS more p-type. However, the clear definition of Cu-poor and Zn-rich is not defined. Thus, it was found that for the higher efficiency solar cell, the CZTS nanoparticles must have a configuration such that Cu/(Zn+Sn) and Zn/Sn should lie in the ranges of 0.7–0.9 and 1.1–1.3, respectively. As confirmed by Fairbrother et al., the shaded area in Figure 11.6 indicates the CZTS nanoparticles with the compositions such that Cu/Sn = 1.7–1.9, Cu/Zn = 1.3–1.7, and Zn/Sn = 1.1–1.3, in the range such that the device efficiency is higher [21]. Figure 11.6 shows the possible range for the chemical potential resulting into the formation of quaternary semiconductor phase of Cu_2ZnSnS_4, which is very narrow and is stable [22].

Thus, it is very difficult to fabricate single-phase Cu_2ZnSnS_4. By the ternary phase equilibrium of $Cu_2S–ZnS–SnS_2$ [23], it was found that with a slight variation in the composition of input precursors, reaction time, and temperature, the formation of binary and ternary compounds is also possible during CZTS synthesis. These secondary phase formations during the synthesis of CZTS may lead to poor and low efficiency device quality [24, 25]. Cu-poor and Zn-rich composition results in ZnS phases many times, having a very high bandgap, resulting in insulator regions in the device, thus reducing the device performances [26].

However, these binary phases may also result in the evolution of CZTS nanocrystals depending on the reaction conditions. Guo et al. described the final formation of CZTS nanoparticles from the spherical shaped Cu_2S nanoparticles to pure rice-like wurtzite phased CZTS nanoparticles. The authors of the chapter also studied the development of kesterite phased rod-shaped CZTS nanoparticles from Cu_2S nanoparticles (shown in Figure 11.7), which involves three steps: (1) The rapid nucleation of Cu_2S, (2) The asynchronous diffusion of zinc (Zn) and tin (Sn) with simultaneous growth and shape transformation, and finally, (3) conversion to pure CZTS NCs was obtained. Due to similar anionic lattice structures of both Cu_2S and CZTS, the inter-diffusion of cations take place at high reaction temperatures [11, 27].

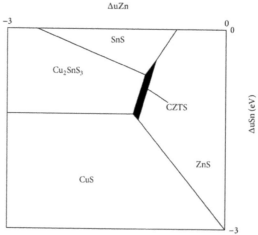

Figure 11.6. The chemical potential phase diagram of Cu_2ZnSnS_4 under Cu-rich conditions [22].

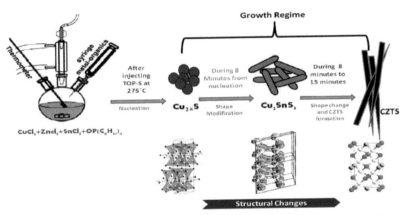

Figure 11.7. Schematic showing the development of CZTS nanorods from Cu$_2$S nanoparticles at a molecular level [27].

However, different reaction conditions, such as temperature, various surfactants, etc. lead to different phases explained further in the chapter. Furthermore, Zhang et al. found that wurtzite phase of CZTS can be formed by partial replacing of Zn sites in ZnS by Cu and Sn atoms [18].

3. Synthesis of CZTS Nanoparticles

CZTS thin films fabrication process can be done by both vacuum and non-vacuum method, but it was observed during the literature survey that the efficiency of solar cells based on CZTS-family (CZTSSe 12.6% using chemical route) [91] thin film synthesized by non-vacuum technique have obtained the highest efficiency. Here, authors are discussing different methods for the fabrication of CZTS thin film for the introduction, however, the chemical route has been discussed in detail (summarized in Figure 11.8).

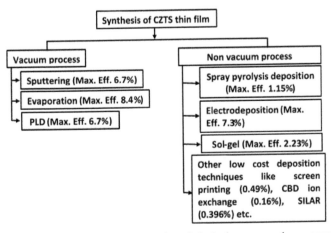

Figure 11.8. The efficiency for CZTS thin film solar cells by both vacuum and non-vacuum methods.

3.1 Vacuum-Based Deposition

The vacuum-based techniques involve the efficient control over the synthesis parameters, such as temperature and pressure, resulting in controlled chemical composition and phase of the samples with good reproducibility. The vacuum-based methods mainly involve sputtering, evaporation, and PLD based techniques. In these techniques, the constituent precursors of the CZTS nanoparticles were deposited under various temperature and pressure conditions.

3.1.1 Sputtering Techniques

This technique involves many methods used by different researchers, such as Ar beam sputtering by Ito and Nakazawa in 1988 [28], DC-magnetron and RF magnetron sputtering [29, 30], ion beam sputtering [31], pulsed laser deposition [32], atom beam sputtering [33], and many more. The first CZTS solar cell with an efficiency of 3.7% based on sputtering method was reported [34]. After that, Chalapathy et al. reported a CZTS-based solar cell with a conversion efficiency of 4.59% by DC sputtering [35]. Katagiri et al. also prepared the solar with CZTS thin films as an acceptor layer with an efficiency of 5.74% using Cu, ZnS, and SnS nanomaterials as precursors [36]. However, the CZTS solar cell with an efficiency of 6.77% was established in 2008 using this technique [37, 38].

3.1.2 Evaporation Techniques

This technique involves co-evaporation, fast evaporation, electron beam (EB), and thermal evaporation techniques for the deposition of CZTS thin films [39, 40, 41, 42, 43, 13]. It involved two different approaches, first the simultaneous deposition or sequential deposition of all precursors, followed by the second step, called sulfurization or annealing.

The first CZTS-based inorganic PV device with PEC 0.66% was designed using EB evaporation method by Katagiri et al. in 1997. However, the device synthesized by sequential deposition of precursors showed a conversion efficiency of 2.62% by the same group in 2001, followed by vapor-phase sulfurization [39]. Kobayashi et al., who followed a similar technique of sulfurization at 520°C, observed an efficiency of 4.53% [40]. Schubert et al. observed a maximum PEC of 4.1% in a device synthesized based on the co-evaporation process [43]. However, Shin et al. observed that the device based on thermal evaporation method showed a maximum efficiency of 8.4% [13].

3.1.3 PLD Technique

This technique has many advantages, such as consistent and uniform deposition on the target, increased crystallinity, clean deposition because of inert atmosphere, simplicity, and flexibility [44, 45]. The quality of the deposited film depends on many factors, such as pulse-rate, distance, and length between substrate and target temperature and substrate orientation. The literature survey shows that the small area deposition (~ 1 cm^2) is the only disadvantage in using this method.

Sekiguchi et al. reported the first PLD-based solar device having CZTS thin films on GaP substrate [46]. Moholkar et al. observed the maximum efficiency of 4.1% for CZTS-based solar cell device using this technique [47, 13].

3.2 Non Vacuum-Based Deposition

The large production of the low cost and vacuum-free production of photovoltaic device involves various techniques for deposition, such as chemical bath deposition [48], hydrazine solution preparation and spin coating [49], electroplating [50], spray pyrolysis [51], ultrasonic spray pyrolysis [USP] setup [52], sol-gel [53], and electrodeposition [54].

The vacuum-based methods are expensive and suffer low materials utilization, and consumption of large energy. However, non-vacuum-based methods are cheap, have less energy consumption, and provide good and uniform thin films over a large area. This technique can also incorporate flexible substrate for thin film deposition. Thus, the successful low cost, non-vacuum deposition techniques, in combination with the non-toxic and earth abundant constituents of CZTS, lead to the unparalleled growth in photovoltaic technology.

3.2.1 Spray Pyrolysis Technique

In this method, the target surface is heated to a high temperature of about 600°C, and then the precursor metal salts solutions are sprayed onto the heated surface, which causes the pyrolysis of the sprayed material, and the formation of a thin film coat on the surface. The structural and optical properties of the film depend highly on the temperature of the target surface, and it was found to have better quality in the range of 500°C–650°C in pyrolysis.

This technique was first used by Nakayama and Ito in 1996 for the stannite-phased CZTS films [55]. Kamoun et al. followed a similar technique for the preparation of kesterite-phased CZTS. However, secondary phases of CuS and Cu_2S were observed, which were removed by post annealing and sulfurization [56]. Furthermore, the same technique can also be used without annealing or post-deposition sulfurization to obtain single-phased CZTS [57].

In spite of advantages, such as low cost and vacuum-free technique, this technique cannot be scaled for industrial applications, which is the biggest disadvantage with this method.

3.2.2 Electrodeposition Technique

It is a non-vacuum low-cost method which can also be used for large scale production in industries. It involves sequential or 1-step electrodeposition of precursors, followed by sulfurization/annealing. Scragg et al. first fabricated the solar cell using CZTS as thin films using this technique, and got an efficiency of 3.2% with some improvements in annealing conditions [58, 59]. However, Ennaoui et al. observed a PEC of 3.4% using the same technique for copper-poor CZTS material as an absorber layer in 2009 [60]. The solar cell developed by electrodeposited CZTS got a record efficiency of 7.3% fabricated by Deligianni et al., which includes metal stacking using electrodeposition technique, followed by low temperature annealing in N_2, and sulfur vapor producing highly crystalline CZTS phase [61].

3.2.3 Sol-gel Deposition Technique

It is also a very simple and low-cost technique based on spin coating. The first device fabricated by Tanaka et al. based on spin-coated CZTS layer shows the conversion efficiency of 1.01%. The same group further improves the conversion efficiency to 2.03% by improving the synthesis condition of CZTS, which involves annealing of CZTS precursor film in sulfur environment at 500°C for 1 hour [62, 63, 64, 53].

3.2.4 Colloidal Route Technique

CZTS can also be synthesized by the colloidal route. This method develops CZTS nanoparticles in the pure kesterite phase with a good yield. The main advantage of this method is that it does not require post synthesis sulfurization/selenization.

The CZTS nanocrystals were synthesized firstly by Guo et al. by the hot injection technique using Olylamine as a solvent containing 1.5 mmol copper (II) acetylacetonate, 0.75 mmol of zinc acetylacetonate, and 0.75 mmol of tin(IV) bis(acetylacetonate) dibromide in oleylamine at 225°C. Xin et al. fabricated a dye sensitized solar cell using CZTS nanoparticles synthesized by the above method, and observed to have a maximum efficiency of 7.37% [65]. Our group also synthesized kesterite-phased CZTS nanoparticles using hot-injection technique using TOPO (Tri-Octyl Phosphine Oxide), and studied the formation of CZTS from Cu_2S nanoparticles [27].

3.2.5 Other Non-Vacuum Technique

Zhou et al. reported a screen printing (SP) approach for the CZTS-based solar cell preparation of 0.49% efficiency [66].

The solvothermal route is another non-vacuum method as introduced by Cao and Shen to synthesize high quality CZTS nanoparticles at 180°C in an autoclave with a Teflon lining [67]. Successive ionic layer adsorption and reaction (SILAR) technique is another good chemical route method, as reported by Shinde et al., for the formation of CZTS thin films by sequential reaction on SLG substrate surface [68]. The maximum efficiency of 1.85% was reported by Mal et al. using wet chemistry (SILAR) technique [69]. Recently, another novel technique known as an open atmosphere chemical vapor deposition (OACVD) was reported by Washio et al. It includes the synthesis of CZTS nanoparticles using oxide precursors. CZTS-based best solar cells yielded an efficiency of 6.03%.

In all the synthesis methods, the morphology and size of CZTS nanoparticles depends on many factors, such as the surfactant used, the reaction time, atomic precursors, etc. It has been found in literature that with change in reaction time, different shapes of CZTS nanoparticles, such as sphere, rhombus, and rice can be obtained due to different kinetical and thermodynamically favorable environments [70]. Here in this chapter, we will focus on hot-injection, colloidal route synthesis method of CZTS nanoparticles. The authors also synthesized various shaped

CZTS nanoparticles by a change in different capping ligands by hot-injection technique [71].

4. Colloidal Route Synthesis

Synthesis of monodispersed nanoparticles is also very important, as the properties of nanoparticles strongly depend upon their size and shape. In colloidal route synthesis, various properties, such as shape, size, and surface of nanoparticles can be tuned or varied depending on different parameters, such as reaction time, capping ligand, temperature, etc. [72]. So, using this non-vacuum, low cost technique, both the quality and properties of the nanoparticles can be varied. However, understanding the formation mechanism, including nucleation and growth of monodisperse nanocrystals, is important.

Figure 11.9 shows a scheme depicting various parameters influencing the reaction, such as composition of precursors, which consists of the constituting elements of nanoparticles, and the capping ligands which control the growth of the nanoparticles by preventing their growth to bulk material and agglomeration, respectively, temperature and duration of synthesis controlling the size and monodispersity of nanoparticles, etc. All the parameters mentioned above highly influence the formation and properties of the obtained nanocrystals.

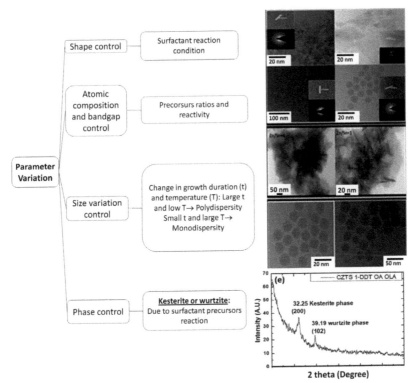

Figure 11.9. Scheme depicting various parameters influencing the synthesis chemical reaction.

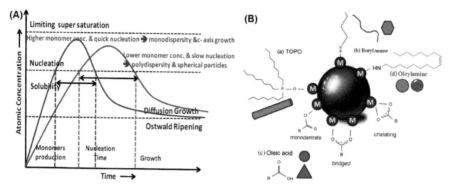

Figure 11.10. (a) LaMer's Plot showing the effect of nucleation and development of monomers on the growth of nanoparticles. (b) Schematic showing the effect of different surfactants producing varying shaped nanoparticles due to their different binding sites [27].

The development mechanism for nanoparticles includes two steps: the nucleation (i.e., a formation of "nuclei or seeds") and the growth of nuclei to nanocrystals. The ligands play an important role in both stages. These surfactant molecules influence the development of "monomers" prior to the nucleation stage. The author studied the effect of various ligands (TOPO, Oleic acid, Butylamine, and oleyl amine) on the morphology and shape of CZTS nanoparticles, and obtained different shaped nanoparticles, as shown in Figure 11.10 [71].

As suggested by Yu et al., and also observed by authors, the nucleation stage controls the monomers' activity by three effects [73, 71]. First, it is via the strong coordination bond between surfactant and monomers, leading to the reduction of their reactivity and their concentration. Next, the length of hydrocarbon chains in ligand molecule also affects the reactivity, as the long hydrocarbon chains decrease the reactivity of monomers.

Lastly, the concentration of surfactant also affects the reactivity, as the increased concentration of surfactant suppresses the reactivity. The length and the concentration of the surfactants also constitute the stearic hindrance effect. The low reactivity of monomer leads to a lower monomer concentration and reduced number of nuclei. The temperature of the reaction affects the stability and constituents of the complex at the nucleation stage. However, in the growth stage, the shape of nanocrystals is affected by the ligand molecules. The higher monomer activity and remaining concentration of monomers, with a low nuclei concentration results in elongated nanocrystals. This stage is known as "defocusing stage".

After the nucleation time, growth time starts influencing the size of nanoparticles. The desired size can be obtained by quenching the reaction at the required and optimized time by reducing the reaction temperature or by separating the as-synthesized nanomaterial from the solution. The surfactant capped nanoparticles will settle down, while the side-products, unreacted precursors, remaining stabilizers, and the solvent remain in the supernatant. Thus, nanoparticles can be used in different applications after the washing, followed by synthesis of the nanoparticles.

In summary, by the optimization of various reaction conditions as mentioned above, materials of desired shape and size with high quality can be produced.

4.1 Separation Between Nucleation and Growth

To obtain the monodisperse particle, the separation of nucleation and growth stage is very important. There are various techniques to separate the nucleation from the growth stages. The seed mediated growth method is used to obtain heterogeneous nucleation [72]. This method uses the already prepared material as nuclei, and the monomer formation takes place on its surface. The concentration of these monomers is kept low to avoid the homogeneous nucleation. However, this method can be used for the formation of homogeneous particles [74] as well as heterogeneous structures [75].

The homogeneous nucleation can be achieved using two methods- "hot-injection" [76] and "heating-up" [77], for the separation of nuclei formation and growth stages during their synthesis [72]. In the case of hot-injection, one of the precursors is in the solution while the other one is added or injected at a higher temperature to achieve the supersaturation. After injection at higher temperature, nucleation starts, which leads to a dramatic decrease in monomer concentration, followed by the growth process. However, this high temperature is the main disadvantage of this method, particularly for the industrial use [72].

However, in the heating-up method, the metal salts are added initially and the temperature is increased. The heating results in the decomposition of the precursors, which increases the monomer concentration till the supersaturation state has been achieved, followed by the growth [78]. The size distribution of as-prepared nanoparticles is similar in both the methods [79]. However, this method is easy and gives a high yield, which is advantageous for large-scale production.

4.2 Shape Control of Nanoparticles

Shape control is an important parameter during the synthesis of nanoparticles, which can be achieved during nucleation stage only. The lattice nuclei structure determines the growth direction that can be altered by a change in synthesis temperature. Due to the large surface to volume ratio of nanoparticles, the shape of the particle depends on the material energies and specific facets [80, 81, 82]. For example, low temperature during synthesis leads to wurtzite-phased rod-shaped cadmium sulphide particles, however, high temperatures results in zincblende structure easing the formation of branched particles [83].

Surface energy is another crucial factor during growth state which results in the different facets with different energy. The facet with high energy grows faster as compared to the one with lower energy. This control is possible by using different stabilizers which can be adsorbed only on certain facets, and thus slows down the growth of that facet [84, 85, 86]. The variation of the stabilizer concentration also leads to diverse shapes [87]. The specific attachment of hexylphosphonic acid to specific surface leads to the growth of rod-shaped CdSe nanoparticles.

However, spherical-shaped CdSe nanoparticles can be formed in the absence of this surfactant [88].

Other than the stabilizers and the temperature control at the nucleation state, the concentration of monomers and temperature required during growth are other important factors to be considered during shape control, as it is a necessary condition for the thermodynamic control. However, the low concentration and high monomer concentration are important for the low temperature and high monomer concentration included in the kinetic control [89]. The effects of both thermodynamic and kinetic control can been seen during the synthesis of truncated cubes and nanorods of PbS, respectively, as shown in [81].

Another important factor that can control the shape of nanoparticles during synthesis process is the diffusion flux of monomers [90]. A higher rate of diffusion leads to one-dimensional nanoparticle, however, lower diffusion rate of monomers leads to three-dimensional growth of nanoparticles, which is also known as the ripening stage [89].

Thus, it can be concluded that any specific shapes of nanoparticles are attained mainly to minimize the surface energy and attain stability. The various factors controlling the shape of nanoparticles are—the lattice structure of the nuclei, growth rate, surface energy, surface-specific molecules, as well as the temperature, and can be controlled during the synthesis.

5. Application of CZTS Nanoparticles in Photovoltaics

5.1 Basic Inorganic Thin Film Solar Cell

The CZTS has been widely explored for second generation inorganic thin film solar cell, in which it has been used as a p-type absorber layer. The schematic of solar cell device fabricated using CZTS has been shown in Figure 11.11. The first layer is molybdenum (Mo) acting as back contact, with a thickness of ~ 1 μm, which is deposited using sputtering on the glass/soda lime glass substrate. Here, Mo is used because of its stability in extreme conditions, such as high temperature and sulfur mixed vapors. The second layer is the absorber layer of CZTS, having thickness in the range 1.0 to 2.0 μm, which was coated over the Mo. The next layer is the CdS thin film, which acts as a buffer layer of thickness 50–100 nm, which has been deposited over CZTS using chemical bath deposition technique to form the p-n junction. It is very difficult to obtain the uniform deposition of CdS over CZTS thin film having pin holes, which can lead to leakage. To prevent this shortage between front and back contact in the device, a layer of intrinsic ZnO (i-ZnO) of thickness

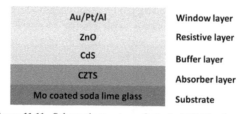

Figure 11.11. Schematic structure of a typical CZTS solar cell.

50 ~ 90 nm is sputter coated on CdS. Finally, the metal grids are deposited for the electrical measurements, i.e., the I–V characteristics of the device [1]. The efficiency of champion cell fabricated using kesterite semiconductors CZTSSe and CZTS have achieved the PCE of 12.6% and 8.4% via non-vacuum and vacuum-based methods, respectively [91, 92]. Recently, KIER reported an efficiency of 13.08% at PVSEC-36 at cell of area 0.181 cm² [93]. However, this value is very less than its Shockley Quisser theoretical limit (n ~ 32%). Researchers are working on this cell configuration, shown in Figure 11.11, by exploring alternative non-toxic buffer layers having low lattice mismatch with CZTS. Several combinations have been tried as the buffer (CdS, ZnO, ZnS and CeO_2)/CZTS interface to reduce the recombination effects [94, 95].

5.2 Hybrid Organic-Inorganic Photovoltaic

Hybrid organic inorganic polymer solar cells contain both organic (conjugated polymer) and inorganic (semiconductor nanostructures) components, as electron donor and electron acceptor, respectively. Researchers are attracted to these solar cells due to their eminent properties, such as light weight, cost-effectiveness, flexibility, and solution-processing. These devices have benefits of both organic and inorganic photovoltaic materials, and provide enhanced stability, electrical and charge transfer capabilities. The conjugate polymers act as donor-absorber material, and generate excitons which get dissociated at the donor-acceptor interfaces, thus transferring the charge particles by injecting the electrons into an inorganic semiconductor material, while the holes remaining in the polymer are collected by their respective collection electrodes. The organic solar cell (3rd generation) have attained the PCE of ~ 10%. Poly(3-hexylthiophene) (P3HT) and phenyl-C61-butyric acid methyl ester (PCBM) are the commonly used donor and acceptor materials, respectively, in OPV with the PCE of about 0.5 to 5.0%. However, they lack in their performance due to high instability. Many other inorganic semiconductor compounds have been incorporated as acceptor material in this configuration, such as metal-oxides (e.g., ZnO and TiO_2) [96, 97] and binary/ternary chalcogenides (e.g., CdS, CdSe, CdTe, PbS, FeS_2, or $CuInS_2$) [98, 97] to increase the stability and performance of the device. However, these materials suffer from many problems, such as wide band gap, toxicity, and costly constituents, as already mentioned in the chapter [70]. Recently, these materials are replaced by CZTS nanoparticles because of their low cost, non-toxic, and high abundant constituents. Researchers have started exploring as very few reports were found in the literature based on application of CZTS nanocrystals in the hybrid solar cell application. The CZTS/organic interface (PCBM) has been explored for the effect of surfactant removal using pyridine for high efficiency [99]. Cheng et al. in 2016 reported the use of solvothermaly synthesized CZTS/P3HT/PCBM blend as an active layer, and got an improved PCE from 3.30% to 3.65%.

Our group explored the kesterite-phased CZTS NPs synthesized by a hot-injection technique using three different S-sources (Thiourea, elemental-S, and 1-dodecanthiol). These NPs were blended with P3HT-PCBM to demonstrate the photovoltaic properties of CZTS NPs obtained using different precursors. The solar cells with a simple geometry of ITO/MoOx/P3HT:PCBM/Al were fabricated

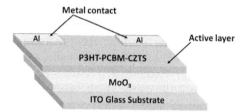

Figure 11.12. Schematic structure of hybrid organic-inorganic CZTS solar cell.

(Figure 11.12), and the overall enhancement of 34% efficiency was observed *(work under communication)*.

5.3 Tandem Solar Cell

Tandem solar cells are constituted of two or more materials having different energy band-gaps, which spans a wide range of the sun's spectra, and have an enhanced efficiency of solar cell devices [100, 101]. Figure 11.13 shows the device having two-junction CZTS/CZTSSe-based tandem structures. It consists of the upper cell with CZTS as an absorber layer and the bottom cell with CZTSSe as an absorber layer of a lower band gap of ~ 1.1 eV. The theoretical investigations for kesterite-based tandem cells assume the ideal tunnel junction with no electrical and optical losses between the cells, and thus, they are of great significance to understand both electrical and optical behavior. In the tandem cell, each cell is studied separately and then the V_{oc} of the complete cell can be estimated by adding individual voltages of constituting cells. The current is same for both the cells by applying the continuity of the current across the interface. Recently, an overall efficiency of ~ 22.5% has been achieved in this configuration [102].

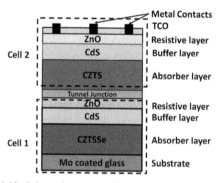

Figure 11.13. Schematic structure of CZTS/CZTSSe-based solar cell.

5.4 Perovskite Solar Cell

$CaTiO_3$ is the first invented perovskite oxide compound available in nature, found in the year 1839. It is used as an absorber layer in a solar device because of its high absorbance. The perovskite fabrication-based thin-film solar cell device is simple as compared to the first-generation silicon-based solar cell, as it requires more accuracy and cost for the fabrication process. The efficiency and stability of these solar cells can be further improved by using Pb-based quantum dots, such as $CsPbCl_3$:Mn. Qian et al. observed the increase of 3.34% in the overall efficiency of the cell [103]. The other category of these materials are the lead halide perovskites, having both organic and inorganic substances, i.e., metal cations and halide anion, respectively.

Indeed, perovskite materials have been explored by the scientific community over the last few years due to their enhanced power conversion efficiency of ~ 23.7% in photovoltaic devices. They have many prominent properties for their applications in photovoltaics, such as high absorbance, easy fabrication, high carrier mobility, and the wide absorption spectrum. However, these compounds suffer from toxicity due to the presence of lead and the long-term instability in the ambient environment because of its hygroscopic amine cations [104]. The stability and efficiency of the perovskite-based device can further be enhanced by incorporating the low-cost, stable inorganic compound CZTS as the hole transport layer in the device. Figure 11.14 shows the schematic of modified CZTS/perovskite cell with Au as top contact, while FTO acts as the bottom layer of the device. The good optical absorptions of both CZTS and perovskite materials increase the overall enhancement of the device efficiency.

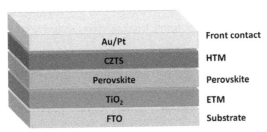

Figure 11.14. Schematic structure of perovskite/CZTS solar cell.

References

[1] Dhawale, D.S., A. Ali and A.C. Lokhande. 2019. Impact of various dopant elements on the properties of kesterite compounds for solar cell applications: a status review. Sustainable Energy Fuels, 3(6): 1365–1383.

[2] Hosenuzzaman, M., N.A. Rahim, J. Selvaraj, M. Hasanuzzaman, A.A. Malek and A. Nahar. 2015. Global prospects, progress, policies, and environmental impact of solar photovoltaic power generation. Renewable and Sustainable Energy Rev., 41: 284–297.

[3] Cotal, H., C. Fetzer, J. Boisvert, G. Kinsey, R. King, P. Hebert, H. Yoon and N. Karam. 2009. III–V multijunction solar cells for concentrating photovoltaics. Energy & Environ. Sci., 2(2): 174–192.

[4] Shockley, W. and H.J. Queisser. 1961. Detailed balance limit of efficiency of p-n junction solar cells. J. Appl. Phys., 32(3): 510–519.

[5] Pal, K., P. Singh, A. Bhaduri and K.B. Thapa. 2019. Current challenges and future prospects for a highly efficient (> 20%) kesterite CZTS solar cell: A review. Sol. Energy Mater. Sol. Cells, 196: 138–156.

[6] De Vos, A. 1980. Detailed balance limit of the efficiency of tandem solar cells. J. Phys. D: Appl. Phys., 13(5): 839.

[7] Khalate, S., R. Kate and R. Deokate. 2018. A review on energy economics and the recent research and development in energy and the Cu2ZnSnS4 (CZTS) solar cells: A focus towards efficiency. Sol. Energy, 169: 616–633.

[8] Kranz, L., C. Gretener, J. Perrenoud, R. Schmitt, F. Pianezzi, F. La Mattina, P. Blösch, E. Cheah, A. Chirilă and C.M. Fella. 2013. Doping of polycrystalline CdTe for high-efficiency solar cells on flexible metal foil. Nat. Commun., 4: 2306.

[9] Green, M.A., K. Emery, Y. Hishikawa and W. Warta. 2010. Solar cell efficiency tables (version 36). Prog. Photovoltaics Res. Appl., 18(5): 346–352.

[10] Mitzi, D.B., O. Gunawan, T.K. Todorov, K. Wang and S. Guha. 2011. The path towards a high-performance solution-processed kesterite solar cell. Sol. Energy Mater. Sol. Cells, 95(6): 1421–1436.

[11] Li, M., W.-H. Zhou, J. Guo, Y.-L. Zhou, Z.-L. Hou, J. Jiao, Z.-J. Zhou, Z.-L. Du and S.-X. Wu. 2012. Synthesis of pure metastable wurtzite CZTS nanocrystals by facile one-pot method. J. Phys. Chem. C, 116(50): 26507–26516.

[12] Pal, K., K.B. Thapa and A. Bhaduri. 2018. A review on the current and future possibilities of copper-zinc tin sulfur thin film solar cell to increase more than 20% efficiency. Advanced Science, Engineering and Medicine, 10(7-8): 645–652.

[13] Suryawanshi, M., G. Agawane, S. Bhosale, S. Shin, P. Patil, J. Kim and A. Moholkar. 2013. CZTS based thin film solar cells: a status review. Mater. Technol., 28(1-2): 98–109.

[14] Touati, R., M.B. Rabeh and M. Kanzari. 2014. Optical constants Cu_2ZnSnS_4 thin films deposited at different substrate temperatures. Green Energy, 2014 International Conference on, IEEE.

[15] Schorr, S. 2007. Structural aspects of adamantine like multinary chalcogenides. Thin Solid Films, 515(15): 5985–5991.

[16] Chen, S., X. Gong, A. Walsh and S.-H. Wei. 2009. Crystal and electronic band structure of $Cu_2ZnSnX4$ (X = S and Se) photovoltaic absorbers: First-principles insights. Appl. Phys. Lett., 94(4): 041903.

[17] Paier, J., R. Asahi, A. Nagoya and G. Kresse. 2009. Cu_2ZnSnS_4 as a potential photovoltaic material: a hybrid Hartree-Fock density functional theory study. Phys. Rev. B: Condens. Matter, 79(11): 115126.

[18] Zhang, Y., X. Yuan, X. Sun, B.-C. Shih, P. Zhang and W. Zhang. 2011. Comparative study of structural and electronic properties of Cu-based multinary semiconductors. Phys. Rev. B: Condens. Matter, 84(7): 075127.

[19] Nagoya, A., R. Asahi and G. Kresse. 2011. First-principles study of Cu_2ZnSnS_4 and the related band offsets for photovoltaic applications. J. Phys.: Condens. Matter, 23(40): 404203.

[20] Schorr, S. 2011. The crystal structure of kesterite type compounds: A neutron and X-ray diffraction study. Sol. Energy Mater. Sol. Cells, 95(6): 1482–1488.

[21] Fairbrother, A., M. Neuschitzer, E. Saucedo and A. Pérez-Rodríguez. 2015. Zn-poor $Cu_2ZnSnSe_4$ thin films and solar cell devices. Phys. Status Solidi A, 212(1): 109–115.

[22] Nagoya, A., R. Asahi, R. Wahl and G. Kresse. 2010. Defect formation and phase stability of Cu_2ZnSnS_4 photovoltaic material. Phys. Rev. B: Condens. Matter, 81(11): 113202.

[23] Olekseyuk, I., I. Dudchak and L. Piskach. 2004. Phase equilibria in the Cu_2S–ZnS–SnS_2 system. J. Alloys Compd., 368(1-2): 135–143.

[24] Nozaki, H., K. Shibata and N. Ohhashi. 1991. Metallic hole conduction in CuS. J. Solid State Chem., 91(2): 306–311.

[25] Wu, C., Z. Hu, C. Wang, H. Sheng, J. Yang and Y. Xie. 2007. Hexagonal Cu_2SnS_3 with metallic character: Another category of conducting sulfides. Appl. Phys. Lett., 91(14): 143104.

[26] Wang, K., B. Shin, K.B. Reuter, T. Todorov, D.B. Mitzi and S. Guha. 2011. Structural and elemental characterization of high efficiency Cu_2ZnSnS_4 solar cells. Appl. Phys. Lett., 98(5): 051912.

[27] Jain, S., D. Singh, N. Vijayan and S.N. Sharma. 2018. Time-controlled synthesis mechanism analysis of kesterite-phased Cu_2ZnSnS_4 nanorods via colloidal route. Appl. Nanosci., 8(3): 435–446.

[28] Ito, K. and T. Nakazawa. 1988. Electrical and optical properties of stannite-type quaternary semiconductor thin films. Jpn. J. Appl. Phys., 27(11R): 2094.

[29] Tanaka, T., D. Kawasaki, M. Nishio, Q. Guo and H. Ogawa. 2006. Fabrication of Cu_2ZnSnS_4 thin films by co-evaporation. Phys. Status Solidi C, 3(8): 2844–2847.

[30] Fernandes, P., P. Salomé and A. Da Cunha. 2009. Precursors' order effect on the properties of sulfurized Cu_2ZnSnS_4 thin films. Semicond. Sci. Technol., 24(10): 105013.

[31] Zhang, J. and L. Shao. 2009. Cu_2ZnSnS_4 thin films prepared by sulfurizing different multilayer metal precursors. Sci. China Ser. E: Technol. Sci., 52(1): 269–272.

[32] Yoo, H. and J. Kim. 2010. Growth of Cu_2ZnSnS_4 thin films using sulfurization of stacked metallic films. Thin Solid Films, 518(22): 6567–6572.

[33] Tanaka, T., T. Nagatomo, D. Kawasaki, M. Nishio, Q. Guo, A. Wakahara, A. Yoshida and H. Ogawa. 2005. Preparation of Cu_2ZnSnS_4 thin films by hybrid sputtering. J. Phys. Chem. Solids, 66(11): 1978–1981.

[34] Momose, N., M.T. Htay, T. Yudasaka, S. Igarashi, T. Seki, S. Iwano, Y. Hashimoto and K. Ito. 2011. Cu_2ZnSnS_4 thin film solar cells utilizing sulfurization of metallic precursor prepared by simultaneous sputtering of metal targets. Jpn. J. Appl. Phys., 50(1S2): 01BG09.

[35] Chalapathy, R., G.S. Jung and B.T. Ahn. 2011. Fabrication of Cu_2ZnSnS_4 films by sulfurization of Cu/ZnSn/Cu precursor layers in sulfur atmosphere for solar cells. Sol. Energy Mater. Sol. Cells, 95(12): 3216–3221.

[36] Jimbo, K., R. Kimura, T. Kamimura, S. Yamada, W.S. Maw, H. Araki, K. Oishi and H. Katagiri. 2007. Cu_2ZnSnS_4-type thin film solar cells using abundant materials. Thin Solid Films, 515(15): 5997–5999.

[37] Katagiri, H., K. Jimbo, S. Yamada, T. Kamimura, W.S. Maw, T. Fukano, T. Ito and T. Motohiro. 2008. Enhanced conversion efficiencies of Cu_2ZnSnS_4-based thin film solar cells by using preferential etching technique. Appl. Phys. Express, 1(4): 041201.

[38] Katagiri, H., K. Jimbo, M. Tahara, H. Araki and K. Oishi. 2009. The influence of the composition ratio on CZTS-based thin film solar cells. MRS Online Proceedings Library Archive, 1165.

[39] Araki, H., A. Mikaduki, Y. Kubo, T. Sato, K. Jimbo, W.S. Maw, H. Katagiri, M. Yamazaki, K. Oishi and A. Takeuchi. 2008. Preparation of Cu_2ZnSnS_4 thin films by sulfurization of stacked metallic layers. Thin Solid Films, 517(4): 1457–1460.

[40] Katagiri, H., K. Saitoh, T. Washio, H. Shinohara, T. Kurumadani and S. Miyajima. 2001. Development of thin film solar cell based on Cu_2ZnSnS_4 thin films. Sol. Energy Mater. Sol. Cells, 65(1-4): 141–148.

[41] Kobayashi, T., K. Jimbo, K. Tsuchida, S. Shinoda, T. Oyanagi and H. Katagiri. 2005. Investigation of Cu_2ZnSnS_4-based thin film solar cells using abundant materials. Jpn. J. Appl. Phys., 44(1S): 783.

[42] Bär, M., B.-A. Schubert, B. Marsen, S. Krause, S. Pookpanratana, T. Unold, L. Weinhardt, C. Heske and H.-W. Schock. 2011. Impact of KCN etching on the chemical and electronic surface structure of Cu_2ZnSnS_4 thin-film solar cell absorbers. Appl. Phys. Lett., 99(15): 152111.

[43] Wang, K., O. Gunawan, T. Todorov, B. Shin, S. Chey, N. Bojarczuk, D. Mitzi and S. Guha. 2010. Thermally evaporated Cu_2ZnSnS_4 solar cells. Appl. Phys. Lett., 97(14): 143508.

[44] Geohegan, D., D. Chrisey and G. Hubler. 1994. Pulsed Laser Deposition of Thin Films. Chrisey and G.K. Hubler (eds.). Wiely, New York: 59–69.

[45] Krebs, H.-U., M. Weisheit, J. Faupel, E. Süske, T. Scharf, C. Fuhse, M. Störmer, K. Sturm, M. Seibt and H. Kijewski. 2003. Pulsed Laser Deposition (PLD)—A Versatile Thin Film Technique. Advances in Solid State Physics, Springer: 505–518.

[46] Sekiguchi, K., K. Tanaka, K. Moriya and H. Uchiki. 2006. Epitaxial growth of Cu_2ZnSnS_4 thin films by pulsed laser deposition. Phys. Status Solidi C, 3(8): 2618–2621.

[47] Moholkar, A., S. Shinde, A. Babar, K.-U. Sim, Y.-b. Kwon, K. Rajpure, P. Patil, C. Bhosale and J. Kim. 2011. Development of CZTS thin films solar cells by pulsed laser deposition: influence of pulse repetition rate. Sol. Energy, 85(7): 1354–1363.

[48] Wangperawong, A., J. King, S. Herron, B. Tran, K. Pangan-Okimoto and S. Bent. 2011. Aqueous bath process for deposition of Cu_2ZnSnS_4 photovoltaic absorbers. Thin Solid Films, 519(8): 2488–2492.

[49] Todorov, T., O. Gunawan, S.J. Chey, T.G. de Monsabert, A. Prabhakar and D.B. Mitzi. 2011. Progress towards marketable earth-abundant chalcogenide solar cells. Thin Solid Films, 519(21): 7378–7381.

[50] Ma, G., T. Minegishi, D. Yokoyama, J. Kubota and K. Domen. 2011. Photoelectrochemical hydrogen production on Cu_2ZnSnS_4/Mo-mesh thin-film electrodes prepared by electroplating. Chem. Phys. Lett., 501(4-6): 619–622.

[51] Yoo, H. and J. Kim. 2011. Comparative study of Cu_2ZnSnS_4 film growth. Sol. Energy Mater. Sol. Cells, 95(1): 239–244.

[52] Prabhakar, T. and N. Jampana. 2011. Effect of sodium diffusion on the structural and electrical properties of Cu_2ZnSnS_4 thin films. Sol. Energy Mater. Sol. Cells, 95(3): 1001–1004.

[53] Tanaka, K., Y. Fukui, N. Moritake and H. Uchiki. 2011. Chemical composition dependence of morphological and optical properties of Cu_2ZnSnS_4 thin films deposited by sol–gel sulfurization and Cu_2ZnSnS_4 thin film solar cell efficiency. Sol. Energy Mater. Sol. Cells, 95(3): 838–842.

[54] Scragg, J.J., P.J. Dale and L.M. Peter. 2008. Towards sustainable materials for solar energy conversion: Preparation and photoelectrochemical characterization of Cu_2ZnSnS_4. Electrochem. Commun., 10(4): 639–642.

[55] Nakayama, N. and K. Ito. 1996. Sprayed films of stannite Cu_2ZnSnS_4. Appl. Surf. Sci., 92: 171–175.

[56] Kamoun, N., H. Bouzouita and B. Rezig. 2007. Fabrication and characterization of Cu_2ZnSnS_4 thin films deposited by spray pyrolysis technique. Thin Solid Films, 515(15): 5949–5952.

[57] Kumar, Y.K., G.S. Babu, P.U. Bhaskar and V.S. Raja. 2009. Preparation and characterization of spray-deposited Cu_2ZnSnS_4 thin films. Sol. Energy Mater. Sol. Cells, 93(8): 1230–1237.

[58] Scragg, J.J., P.J. Dale, L.M. Peter, G. Zoppi and I. Forbes. 2008. New routes to sustainable photovoltaics: evaluation of Cu_2ZnSnS_4 as an alternative absorber material. Phys. Status Solidi B, 245(9): 1772–1778.

[59] Scragg, J.J., D.M. Berg and P.J. Dale. 2010. A 3.2% efficient Kesterite device from electrodeposited stacked elemental layers. J. Electroanal. Chem., 646(1-2): 52–59.

[60] Ennaoui, A., M. Lux-Steiner, A. Weber, D. Abou-Ras, I. Kotschau, H.-W. Schock, R. Schurr, A. Hölzing, S. Jost and R. Hock. 2009. Cu_2ZnSnS_4 thin film solar cells from electroplated precursors: Novel low-cost perspective. Thin Solid Films, 517: 2511–2514.

[61] Ahmed, S., K.B. Reuter, O. Gunawan, L. Guo, L.T. Romankiw and H. Deligianni. 2012. A high efficiency electrodeposited Cu_2ZnSnS_4 solar cell. Advanced Energy Materials, 2(2): 253–259.

[62] Moritake, N., Y. Fukui, M. Oonuki, K. Tanaka and H. Uchiki. 2009. Preparation of Cu_2ZnSnS_4 thin film solar cells under non-vacuum condition. Phys. Status Solidi C, 6(5): 1233–1236.

[63] Tanaka, K., N. Moritake and H. Uchiki. 2007. Preparation of Cu_2ZnSnS_4 thin films by sulfurizing sol–gel deposited precursors. Sol. Energy Mater. Sol. Cells, 91(13): 1199–1201.

[64] Tanaka, K., M. Oonuki, N. Moritake and H. Uchiki. 2009. Cu_2ZnSnS_4 thin film solar cells prepared by non-vacuum processing. Sol. Energy Mater. Sol. Cells, 93(5): 583–587.

[65] Xin, X., M. He, W. Han, J. Jung and Z. Lin. 2011. Low-cost copper zinc tin sulfide counter electrodes for high-efficiency dye-sensitized solar cells. Angew. Chem. Int. Ed., 50(49): 11739–11742.

[66] Zhou, Z., Y. Wang, D. Xu and Y. Zhang. 2010. Fabrication of Cu_2ZnSnS_4 screen printed layers for solar cells. Sol. Energy Mater. Sol. Cells, 94(12): 2042–2045.

[67] Cao, M. and Y. Shen. 2011. A mild solvothermal route to kesterite quaternary Cu_2ZnSnS_4 nanoparticles. J. Cryst. Growth, 318(1): 1117–1120.

[68] Shinde, N., D. Dubal, D. Dhawale, C. Lokhande, J. Kim and J. Moon. 2012. Room temperature novel chemical synthesis of Cu_2ZnSnS_4 (CZTS) absorbing layer for photovoltaic application. Mater. Res. Bull., 47(2): 302–307.

[69] Mali, S.S., P.S. Shinde, C.A. Betty, P.N. Bhosale, Y.W. Oh and P.S. Patil. 2012. Synthesis and characterization of Cu_2ZnSnS_4 thin films by SILAR method. J. Phys. Chem. Solids, 73(6): 735–740.

[70] Washio, T., T. Shinji, S. Tajima, T. Fukano, T. Motohiro, K. Jimbo and H. Katagiri. 2012. 6% Efficiency Cu_2ZnSnS_4-based thin film solar cells using oxide precursors by open atmosphere type CVD. J. Mater. Chem., 22(9): 4021–4024.

[71] Jain, S., P. Chawla, S.N. Sharma, D. Singh and N. Vijayan. 2018. Efficient colloidal route to pure phase kesterite Cu_2ZnSnS_4 (CZTS) nanocrystals with controlled shape and structure. Superlattices Microstruct., 119: 59–71.

[72] Park, J., J. Joo, S.G. Kwon, Y. Jang and T. Hyeon. 2007. Synthese monodisperser sphärischer Nanokristalle. Angew. Chem. Int. Ed., 119(25): 4714–4745.

[73] Yu, W.W., Y.A. Wang and X. Peng. 2003. Formation and stability of size-, shape-, and structure-controlled CdTe nanocrystals: ligand effects on monomers and nanocrystals. Chem. Mater., 5(22): 4300–4308.

[74] Gole, A. and C.J. Murphy. 2004. Seed-mediated synthesis of gold nanorods: role of the size and nature of the seed. Chem. Mater., 16(19): 3633–3640.

[75] Shi, W., H. Zeng, Y. Sahoo, T.Y. Ohulchanskyy, Y. Ding, Z.L. Wang, M. Swihart and P.N. Prasad. 2006. A general approach to binary and ternary hybrid nanocrystals. Nano Lett., 6(4): 875–881.

[76] Talapin, D.V., A.L. Rogach, A. Kornowski, M. Haase and H. Weller. 2001. Highly luminescent monodisperse CdSe and CdSe/ZnS nanocrystals synthesized in a hexadecylamine–trioctylphosphine oxide–trioctylphospine mixture. Nano Lett., 1(4): 207–211.

[77] Park, J., K. An, Y. Hwang, J.-G. Park, H.-J. Noh, J.-Y. Kim, J.-H. Park, N.-M. Hwang and T. Hyeon. 2004. Ultra-large-scale syntheses of monodisperse nanocrystals. Nat. Mater., 3(12): 891.

[78] Cozzoli, P.D., T. Pellegrino and L. Manna. 2006. Synthesis, properties and perspectives of hybrid nanocrystal structures. Chem. Soc. Rev., 35(11): 1195–1208.

[79] Seo, W.S., H.H. Jo, K. Lee and J.T. Park. 2003. Preparation and optical properties of highly crystalline, colloidal, and size-controlled indium oxide nanoparticles. J. Adv. Mater., 15(10): 795–797.

[80] Lee, S.-M., Y.-w. Jun, S.-N. Cho and J. Cheon. 2002. Single-crystalline star-shaped nanocrystals and their evolution: programming the geometry of nano-building blocks. J. Am. Chem. Soc., 124(38): 11244–11245.

[81] Kim, Y.-H., Y.-w. Jun, B.-H. Jun, S.-M. Lee and J. Cheon. 2002. Sterically induced shape and crystalline phase control of GaP nanocrystals. J. Am. Chem. Soc., 124(46): 13656–13657.

[82] Hull, K.L., J.W. Grebinski, T.H. Kosel and M. Kuno. 2005. Induced branching in confined PbSe nanowires. Chem. Mater., 17(17): 4416–4425.

[83] Jun, Y.-w., S.-M. Lee, N.-J. Kang and J. Cheon. 2001. Controlled synthesis of multi-armed CdS nanorod architectures using monosurfactant system. J. Am. Chem. Soc., 123(21): 5150–5151.

[84] Murray, C.B., S. Sun, W. Gaschler, H. Doyle, T.A. Betley and C.R. Kagan. 2001. Colloidal synthesis of nanocrystals and nanocrystal superlattices. IBM J. Res. Dev., 45(1): 47–56.

[85] Zhong, X., Y. Feng, W. Knoll and M. Han. 2003. Alloyed $Zn_xCd_{1-x}S$ nanocrystals with highly narrow luminescence spectral width. J. Am. Chem. Soc., 125(44): 13559–13563.

[86] Ghezelbash, A. and B.A. Korgel. 2005. Nickel sulfide and copper sulfide nanocrystal synthesis and polymorphism. Langmuir, 21(21): 9451–9456.

[87] Manna, L., E.C. Scher and A.P. Alivisatos. 2000. Synthesis of soluble and processable rod, arrow-, teardrop-, and tetrapod-shaped CdSe nanocrystals. J. Am. Chem. Soc., 122(51): 12700–12706.

[88] Peng, X., L. Manna, W. Yang, J. Wickham, E. Scher, A. Kadavanich and A.P. Alivisatos. 2000. Shape control of CdSe nanocrystals. Nature, 404(6773): 59.

[89] Jun, Y.w., J.s. Choi and J. Cheon. 2006. Formkontrolle von Halbleiter-und Metalloxid-Nanokristallen durch nichthydrolytische Kolloidverfahren. Angew. Chem. Int. Ed., 118(21): 3492–3517.

[90] Peng, Z.A. and X. Peng. 2001. Mechanisms of the shape evolution of CdSe nanocrystals. J. Am. Chem. Soc., 123(7): 1389–1395.

[91] Wang, W., M.T. Winkler, O. Gunawan, T. Gokmen, T.K. Todorov, Y. Zhu and D.B. Mitzi. 2014. Device characteristics of CZTSSe thin-film solar cells with 12.6% efficiency. Adv. Energy Mater., 4(7): 1301465.

[92] Shin, B., O. Gunawan, Y. Zhu, N.A. Bojarczuk, S.J. Chey and S. Guha. 2013. Thin film solar cell with 8.4% power conversion efficiency using an earth-abundant Cu_2ZnSnS_4 absorber. Prog. Photovoltaics Res. Appl., 21(1): 72–76.

[93] Wang, D., W. Zhao, Y. Zhang and S.F. Liu. 2018. Path towards high-efficient kesterite solar cells. J. Energy Chem., 27(4): 1040–1053.

[94] Olopade, M., O. Oyebola and B. Adeleke. 2012. Investigation of some materials as buffer layer in copper zinc tin sulphide (Cu_2ZnSnS_4) solar cells by SCAPS-1D. Advances in Applied Science Research, 3(6): 3396–3400.

[95] Crovetto, A., C. Yan, B. Iandolo, F. Zhou, J. Stride, J. Schou, X. Hao and O. Hansen. 2016. Lattice-matched Cu_2ZnSnS_4/CeO_2 solar cell with open circuit voltage boost. Appl. Phys. Lett., 109(23): 233904.

[96] Han, J., Z. Liu, X. Zheng, K. Guo, X. Zhang, T. Hong, B. Wang and J. Liu. 2014. Trilaminar $ZnO/ZnS/Sb_2S_3$ nanotube arrays for efficient inorganic–organic hybrid solar cells. RSC Adv., 4(45): 23807–23814.

[97] Liu, C., Z. Qiu, F. Li, W. Meng, W. Yue, F. Zhang, Q. Qiao and M. Wang. 2015. From binary to multicomponent photoactive layer: a promising complementary strategy to efficient hybrid solar cells. Nano Energy, 12: 686–697.

[98] Wang, L., H.Y. Wang, H.T. Wei, H. Zhang, Q.D. Chen, H.L. Xu, W. Han, B. Yang and H.B. Sun. 2014. Unraveling charge separation and transport mechanisms in aqueous-processed polymer/CdTe nanocrystal hybrid solar cells. Adv. Energy Mater., 4(9).

[99] Engberg, S., K. Agersted, A. Crovetto, O. Hansen, Y.M. Lam and J. Schou. 2017. Investigation of Cu_2ZnSnS_4 nanoparticles for thin-film solar cell applications. Thin Solid Films, 628: 163–169.

[100] Saha, U. and M.K. Alam. 2017. Proposition and computational analysis of a kesterite/kesterite tandem solar cell with enhanced efficiency. RSC Adv., 7(8): 4806–4814.

[101] Todorov, T., T. Gershon, O. Gunawan, Y.S. Lee, C. Sturdevant, L.Y. Chang and S. Guha. 2015. Monolithic perovskite-CIGS tandem solar cells via *in situ* band gap engineering. Adv. Energy Mater., 5(23): 1500799.

[102] Ferhati, H. and F. Djeffal. 2019. An efficient analytical model for tandem solar cells. Mater. Res. Express, 6(7): 076424.

[103] Ravindiran, M. and C. Praveenkumar. 2018. Status review and the future prospects of CZTS based solar cell—A novel approach on the device structure and material modeling for CZTS based photovoltaic device. Renewable and Sustainable Energy Rev., 94: 317–329.

[104] Sanad, M., A. Elseman, M. Elsenety, M. Rashad and B. Elsayed. 2019. Facile synthesis of sulfide-based chalcogenide as hole-transporting materials for cost-effective efficient perovskite solar cells. Journal of Materials Science: Materials in Electronics, 30(7): 6868–6875.

CHAPTER 12

Surface Modification of Glass Nanofillers and Their Reinforcing Effect in Epoxy-Based Nanocomposites

Lifeng Zhang[1],* and *Demei Yu*[2],*

1. Introduction

Polymer composite is a multi-phase material in which reinforcing fillers are integrated with polymer matrix and result in synergistic mechanical properties that cannot be achieved from either component alone [1]. The "light and strong" characteristic of polymer composite material, in which fiber and particle are two common reinforcing agents, makes it very popular in our daily life as engineering materials for automobile, aerospace, sports utilities, construction, etc.

With widespread interest in nanomaterials recently, nanometer scale fillers for polymer composite materials have attracted growing attention. Polymer nanocomposite (PNC) has become not only a hot term, but also a boosting market in recent years [2–5]. Generally, PNC is a polymer-based composite material in which at least one phase has one, two, or three dimensions that fall into nanometer scale. More commonly, it means a bulk polymer matrix with solid nanofillers dispersed therein. Compared to conventional polymer composites with micrometer scale fillers, PNCs have demonstrated substantial mechanical property enhancements at much lower filler loading, which facilitates to reduce specific weight and simplify processing [6]. Moreover, due to nanometer size feature and concomitant ultra-high specific surface area of the dispersed nanofillers, there is substantially stronger interfacial bonding

[1] Department of Nanoengineering, Joint School of Nanoscience and Nanoengineering, North Carolina A&T State University, Greensboro, NC 27401, United States.
[2] Department of Chemistry, School of Science, State Key Laboratory of Electrical Insulation and Power Equipment, Xi'an Jiaotong University, Xi'an, Shaanxi 710049, China.
* Corresponding authors: lzhang@ncat.edu, dmyu@mail.xjtu.edu.cn

between the nanofillers and polymer matrix, which results in significant mechanical property improvement typically not shared by their conventional counterparts. To date, nanofillers have shown a bright prospect in developing high-performance and lightweight composite materials with inherent processability [7].

Epoxy resin is the most widely and commonly used thermoset polymer in various composite-related industries, including automobile, aerospace, windmills, energy storage, packaging, etc. [8]. Epoxy resin is extensively used as a matrix in fiber-reinforced polymer composites as well as in engineering adhesives due to its excellent properties, such as thermal stability, high adhesion strength, good processability, and mechanical properties [9]. Uncured epoxy is normally a liquid resin, and curing of epoxy resin with hardener produces a cross-linked three-dimensional molecular structure, which results in high modulus, strength, and toughness. The main unsatisfactory property of cured epoxy that restricts its application is low fracture toughness, caused by high cross-linking density and internal stress in the curing process [10, 11]. The limitation of low fracture energy of cured epoxy can be rectified by including a variety of nanoscale fillers, such as carbon nanotubes, graphene, nanoclay, glass nanoparticles, rubber nanoparticles, etc. as secondary phase [10–12]. The performance of epoxy-based composite materials can be uniquely engineered by selecting appropriate nanoscale fillers for advanced composite applications [13].

2. Glass Nanofillers

Glass (SiO_2, silica) is an abundant compound over the earth's crust, and glass nanoparticles (GNPs) have been considered as one of the most important types of nanofillers at the forefront of polymer composite materials [14, 15]. The most advantageous feature of GNPs is their low production cost in combination with ultra-high surface area, strong adsorption, good dispersal ability, high chemical purity, and excellent stability [16]. GNPs that are obtained from natural resources contain impurities, and thus are not favorable for scientific research and industrial applications. This has led to the synthesis of GNPs, including colloidal silica, silica gels, pyrogenic silica, and precipitated silica that produces pure amorphous silica powder, while natural mineral silica is crystalline (quartz, cristobalite).

Recently, electrospun nanofibers have been explored as a new promising reinforcing filler in PNCs [17]. Recent studies have indicated that continuous SiO_2 nanofibers with diameters of ~ 400 nm can be prepared through electrospinning a spin dope containing an alkoxide precursor of SiO_2 followed by pyrolysis at high temperature [18, 19]. These electrospun SiO_2 nanofibers are morphologically uniform and structurally amorphous, and they can retain their fiber morphology when subjected to vigorous ultrasonication; therefore, electrospun SiO_2 nanofibers are nanoscaled glass fibers, and thus termed as electrospun glass nanofibers (EGNFs). Compared to GNPs, very limited research endeavors have been devoted to EGNF-reinforced PNCs.

3. Surface Modification of Glass Nanofillers

It has been observed that mechanical properties of the GNP-reinforced epoxy nanocomposite increase with GNP loading until a certain GNP weight percentage, beyond which mechanical properties of the composite start to degrade. This is caused by GNP agglomeration [20]. The agglomeration of GNPs at high loading level is mainly due to their high surface energy and poor dispersion [11]. Furthermore, the difference of surface property between inorganic GNPs and organic epoxy matrix could lead to insufficient wetting of GNPs by the epoxy resin. When GNP-epoxy interaction is weak, the silanol groups on the surface of GNPs could form hydrogen bonds with those on adjacent GNPs, leading to GNP agglomeration even under high shear mixing condition [14, 21]. Accordingly, various dispersing methods have been investigated to achieve a uniform distribution of GNPs in the epoxy matrix, such as high shear mixing and ultrasonic treatment. It is noteworthy that when high loading of GNPs is employed, uniform dispersion of GNPs in epoxy matrix becomes extremely difficult because the overall cohesive force between GNPs is more than the overall adhesive force between GNPs and epoxy matrix [22].

A good dispersion of GNPs in epoxy matrix can be achieved by modifying the surface of GNPs. Surface modification of GNPs, i.e., changing GNPs' surface from hydrophilic to hydrophobic or lipophilic, could help GNPs' dispersion in hydrophobic epoxy resin. Furthermore, introducing charges or barrier molecules to GNPs' surface could stabilize GNPs in epoxy matrix. In the meantime, GNP surface modification could also introduce certain functional groups that can react with epoxide functional groups in epoxy and improve interfacial bonding between GNPs and epoxy matrix, which consequently improves mechanical properties of the resultant epoxy nanocomposite [21]. EGNFs basically have the same surface and similar situation as GNPs when they are engaged in epoxy-based nanocomposite.

3.1 Surface of Glass Nanofillers

The surface of glass nanofillers (GNPs or EGNFs) consists of –Si–OH (silanol) and –Si–O–Si– groups, which makes them hydrophilic in nature. The available amount of silanol groups on glass nanofillers' surface determines their degree of hydrophilicity. Higher the number of silanol groups available on surface of glass nanofillers, higher the hydrophilicity of the glass nanofillers' surface. Typically, there are three possible silanol types: isolated silanols, vicinal silanols, and germinal silanols on surface of glass nanofillers, as shown in Figure 12.1. The silanol groups that are available on nanofillers' surface can form hydrogen bonding with other silanol groups on adjacent nanofiller surface, leading to glass nanofiller agglomeration that remains intact even with high shear mixing conditions if the filler-matrix interaction is weak [14]. Surface modification of glass nanofillers has been used to promote compatibility between epoxy matrix and glass nanofillers, and can be conducted through chemical or physical methods [23].

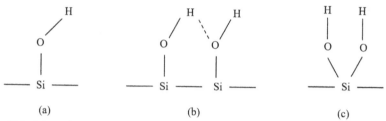

Figure 12.1. Schematic representation of three silanol types: (a) isolated silanol, (b) vicinal silanol, and (c) germinal silanol.

3.2 Surface Modification

Surface modification of glass nanofillers can be carried out physically by adsorbing low molecular weight surfactant or macromolecule onto the surface of glass nanofillers. The principle involved in this case is physical interaction, such as ionic bonding, hydrogen bonding, or dipole-dipole interaction between the molecules adsorbed, and the silanol groups on glass nanofiller surface. This physical coating on glass nanofiller surface could prevent interactions between glass nanofillers, reduce their agglomeration, and improve performance of the resulting nanocomposite.

Compared to physical surface modification, chemical surface modification is more significant because a stronger covalent bond is formed between glass nanofillers and surface modifying agent [23]. Chemical modification leads to much stronger bonding between the surface modifying agent and glass nanofillers, and can avoid desorption of surface modifying agent from the nanofiller surface in the long run. Chemical surface modification also improves interaction and dispersion of glass nanofillers in polymer matrix through chemical reaction between functional groups on glass surface and the matrix [24–26]. The most popular and easiest chemical surface modification technique for glass nanofillers is the use of silane coupling agent. Silane coupling agents can be represented as $R(CH_2)_n SiX_3$, where X represents hydrolyzable group and R represents specific functional group. Typically, X group can be chloro-, methoxy-, or ethoxy- groups, and R groups can be chosen to meet the requirement based on polymer matrix. The principle of the surface modification of glass nanofillers through silane coupling agent is that the X group reacts with hydroxyl group on surface of glass nanofillers, and the R group from the coupling agent reacts with polymer matrix. In the case of epoxy nanocomposite, generally R group contains functional groups that can react with epoxide functional groups. Figure 12.2 demonstrates the reaction between GNPs/EGNFs and 3-Aminopropyltriethoxysilane (APTES), as well as the following reaction between the surface-modified GNPs/EGNFs and epoxy resin. In addition to APTES, there are many other different silane coupling agents available from commercial sources, such as 3-Glycidyloxypropyltrimethoxysilane (GPTMS), Methacryloxypropyltriethoxysilane (MAPTES), Chloropropyltriethoxysilane (CPTES), (3-Mercaptopropyl)trimethoxysilane (MPTMS), etc.

The chemical surface modification of glass nanofillers with silane coupling agents is conducted normally via aqueous or non-aqueous systems. In non-aqueous system, molecules of silane coupling agent are attached to the nanofiller

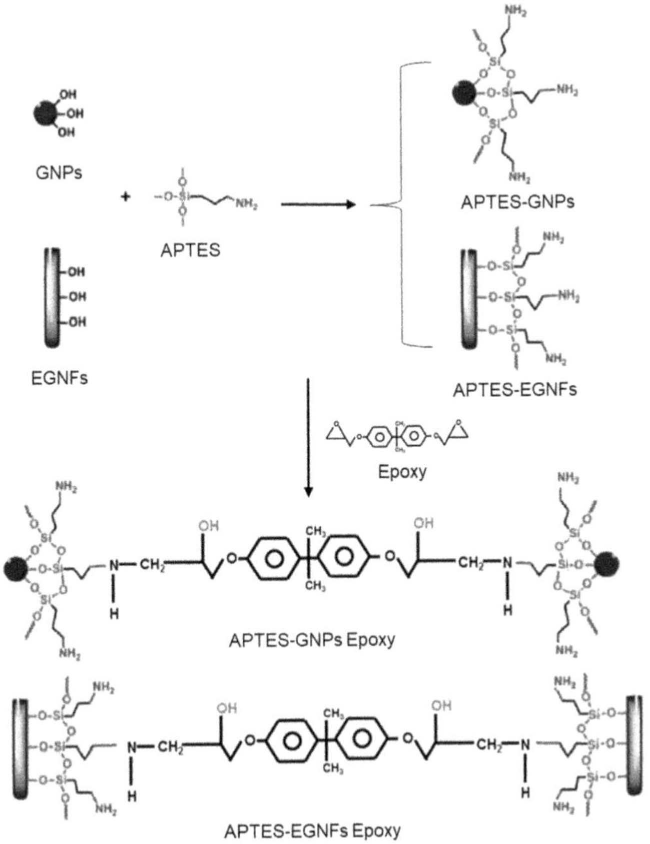

Figure 12.2. Schematic diagram of the reaction between a typical silane coupling agent APTES and glass nanofillers, and the reaction between surface-modified glass nanofillers and epoxy matrix.

surface through direct condensation reaction without the hydrolysis step, which is usually conducted under reflux condition. Silane coupling agents can go through uncontrollable hydrolysis and polycondensation reaction in an aqueous system. Hence, organic solvents are preferred in the surface modification process so that a better control in reaction may be achieved [27]. However, an aqueous system is good for large-scale production of surface-modified GNPs. In this case, silane molecules go through hydrolysis and condensation reaction prior to deposition on the surface of GNPs, as shown in Figure 12.3.

A co-condensation approach has been used to incorporate organic functional groups homogenously to both interior and exterior of GNPs [27]. For example, amino-functionalized monodispersed GNPs can be produced through a modified Stöber method using tetraethyl orthosilicate (TEOS) and APTES in ethanol [28]. The co-condensation approach has been used to synthesize monodispersed mesoporous GNPs with tetramethoxysilane (TMOS) and organic trimethoxysilane (R-TMS) including 3-aminopropyl trimethoxysilane (APTMS), [3-(2-aminoethylamino)

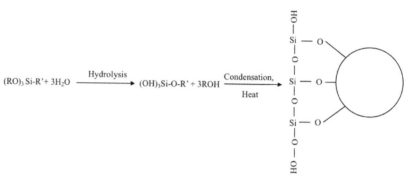

Figure 12.3. Chemical surface modification of GNPs in an aqueous system.

propyl] trimethoxysilane (AEAP-TMS) or 3-[2-(2-aminoethylamino) ethylamino] propyl trimethoxysilane (AEAEAP-TMS) [28–30].

In regular surface modification, premade GNPs are used so that only the exterior of GNPs are modified. In co-condensation synthesis, GNPs are prepared with *in situ* surface modification, so that the resulting GNPs have homogeneous functional groups both inside GNPs and on the surface. When comparing regular surface modification with *in situ* surface modification, the regular post-modification does not affect the size and size distribution of GNPs very much, while co-condensation produces bigger size GNPs with low aggregation due to high rate of hydrolysis that induces GNP growth [27]. However, a disadvantage from co-condensation is that special care must be taken so that organic functionality can be kept during the purification process, and thus calcination is not suitable [31]. Therefore, the post-modification method is the most used method in modifying SNPs with common silane coupling agents, such as APTES, GPTMS, MPTMS, etc. for composite purpose [32–35].

4. Effect of Surface-Modified Glass Nanofillers on Mechanical Properties of the Resultant Epoxy Nanocomposites

4.1 Surface-Modified GNPs

Numerous studies have proved that surface modification of GNPs improve mechanical properties of the cured epoxy resin. Selection of surface-modification agent is dependent on the end application. Specifically, Zheng et al. [36] prepared silane coupling agent-pretreated GNPs, and added them to CYD-128 epoxy at 3 wt.% loading. The comparison indicated that surface treatment of GNPs improved elastic modulus, tensile strength, and impact strength of the cured epoxy resin by 1.3%, 7.0%, and 9.8%. Yu et al. [32] used three types of silane coupling agents APTES, GPTMS, and methacryloxypropyl trimethoxysilane (MAPTMS) to surface-modify GNPs. When surface-modified GNPs with APTES, GPTMS, or MAPTMS were integrated with epoxy resin at 4 wt.% loading, the flexural modulus of the cured epoxy resin was improved by 8.4%, 12.6%, and 6.1%, respectively, compared to that of the one with pristine SNPs. The ultimate flexural strength of cured epoxy resin was correspondingly improved by 10.7%, 29.3%, and 8.4% with APTES, GPTMS, and MAPTMS surface-modified GNPs. Jiang et al. [33] modified GNPs surface with

APTES (SNP+APTES), and further modified GNP+APTES with graphene oxide (GNP+APTES+GO). The prepared GO, GNP+APTES, and GNP+APTES+GO were incorporated in a bisphenol A diglycidyl ether (DGEBA) type epoxy YD-128 with a polyether diamine system at loadings of 0.1 wt.%, 0.25 wt.%, 0.5 wt.%, 1 wt.%, and 3 wt.%. GNP+APTES+GO showed better reinforcing performance than GNP+APTES at all loadings. The results were attributed to the large surface area, multiple interactions between the functional groups of matrix and filler, and high aspect ratio of GNP+APTES+GO, as well as a good dispersion and interfacial interaction of the GNP+APTES+GO with the epoxy. Similarly, Chen et al. [35] modified the surface of GNPs with APTES (GNP+APTES), and further modified GNP+APTES with GO (SNP+APTES+GO). Various larger loadings (5 wt.%, 10 wt.%, 15 wt.%, and 20 wt.%) of GNP+APTES+GO were used in an epoxy system of EPON 828 with a diamine curing agent. It was shown that as loading of GNP+APTES+GO increased, modulus of the resulting nanocomposite kept increasing, but the tensile strength of the nanocomposite first increased with GNP+APTES+GO loadings up to 10 wt.%, and then started to reduce. At 10 wt.% loading, GNP+APTES+GO outperformed pristine GNPs and GNP+APTES. Compared to those from the epoxy system with pristine GNPs, 10 wt.% GNP+APTES led to increases of modulus and tensile strength of the resulting epoxy nanocomposite by 5.2% and 11.6%, respectively, while 10 wt.% GNP+APTES+GO resulted in corresponding improvements by 16.2% and 36.3%, respectively.

4.2 Surface-Modified EGNFs

Although EGNFs have been successfully prepared through electrospinning a spin dope containing glass precursor followed by sol-gel processing and subsequent pyrolysis, there are only a few research reports regarding EGNFs for reinforcing purpose in polymer nanocomposites. For the first time, Fong et al. dedicated their effort to the usage of EGNFs as a reinforcing filler in 2,2'-bis[4-(methacryloxypropoxy)-phenyl]-propane (Bis-GMA)/triethylene glycol dimethacrylate (TEGDMA) dental composites [37]. EGNFs were employed mainly because they could offer the resultant dental composite with satisfying mechanical performance as well as translucent appearance, which is important for dental restorations. In that study, EGNFs with diameters of approximately 500 nm were prepared through electrospinning a solution composed of TEOS (SiO_2 precursor) and poly vinyl pyrrolidone (PVP, a carrying polymer for good nanofiber formation), followed by pyrolysis at a high temperature (800°C). The obtained continuous glass nanofibers were subjected to ultrasonic vibration, and converted to short fibers with overall fiber morphology being retained, and average aspect ratio larger than 100. Prior to incorporating them into dental resin, EGNFs were surface-modified by 3-methacryloxypropyltrimethoxy silane and n-propylamine in order to improve nanofiller-matrix interfacial adhesion. The results indicated that 7.5 wt.% substitution of conventional dental filler, i.e., glass powder, with ECNFs in Bis-GMA/TEGDMA dental composite brought about considerable improvement in flexural strength, modulus, and work of fracture by as much as 44%, 29%, and 66%, respectively. Fong and Zhang extended the application of short EGNFs to epoxy resin matrix (SC-15A) [38]. The effects of surface modification of

EGNFs by two types of silane coupling agents with respective amine end groups (APTES) and epoxide end groups (GPTMS), as well as mass fraction of EGNFs in PNCs on overall mechanical properties of corresponding epoxy PNCs were carefully studied. It was found that EGNFs (ca. 400 nm diameter) remarkably outperformed conventional glass fibers (CGFs, ca. 10 μm diameter) in both tensile and impact tests, and yielded simultaneous enhancement in strength, stiffness, and toughness of the epoxy-based PNCs at small mass fractions of 0.5 wt.% and 1 wt.%. Tensile strength, Young's modulus, work of fracture, and impact strength of the EGNFs reinforced epoxy composite were increased by up to 40%, 201%, 67%, and 363%, respectively. Saline treatment played an important role in mechanical performance improvement. EGNFs with epoxide end groups (G-EGNFs) had a better toughening effect, while the one with amine end groups (A-EGNFs) exhibited better reinforcing effect.

4.3 Surface-Modified GNPs vs Surface-Modified EGNFs

To further explicate the merit of EGNFs as a reinforcing/toughening agent, Zhang and Yu conducted a side-by-side direct comparison of nanoscale glass filler reinforced epoxy composites on the basis of EGNFs (with diameter of ~ 300 nm and length of about a few tens of micometers) and glass nanoparticles (GNPs, with diameter of approximately ~ 20 nm) [39]. No appreciable difference on morphology was identified for both EGNFs and GNPs after APTES surface modification, except for some denser packing (Figure 12.4). FTIR was used to characterize surface modification of these glass nanofillers. As shown in Figure 12.5a, FT-IR spectra of both GNPs and EGNFs prior to silane treatments exhibited a broad peak centered at 3,410 cm^{-1} that was attributed to Si-OH groups. Peaks at 1,070 cm^{-1}, 960 cm^{-1}, and 800 cm^{-1} in the IR spectra could be assigned to stretching or bending vibrations of Si-O-Si groups in SiO$_2$. Compared to pristine GNPs and EGNFs, the IR spectra of both APTES-GNPs and APTES-EGNFs exhibited typical C-H stretching vibration of -CH$_2$- groups at 2,925 cm^{-1} and 2,850 cm^{-1}, which indicated successful attachment of APTES to SiO$_2$ surface. This was further confirmed in both FTIR spectra of APTES-

Figure 12.4. SEM images of silanized glass nanofillers: (a) APTES-EGNFs, (b) APTES-GNPs. The right column showed filler morphology at a higher magnification [39].

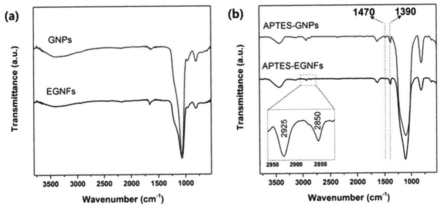

Figure 12.5. FT-IR spectra of glass nanofillers: (a) pristine GNPs and EGNFs, (b) APTES-GNPs and APTES-EGNFs [39].

GNPs and APTES-EGNFs by additional peaks at ~ 1,470 cm^{-1} and ~ 1,390 cm^{-1}, which was associated with the -CH$_2$- and -CH$_3$ groups of APTES coupling agent. Surface modification of the glass nanofillers by APTES brought in mechanical improvement of the resultant epoxy nanocomposites. Within the surveyed filler loading scope ($\leq$ 0.5 wt.%), EGNFs outperformed GNPs in terms of reinforcing and toughening effect. Incorporation of 0.5 wt.% amino-functionalized EGNFs simultaneously increased tensile strength, elongation at break, and work of fracture of the resultant epoxy PNC by ~ 20%, ~ 20%, and ~ 50%, respectively, along with negligible change of Young's modulus. Given the much lower specific surface area of EGNFs than that of GNPs, the better performance of EGNFs was ascribed to larger aspect ratio (shape factor) that aroused "fiber bridging" mechanism as well as crack deflection, which were not shared by GNP counterparts. The reinforcing/toughening mechanisms of EGNFs in PNCs, however, remain to be further understood in order to take full advantage of EGNFs in designing next-generation PNC materials.

5. Conclusion

As reinforcing agents for epoxy-based nanocomposites, glass nanoparticles (GNPs) have been in the spotlight for a long time, while electrospun glass nanofibers (EGNFs) demonstrated their great potential in recent years. Appropriate surface modification of these glass nanofillers is definitely playing a dominant role in epoxy-based nanocomposites from the point of view of processing, as well as mechanical properties of the resultant nanocomposites. The surface modification will (1) reduce agglomeration of these glass nanofillers and consequently lead to uniform dispersion of these nanofillers in epoxy matrix; (2) introduce certain functional groups on the glass surface that can react with epoxide functional groups in epoxy resin and improve interfacial bonding between these glass nanofillers and epoxy matrix, which consequently improves mechanical properties of the resulting epoxy nanocomposite. It is envisioned that glass nanofillers, especially EGNFs, are going to shine with greater luster as reinforcing agents in the field of polymer nanocomposites.

Acknowledgments

This work was performed at the Joint School of Nanoscience and Nanoengineering, a member of Southeastern Nanotechnology Infrastructure Corridor (SENIC) and National Nanotechnology Coordinated Infrastructure (NNCI), which is supported by the National Science Foundation (ECCS-1542174).

References

[1] Chawla, K. 2013. Composite Materials: Science and Engineering (3rd Edition). New York: Springer, p. 1–2.

[2] Zou, H., S. Wu and J. Shen. 2008. Polymer/silica nanocomposites: preparation, characterization, properties, and applications. Chem. Rev., 108: f3893–3957.

[3] Sahoo, N.G., S. Rana, J.W. Cho, L. Li and S.H. Chan. 2010. Polymer nanocomposites based on functionalized carbon nanotubes. Prog. Polym. Sci., 35: 837–867.

[4] Potts, J.R., D.R. Dreyer, C.W. Bielawski and R.S. Ruoff. 2011. Graphene-based polymer nanocomposites. Polymer, 52: 5–25.

[5] Dang, Z.-M., J.-K. Yuan, S.-H. Yao and R.-J. Liao. 2013. Flexible nanodielectric materials with high permittivity for power energy storage. Adv. Mater., 25: 6334–6365.

[6] Winey, K.I. and R.A. Vaia. 2007. Polymer nanocomposites. MRS Bull., 32: 314–322.

[7] Giannees, E.P. 1996. Polymer layered silicate nanocomposites. Adv. Mater., 8: 29–35.

[8] Bilyeu, B., W. Brostow and K.P. Menard. 1999. Epoxy thermosets and their applications I: chemical structures and applications. J. Mater. Educ., 21: 281–286.

[9] Ueki, T., S. Nishijima and Y. Izumi. 2005. Designing of epoxy resin systems for cryogenic use. Cryogenics, 45: 141–148.

[10] Zhao, Y., Z.-K. Chen, Y. Liu, H.-M. Xiao, Q.-P. Feng and S.-Y. Fu. 2013. Simultaneously enhanced cryogenic tensile strength and fracture toughness of epoxy resins by carboxylic nitrile-butadiene nano-rubber. Compos. Part A-Appl. S., 55: 178–187.

[11] Domun, N., H. Hadavinia, T. Zhang, T. Sainsbury, G.H. Liaghat and S. Vahid. 2015. Improving the fracture toughness and the strength of epoxy using nanomaterials—a review of the current status. Nanoscale, 7: 10294–10329.

[12] Martone, A., C. Formicola, M. Giordano and M. Zarrelli. 2010. Reinforcement efficiency of multi-walled carbon nanotube/epoxy nanocomposite. Compos. Sci. Technol., 70: 1154–1160.

[13] Sprenger, S. 2015. Improving mechanical properties of fiber-reinforced composites based on epoxy resins containing industrial surface-modified silica nanoparticles: review and outlook. J. Compos. Mater., 49: 53–63.

[14] Zou, H., S. Wu and J. Shen. 2008. Polymer/silica nanocomposites: preparation, characterization, properties, and applications. Chem. Rev., 108: 3893–3957.

[15] Nazari, A. and S. Riahi. 2011. The effects of SiO_2 nanoparticles on physical and mechanical properties of high strength compacting concrete. Compos. Part B-Eng., 42: 570–578.

[16] Khajeh, M., S. Laurent and K. Dastafkan. 2013. Nanoadsorbents: classification, preparation, and applications (with emphasis on aqueous media). Chem. Rev., 113: 7728–7768.

[17] Zucchelli, A., M.L. Focarete, C. Gualandi and S. Ramakrishna. 2011. Electrospun nanofibers for enhancing structural performance of composite materials. Polym. Adv. Technol., 22: 339–349.

[18] Liu, Y., S. Sagi, R. Chandrasekar, L. Zhang, N.E. Hedin and H. Fong. 2008. Preparation and characterization of electrospun SiO_2 nanofibers. J. Nanosci. Nanotechnol., 8: 1528–1536.

[19] Wen, S., L. Liu, L. Zhang, Q. Chen, L. Zhang and H. Fong. 2010. Hierarchical electrospun SiO_2 nanofibers containing SiO_2 nanoparticles with controllable surface-roughness and/or porosity. Mater. Lett., 64: 1517–1520.

[20] Chen, C., R.S. Justice, D.W. Schaefer and J.W. Baur. 2008. Highly dispersed nanosilica–epoxy resins with enhanced mechanical properties. Polymer, 49: 3805–3815.

[21] Schadler, L.S., S.K. Kumar, B.C. Benicewicz, S.L. Lewis and S.E. Harton. 2007. Designed interfaces in polymer nanocomposites: a fundamental viewpoint. MRS Bulletin, 32: 335–340.

[22] Tanahashi, M. 2010. Development of fabrication methods of filler/polymer nanocomposites: With focus on simple melt-compounding-based approach without surface modification of nanofillers. Materials, 3: 1593–1619.

[23] Rong, M.Z., M.Q. Zhang and W.H. Ruan. 2006. Surface modification of nanoscale fillers for improving properties of polymer nanocomposites: a review. Mater. Sci. Technol., 22: 787–796.

[24] Kickelbick, G. 2003. Concepts for the incorporation of inorganic building blocks into organic polymers on a nanoscale. Prog. Polym. Sci., 28: 83–114.

[25] Shu, H.-H., X.-H. Li and Z.-J. Zhang. 2008. Surface modified nano-silica and its action on polymer. Prog. Chem., 20: 1509–1514.

[26] Bailly, M., M. Kontopoulou and K. El Mabrouk. 2010. Effect of polymer/filler interactions on the structure and rheological properties of ethylene-octene copolymer/nanosilica composites. Polymer, 51: 5506–5515.

[27] Rahman, I.A. and V. Padavettan. 2012. Synthesis of silica nanoparticles by sol-gel: size-dependent properties, surface modification, and applications *in silica*-polymer nanocomposites-a review. J. Nanomater., 2012: 132424.

[28] Branda, F., B. Silvestri, G. Luciani and A. Costantini. 2007. The effect of mixing alkoxides on the Stöber particles size. Colloids Surface A, 299: 252–255.

[29] Kobler, J. and T. Bein. 2008. Porous thin films of functionalized mesoporous silica nanoparticles. ACS Nano, 2: 2324–2330.

[30] Suzuki, T.M., T. Nakamura, K. Fukumoto, M. Yamamoto, Y. Akimoto and K. Yano. 2008. Direct synthesis of amino-functionalized monodispersed mesoporous silica spheres and their catalytic activity for nitroaldol condensation. J. Mol. Catal. A-Chem., 280: 224–232.

[31] Hoffmann, F., M. Cornelius, J. Morell and M. Fröba. 2006. Cover picture: silica-based mesoporous organic–inorganic hybrid materials. Angewandte Chemie International Edition, 45: 3187.

[32] Yu, Z.Q., S.L. You and H. Baier. 2012. Effect of organosilane coupling agents on microstructure and properties of nanosilica/epoxy composites. Polym. Compos., 33: 1516–1524.

[33] Jiang, T., T. Kuila, N.H. Kim, B.-C. Ku and J.H. Lee. 2013. Enhanced mechanical properties of silanized silica nanoparticle attached graphene oxide/epoxy composites. Compos. Sci. Technol., 79: 115–125.

[34] Chen, L., S. Chai, K. Liu, N. Ning, J. Gao, Q. Liu, F. Chen and Q. Fu. 2012. Enhanced epoxy/silica composites mechanical properties by introducing graphene oxide to the interface. ACS Appl. Mater. Interfaces, 4: 4398–4404.

[35] Işın, D., N. Kayaman-Apohan and A. Güngör. 2009. Preparation and characterization of UV-curable epoxy/silica nanocomposite coatings. Prog. Org. Coat., 65: 477–483.

[36] Zheng, Y., Y. Zheng and R. Ning. 2003. Effects of nanoparticles SiO_2 on the performance of nanocomposites. Mater. Lett., 57: 2940–2944.

[37] Gao, Y., S. Sagi, L. Zhang, Y. Liao, D.M. Cowles, Y. Sun and H. Fong. 2008. Electrospun nano-scaled glass fiber reinforcement of Bis-GMA/TEGDMA dental composites. J. Appl. Polym. Sci., 110: 2063–2070.

[38] Chen, Q., L. Zhang, M.-K. Yoon, X.-F. Wu, R.H. Arefin and H. Fong. 2012. Preparation and evaluation of nano-epoxy composite resins containing electrospun glass nanofibers. J. Appl. Polym. Sci., 124: 444–451.

[39] Wang, G., D. Yu, R.V. Mohan, S. Gbewonyo and L. Zhang. 2016. A comparative study of nanoscale glass filler reinforced epoxy composites: Electrospun nanofiber vs nanoparticle. Compos. Sci. Technol., 129: 19–29.

CHAPTER 13

Gas Sensor Application of Zinc Oxide

Bharat R. Pant and *Ahalapitiya H. Jayatissa**

1. Introduction

Zinc oxide is an n-type semiconductor having a large band gap of 3.37 eV, and it belongs to group II–VI compounds. Due to its large bandgap, ZnO provides many benefits, such as high temperature and high power operation, higher breakdown voltages, lower electronic noise, ability to sustain large electrical fields, and non-toxicity [1, 2]. ZnO has many applications in different fields, such as optoelectronic devices, gas sensors, humidity sensors, surface acoustic wave (SAW) devices, and piezoelectric devices. ZnO is soluble in most of the acids, but not soluble in water, and therefore, this material can be synthesized and patterned to interstate on traditional VLSI processes. Among many applications of ZnO, the most important application is its use in the gas sensors. ZnO is an ultraviolet semiconductor with outstanding radiation hardness and high electron mobility. Also, high stability, selectivity, excellent surface to volume ratio in nanoscale, and variety of surface structures make ZnO an exceptional material for gas sensor applications. ZnO can be used as a gas sensor to detect different gases, such as ammonia, hydrogen, methane, carbon monoxide, nitric oxide, and organic vapors. ZnO can be obtained in 1D forms, such as nanorods, nanowires, and nanorings. Also, ZnO can be fabricated in 2D forms of nanostructures, such as nanobelts, and 3D forms of nanostructures, such as nanoparticles. The gas sensing properties of all these nanostructures have been investigated in recent years.

2. Theoretical Background

Since this chapter is devoted to the application of ZnO in gas sensors, the gas sensor mechanism related to an n-type metal oxide is described. Analogous to the

Nanotechnology & MEMS Laboratory, Mechanical, Industrial, and Manufacturing Engineering (MIME) Department, The University of Toledo, Oh 43606, USA.
* Corresponding author: ahalapitiya.jayatissa@utoledo.edu

grain boundary trapping model, the free electrons can be trapped by surface states, as illustrated in Figure 13.1a. The trapped electrons can also be released from the occupied surface states to the conduction band. The occupied surface states, i.e., the trapped electrons, can create a repulsive potential barrier at the surface to prevent further electron trapping. Given enough time, thermodynamic equilibrium can be established between the occupied surface states and the electrons in the conduction band. In a conductance measurement, an electrical potential is applied to a thin film semiconductor gas sensor. The potential drop occurs primarily across the grain boundaries, as illustrated in Figure 13.1b. Conduction electrons must overcome these potential barriers to conduct an electrical current. Therefore, the conductance of metal-oxide (G) is determined by the potential barrier (V_s), as described by the following relationship [3, 4]:

$$G = gq\ \mu_s\ N_d \exp\left(-\frac{qV_s}{kT}\right) = G_o \exp\left(-\frac{qV_s}{kT}\right), \tag{1}$$

where g is a constant determined by the semiconductor geometry and N_d is the donor density. Here, q and μ_s are the charge and the mobility of electrons, respectively. Although G_0 is a temperature-dependent parameter since $\mu_s\ \alpha\ T^{-3/2}$, it is not as sensitive to the temperature change as the exponential part [5]. Thus, G_0 is often regarded as a temperature-independent parameter. The potential barrier (V_s) depends on the occupied surface states (N_s), and can be described as:

$$V_s = \frac{q^2 N_s^2}{2\varepsilon\varepsilon_o N_d} \tag{2}$$

In order to explain baseline drift, a constant G_c is added into equation (2), so that

$$G = G_o \exp\left(-\frac{q^2 N_s^2}{2\varepsilon\varepsilon_o N_d kT}\right) + G_c \tag{3}$$

G_0 and G_c can be estimated from a dataset, while other parameters given in equation (3) depend on the MOS layer.

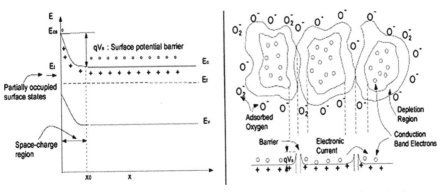

Figure 13.1. (a) Band diagram at the n-type metal-oxide grain boundaries, and (b) schematic of oxygen trapping states in a thin film of metal-oxide grains [3, 4].

The extrinsic surface state-trapping model describes the exchange of conduction electrons between the conduction band and the extrinsic surface states at non-equilibrium conditions. This situation can be applied in ambient because adsorbed oxygen contributes to the negatively charged surface states. As considered before, the ionized oxygen (O^-) occupies the surface state in the metal-oxide films, whereas the chemisorbed (not ionized) oxygen can be considered as the unoccupied surface state.

The gas sensing mechanism depends on the type of careers in the MOS layer. If we consider an n-type MOS material, the sensing mechanism can be expressed as follows. When a reducing gas is adsorbed into the grain boundary region, the trapped oxygen reacts with the surface, removing the adsorbed oxygen, such that the electron density is increased in the grain boundary region, whereas an oxidizing gas causes a decrease in the electron density. In general, the electrical conductivity is increased when a reducing gas is in contact, or the conductivity is decreased when an oxidizing gas is in contact with an n-type MOS. The opposite sensing mechanism can be seen for the p-type MOS [6, 7].

Figure 13.1 illustrates the experimental results of oxygen species adsorbed on the SnO_2 surfaces [19]. It can be found that at a relatively low temperature range, the molecule-ion species (O_2^- or O_2^{2-}) dominated the surface of the metal oxide. At a relatively high-temperature range, the atomic-ion species (O^- or O^{2-}) dominated the metal oxide surface. Yamazoe et al. [8] reported that four kinds of oxygen species are formed on the SnO_2 surface, such as O_2 at 80°C, O_2^- at 150°C, O^- or O^{2-} at 560°C, and a part of lattice oxygen at above 600°C. The O^- (300–500°C) is found in the most reactive oxygen species with gases. Therefore, the most MOS gas sensors are operated in this temperature range.

The sensing properties of MOS, including ZnO, depend on a number of factors, such as operation temperature, porosity of MOS layer, grain size, physical structure, and dopants. All these factors affect the surface states of MOS layer, which vary the sensing mechanism of metal oxides. Thus, ZnO gas sensor characteristics can be tuned in numerous ways. However, in this chapter, we describe the change of gas sensor properties depending on the experimental conditions of gas sensor preparation.

3. Thin Films and Thick Films

Metal oxide films can be categorized into two types—thin films and thick films. The main difference between thick film and thin films is the thickness of the film. Thin films have a thickness below 100 nm, whereas thick film sensors have a typical thickness in the 20–500 μm range. The deposition techniques for thick and thin films are different. To deposit a thin film, vacuum coating technology, such as physical vapor deposition and sputtering, and chemical vapor deposition, solution-based sol-gel method, spray pyrolysis, etc. are used. To coat thick films, techniques such as screen printing, spray coating, mist coating, and dip coating are used. Thin films' coating process is more expensive than thick film coating process, but the thin film is of a very high quality, uniform, precise, and stable.

Kalyamwar et al. [9] synthesized ZnO thick film using the hydrothermal method and deposited on a glass substrate using screen printing for the detection of H_2S gas. The sensor was able to detect 25 ppm of H_2S at room temperature with good sensitivity. The sample that was sonicated for 90 minutes showed the highest sensitivity, whereas the sample that was sonicated for 30 minutes showed the lowest sensitivity. Hou et al. [10] used the sol-gel method to coat Al-doped ZnO thin film and tested for H_2, NH_3, and CH_4 gas. The Al-ZnO properties of the films were modified using a pulsed laser system. The irradiated Al-ZnO sample showed higher sensing performance than as-deposited Al-ZnO samples.

4. Dependence of Sensitivity on Temperature

The sensitivity of the metal oxides gas sensor depends upon operating temperature. The optimal operating temperature is that temperature at which metal oxide shows the highest sensitivity. Sensors operating at a higher temperature consume high power. The sensors that can be operated at a lower temperature with good sensitivity are desirable because of low operation cost. Usually, the sensitivity of the metal oxide increases with increasing temperature until a certain temperature. However, the sensitivity decreases when the temperature is further increased. The cause of the change in the resistance of the metal is due to the presence of oxygen species, such as O^{2-}, O^-, and O_2^-. These oxygen species are absorbed on the surface of the film. They gain electrons from conduction band and become negatively charged. The stability of these species depends upon the temperature. O_2^- is stable below 100°C, O^- is stable between 100°C to 300°C, and O^{2-} is stable above 300°C [11]. When the target gas is exposed to the surface of metal oxide, the reaction occurs between oxygen species and the target. As a result of the reaction, electrons are injected back to the grain of the film. In this way, the resistance of the film is changed.

Wagh et al. [12] explained the NH_3 sensing mechanism of ZnO thick films, and the effect of temperature on the sensor's response. On the surface of the film, ammonia gas reacts with adsorbed oxygen species, and the following reactions happen at the surface:

$$2NH_3 + 3O^-_{ZnO} \rightarrow 3H_2O + N_2 + 3e^- \tag{4}$$

The equation (4) requires some activation energy to proceed. The activation energy is provided in the form of heat energy. When the temperature increases, the activation energy also increases, resulting in an increase in the response (sensitivity) of the gas sensor. The temperature at which sensitivity reaches maximum is the point at which the maximum thermal energy is needed. When the temperature is further increased, sensitivity starts to decrease, because oxygen species are desorbed from the surface at higher rates. Additionally, at a higher temperature, due to intrinsic thermal excitation, carrier concentration increases, and Debye length decreases. As a result, sensitivity decreases.

Figure 13.2 shows the effect of temperature on the resistance and response of the sensor. From Figure 13.2a, it can be seen that the resistance of the film decreases with increasing temperature. The oxygen is adsorbed on the surface of the film and the oxygen gains electrons from the conduction band, resulting in the depletion of electrons. For n-type films, such as ZnO, the depletion of electrons in the conduction band leads to increase in the resistance. However, the intrinsic resistance decreases

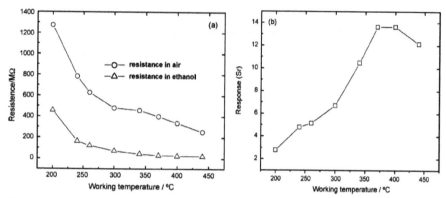

Figure 13.2. (a) Resistance of ZnO nanopillars in the air and in ethanol at a different temperature. (b) Response of ZnO nanopillar sensor to 50 ppm of ethanol at different operating temperatures [13].

with an increase in temperature. The overall change in resistance causes a change in sensitivity. Figure 13.2b shows the sensitivity of the film to 50 ppm of ethanol at different operating temperatures. It can be seen that sensitivity increases with increasing operating temperatures until it reaches 350°C, and remains constant until 400°C. However, above 400°C, the sensitivity starts to decrease. This might be attributed to the increase in desorption rate of adsorbed oxygen.

It can be seen from Figure 13.3 that the sensitivity of 65 nm and 188.5 nm thin films increases rapidly with increasing operating temperature until 400°C, but above 400°C, the sensitivity decreases rapidly. For 280 nm and 390 nm thin film, the sensitivity increases with increasing temperatures, and the sensitivity does not decrease above 400°C. The sensitivity seems to be dependent upon both temperature and thickness of the films. At a lower temperature, lower sensitivity is expected, because the metal oxide surfaces do not have enough activation energy to overcome the energy barrier.

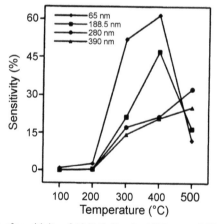

Figure 13.3. Dependence of sensitivity of Al-ZnO film for 1000 ppm of CO gas sensing on operating temperatures [14].

5. Dependence of Sensitivity on Grain Size

Grain size refers to the diameter of the grains present in the films. The films are composed of grains of various diameters, and within the same film, there might be grains of various sizes. The grain size is different from crystal size, which is the size of crystal present within the grain [15]. The grain is composed of many crystals.

The sensitivity of the gas sensor strongly depends on the grain size. Smaller grain size increases the surface area of the film, which allows more interaction of gas molecules with films. As a result, sensitivity increases. Larger grain size means a smaller surface area of interaction with gas molecules. As a result, the sensitivity of the sensor decreases. However, the sensitivity does not always increase with decreasing grain size. For every gas sensor, there is an optimum size of the grain at which it displays maximum sensitivity. When the grain size becomes too small, the porosity of the film decreases. Due to low porosity of the grains, the target gas cannot penetrate deeper into the films, hence sensitivity decreases.

The effects of grain size and film thickness on the sensitivity of the aluminum-doped zinc oxide gas sensor have been investigated [14]. Four AZO films having a thickness of 65 nm, 188.5 nm, 280 nm, and 390 nm were coated on SiO_2/Si substrate using RF magnetron sputtering methods. It was found that the grain size increases with increasing film thickness, and the film having 65 nm thickness has the smallest grain size. From Figure 13.4, it can be seen that sensitivity strongly depends on grain size. The film with a thickness of 65 nm has the smallest grain size and shows the highest sensitivity for 1000 ppm CO gas at 300°C. The lowest sensitivity was found for films having 390 nm thickness and the largest grain size. There is a strong relationship between grain size and the surface area of the films. As the grain size decreases, the surface area increases. More surface area means more interaction of gas molecules with materials, and hence the sensitivity increases.

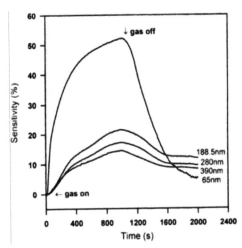

Figure 13.4. Effect of grain size on the sensitivity of the Al-ZnO sample at 300°C under 1000 ppm of CO gas [14].

6. Dependence of Sensitivity on Film Thickness

The gas sensing properties of the films strongly depend on the thickness of the film. The thickness of the film affects sensitivity in two different ways. First, the thickness is related to the grain size of the film. Thicker films have a large grain size, whereas thin films have a smaller grain size. Films having large grain size have lesser surface area than films having small grains. The large surface area leads to more surface reactions between target gas and grains, and hence sensitivity increases. Second, thin films tend to be more porous than thick films. Fabrication of hierarchically three-dimensional (3D) porous ZnO architectures by a template-free, economical hydrothermal method combined with subsequent calcination has been reported [17]. They have prepared these 3D materials by a reaction between zinc carbonate and water at 300°C in a pressurized autoclave. The material has been successfully used to detect ethanol and methanol, which have OH-groups.

   To study the effect of film thickness on gas sensing properties, the ZnO films have been fabricated on SiO_2/Si substrates using electron beam evaporation method [18]. The samples having thickness of 100 nm, 150 nm, and 200 nm were obtained, and these samples were tested for low ppm of H_2 gas (40 ppm). From Figure 13.5, it can be seen that sensitivity decreases with increasing thickness of the film. The film having thickness of 100 nm displayed the highest sensitivity. It might be attributed to the deeper penetration of oxygen molecules into low thickness film, because the smaller grain sizes and wider voids resulted in lower carrier concentration and an increase in sensitivity [18].

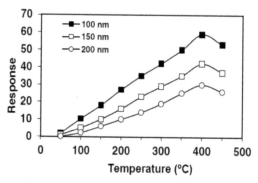

Figure 13.5. Response of the ZnO to H_2 gas as a function of operating temperature and film thickness [12].

7. Dependence of Sensitivity on Doping

Impurity doping of ZnO has attracted much attention for tuning carrier concentration of ZnO. Common doping includes Al and Ga. Metal oxide gas sensors have shown promising results for their applications as gas sensors. Some of the good characteristics of metal oxides are low cost, thermally and chemically stable, high response and capacity to detect a wide variety of gases. Besides good characteristics, there are some limitations associated with the metal oxides gas sensor, such as higher operating temperature, low response, and higher response and recovery time. To overcome these limitations, many efforts have been done. One of the most important efforts is doping of metal oxides with a noble metal, such as Pt, Al, and Pd, graphene, and other metal oxides. It has been reported that doping of the metal oxides results in better performance of ZnO gas sensors [16]. The doping metals, such as Pt, Pd, and Al act as a catalyst and improve the reaction between adsorbed oxygen and target gas, and also decrease the energy needed to split O_2 or target gas molecules [17]. Wagh et al. [12] doped ZnO with ruthenium oxide to see the effect of doping on the gas sensing properties. It was found that the sensitivity of doped ZnO was higher than the pure ZnO. Also, the operating temperature decreased from 350°C to 250°C after doping. The Ru species effectively promote the catalytic reaction between the surface material and target gas, resulting in an increase in sensitivity of the sensor.

Al-doped zinc oxide thin films have been used to detect liquid petroleum gases (LPG) [19]. Chemical spray pyrolysis technique was used to coat the films, and the concentration of the Al was varied from 0 to 1.5%. Figure 13.6 shows the response of pure zinc oxide and Al-doped zinc oxide with three different concentration of Al (0.5, 0.75, and 1%) as a function of operating temperatures. The sensor was tested at three concentrations of LPG, namely, 0.5 vol. %, 0.75 vol. %, and 1.0 vol. %. Compared with undoped ZnO, the Al-doped ZnO films showed an improved response for LPG. The undoped ZnO displayed the highest sensitivity of 40% for 1 vol. % of LPG gas

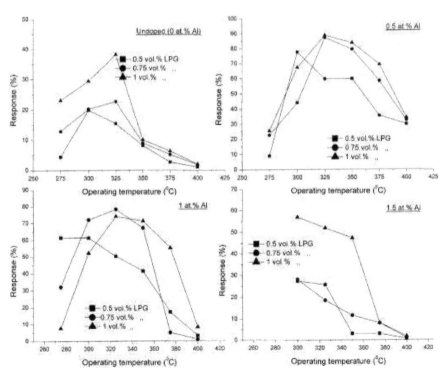

Figure 13.6. Response of Al-doped (0 to 1.5 at %) zinc oxide to LPG as a function of operating temperatures [19].

at 325, while 0.5 at % Al-doped zinc showed the sensitivity of 89% under same conditions. Also, the 0.5 at % Al-doped zinc showed maximum sensitivity when compared to other samples. It might be attributed to the smallest crystal size of this film (103 nm) and the largest surface area of the film [19].

Hou et al. [10] also reported that the Al doping enhanced the sensitivity and selectivity of ZnO prepared by a sol-gel process. In this case, Al has been doped by mixing Al-precursor solution with sol-gel mixture for ZnO coating. They have reported that Al-doped ZnO can be used to fabricate low resistive gas sensors, which do not need a high impedance peripherical electronic circuitry to operate the gas sensors. Figure 13.7a shows the sensitivity for H_2 in the presence of NH_3. The selectivity of H_2 versus NH_3 was tested under the same conditions used for H_2 versus the air. In this case, 15–100 ppm of NH_3 was used because a hazardous level of NH_3 was established within this range [10]. The sensor response for 75 ppm NH_3 in the air is around 40%. The sensitivity reached 70% while passing 2000 ppm H_2 to 75 ppm NH_3 containing air (Figure 13.7a). Figure 13.7b shows the gas sensitivity for CH_4 in the presence of 75 ppm NH_3 in the air. The results shown in Figure 13.7b indicated that the selectivity of NH_3 versus CH_4 was excellent. The results indicated that the selectivity of these reducing gases was in the order of $NH_3 > H_2 > CH_4$.

Doping of ZnO with transition metals and other elements has been also investigated for gas sensor applications. The primary purpose of this research is to enhance low temperature operation capability and selectivity of sensor for a particular gas. Paraguay et al. [20] used spray pyrolysis technique to deposit In, Cu, Fe, and Sn-doped ZnO gas sensor for the detection of ethanol vapor. It was found that the types of dopant affect the microstructure and surface morphology of the films. The sensors were tested as a function of ethanol concentrations at a fixed temperature, and a function of temperatures at fixed ethanol concentration. It can be seen from Figure 13.8 that the sensitivity of doped ZnO films is higher than undoped zinc oxide, except Cu-doped zinc oxide. The highest sensitivity was shown by the Sn-doped (0.4 at %) sample. The second highest sensitivity was shown by Al-doped zinc oxide (1.8 at %).

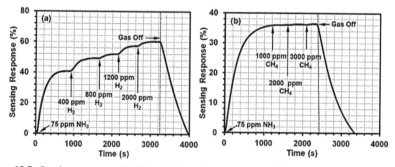

Figure 13.7. Sensing response of 3.0% of Al-doped sensor prepared by precursor-1 for (a) H_2 in the presence of NH_3, and (b) CH_4 in the presence of NH_3 [12].

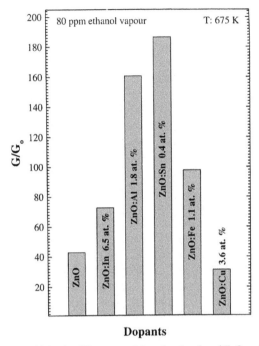

Figure 13.8. Maximum sensitivity for different metal doped and undoped ZnO under 80 ppm of ethanol and at a fixed temperature of 402°C [20].

8. Tuning of Gas Sensor Characteristics by Surface Modifications

It is now obvious that the sensitivity of the metal oxide gas sensor strongly depends upon temperature, grain size, film thickness, and doping. There is another way that can be used to improve the sensitivity, which is surface modifications. Surface modifications are done to enhance the gas sensing performance of the film. After surface modification, properties such as grain sizes, transmittance, conductance, and porosity are changed. These changes directly affect the performance of the sensor. For example, Kakati et al. [21] coated zinc oxide nanorods on alumina substrate using the hydrothermal method. The surface modification of the nanorods was done by coating a thin layer of indium antimonide (InSb) on top of ZnO by using the thermal evaporation method. The activation energy of the InSb deposited ZnO nanorods was found to be very small compared to as-deposited ZnO. Also, the sensitivity of InSb deposited ZnO towards acetone was very high compared to that of pure ZnO nanorods.

Hou et al. [10] deposited Al-doped zinc oxide thin film on a glass substrate using sol-gel methods. Surface modification of the zinc oxide films was done by irradiating the surface with a pulsed laser system having a wavelength of 532 nm, pulse duration of 8 ns, pulse frequency of 5 KHz, and laser fluence in the range of 1.06–3.58 J/cm². The electrical property, optical property, grain size, etc. were found to be changed after laser irradiation. The low-intensity laser formed highly crystalline films with low conductivity and large grains, but high-intensity laser formed low crystalline films with high conductivity and low grain sizes.

From Figure 13.9a, it can be seen that the conductivity of the films increased with increasing fluence of the laser. The conductivity of 3.58 J/cm² irradiated film was the highest, while as-deposited showed the lowest conductivity. This might be attributed to the higher crystallinity and lesser number of charge carrier caused by lower laser fluence irradiation and vice versa [10]. Figure 13.9b shows the sensitivity of 2-layer as-deposited and laser-irradiated ZnO samples under various concentrations of H_2 gas. The highest conductivity was shown by 2.92 J/cm² laser-irradiated sample. Compared to as-deposited ZnO, laser-irradiated films show higher sensitivity, except for 3.58 J/cm². The conductivity decreased when the laser fluence was increased from 2.92 J/cm² to 3.58 J/cm². It can be concluded that the sensitivity of ZnO film can be improved with optimal laser fluence.

Aslam et al. [22] reported the fabrication of RuO_2-doped ZnO films for the detection of ammonia gas. The percentage of RuO_2 was varied from 0.17 at. % to 2.74 at. %. Properties such as electrical, surface morphology and grain size were changed after doping. Also, the sensitivity of the doped film was dramatically increased after doping. The response time for pure ZnO and RuO_2-doped ZnO was found to be 20 seconds and 10 seconds, respectively. It might be attributed to the fact that Ru species enhances fast electron transfer between substrate and absorbate by catalyzing the surface reaction due to the formation of surface states just below the conduction band [22]. They have reported that the Ru-doped ZnO is 400 times more sensitive than undoped ZnO for ammonia.

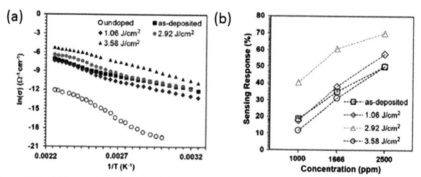

Figure 13.9. (a) Logarithm of conductivity versus the inverse of absolute temperature graph of as-deposited and laser-irradiated 5-layer ZnO samples. (b) The sensitivity of as-deposited and pulsed laser-irradiated 2-layer ZnO towards H_2 gas at 130°C [4].

9. Conclusion

Zinc oxide is a very popular material for application in the gas sensor. There are two types of zinc oxide films, namely, thin film and thick films. Some good characteristics of ZnO are high stability, high sensitivity, low response and recovery time, and ability to detect a wide range of gases. These characteristics can be improved by using many techniques, such as doping with different metal (Pt, Pd, Ru, and Al) and different metal oxides (RuO$_2$, Al$_2$O$_3$), and by doing surface modifications. The sensitivity of the ZnO sensor strongly depends upon film thickness, doping, grain size, and operating temperatures. There is always an optimum temperature at which the sensor displays maximum sensitivity. If the temperature is increased above or decreased below the optimum temperature, sensitivity decreases. There is also an optimum thickness and grain size at which films show maximum sensitivity. Usually, sensitivity increases if the grain size decreases. Surface modification can be done to improve the performance of the sensor. The surface modification can be done by using many methods, such as laser irradiation on the surface and coating another material on top of the ZnO surface.

References

[1] Greenwood, N.N. and A. Earnshaw. 1997. Chemistry of the Elements (2nd ed.). Butterworth-Heinemann. ISBN 0-08-037941-9.
[2] Jayatissa, A.H., A.M. Soleimanpour and Y. Hao. 2012. Manufacturing of multifunctional nanocrystalline ZnO thin films. Adv. Mat. Res., 383: 4073.
[3] Morrison, S.R. 1977. The Chemical Physics of Surfaces. 1st Edition, Plenum Press, New York, NY.
[4] Ding, J., T.J. McAvoy and R.E. Cavicchi. 2001. Surface state trapping models for SnO$_2$-based microhotplate sensors. Sens. Actuators B: Chem., 77: 597.
[5] Madou, M.J. and S.R. Morrison. 1989. Chemical Sensing with Solid State Devices. Academic Press, Boston.
[6] Jayatissa, A.H. and T. Gupta. 2018. Metal-Oxide-Semiconductor (MOS) Gas Sensors. In Advances in Sensors, Chemical Sensor and Biosensor (Edited by Sergey Y. Yurish), 6: 50.
[7] Barsan, N. and U. Weimar. 2001. Conduction model of metal oxide gas sensors. J. Electroceramics, 7: 143.
[8] Yamazoe, N., J. Fuchigami, M. Kishikawa and T. Seiyama. 1979. Interactions of tin oxide surface with O$_2$, H$_2$O, and H$_2$. Surf. Sci., 86: 335.
[9] Kalyamwar, V.S., F.C. Raghuwanshi, N.L. Jadhao and A.J. Gadewar. 2013. Zinc oxide nanostructure thick films as H$_2$S gas sensors at room temperature. J. Sens. Tech., 3: 31.
[10] Hou, Y. and A.H. Jayatissa. 2014. Effect of laser irradiation on gas sensing properties of sol-gel derived nanocrystalline Al-doped ZnO thin films. Thin Solid Films, 562: 585.
[11] Soleimanpour, A.M., A.H. Jayatissa and G. Sumanasekera. 2013. Surface and gas sensing properties of nanocrystalline nickel oxide thin films. Appl. Surf. Sci., 276: 291.
[12] Wagh, M.S., G.H. Jain, D.R. Patil, S.A. PatilL and A. Patilet. 2006. Modified zinc oxide thick film resistors as NH$_3$ gas sensor. Sens. Actuator B, 115: 128.
[13] Bie, L.-J., X.N. Yan, J. Yin, Y.Q. Duan and Z.H. Yuan. 2007. Nanopillar ZnO gas sensor for hydrogen and ethanol. Sens. Actuators B, 126: 604.
[14] Chang, J.F., H.H. Kuo, I.C. Leu and M.H. Hon. 2002. The effects of thickness and operation temperature on ZnO:Al thin film CO gas sensor. Sens. Actuators B, 84: 258.
[15] Hou, Y. and A.H. Jayatissa. 2014. Low resistive gallium doped nanocrystalline zinc oxide for gas sensor application via sol–gel process. Sens. Actuators B: Chem., 204: 310.
[16] Xie, H., K. Wang, Z. Zhang, X. Zhao, F. Liu and H. Mu. 2015. Temperature and thickness dependence of the sensitivity of nitrogen dioxide graphene gas sensors modified by the atomic layer deposited zinc oxide films. RSC Adv., 5: 28030.
[17] Zhang, J., S. Wang, M. Xu, Y. Wang, B. Zhu, S. Zhang, W. Huang and S. Wu. 2009. Hierarchically porous ZnO architectures for gas sensor application. Cryst. Growth & Design, 9: 3533.
[18] Teimoori, F., K. Khojier and N.Z. Dehnavi. 2015. On the dependence of H$_2$ gas sensitivity of ZnO thin films on film thickness. Procedia Mat. Sci., 11: 474.
[19] Sahay, P.P. and R.K. Nath. 2008. Al-doped zinc oxide thin films for liquid petroleum gas (LPG) sensors. Sens. Actuators B, 133: 222.

[20] Paraguay, F., D. Miki-Yoshida, J. Morales, J. Solis and W. Estrada. 2000. Influence of Al, In, Cu, Fe and Sn dopants on the response of thin film ZnO gas sensor to ethanol vapour. Thin Solid Films, 373: 137.

[21] Kakati, N., S.H. Jee, S.H. Kim, H.K. Lee and Y.S. Yoonet. 2009. Sensitivity enhancement of ZnO nanorod gas sensors with surface modification by an InSb thin film. Jpn. J. Appl. Phys., 48: 105002.

[22] Aslam, M., V.A. Chaudhary, S. Mulla, S.R. Sainkar, A.B. Mandale, A.A. Belhekar and K. Vijayamohanan. 1999. A highly selective ammonia gas sensor using surface-ruthenated zinc oxide. Sens. Actuators, 75: 162.

CHAPTER 14

Titanium Dioxide as A Photo Catalyst Material

A Review

Yogesh Singh,[1,2] *Sanju Rani,*[1,2] *Manoj Kumar,*[1,2] *Rahul Kumar*[1,2] and
V.N. Singh[1,2,*]

1. Introduction

The biggest problem the world is facing today is environmental pollution. It can only be solved by using clean and green energy [1–4]. The world population has increased enormously over the past few years, which has given much pace to the environmental pollution and its outcomes, such as toxic air, toxic water, and various diseases. That is why the world today needs an energy source which is clean, green, sustainable, and globally available. Hydrogen is considered the cleanest source of energy, and it is available in abundance in the form of water. Its highly combustible characteristics enable it to be a potential energy vector in the future. The only challenge is to separate hydrogen from oxygen in water molecule. Researchers have made significant progress in splitting water using titanium dioxide. Many other semiconductor materials, such as TiO_2, ZnO, Fe_2O_3, CdS, and WO_3 also show photocatalytic activity in the presence of sunlight, but TiO_2 is the most widely used because of its easy availability, non-toxicity, high comparative yield, and high stability [5–9]. In spite of all these qualities, there is a band gap bottleneck in it for being used on a commercial level. It has a wide band gap, which means only a small portion of sunlight (which is UV) can contribute to photocatalytic activity, which accounts for nearly 5% of all sunlight that falls on it. The biggest part of solar spectrum which lies in the visible region does not take part in photocatalytic activity. The reason behind it is that the recombination rate of photo generated electron-hole pair is very rapid, having a lifespan of 10 ns [10]. So, the electron and hole produced are recombined before they break the hydrogen from water molecule. Researchers have even found the way to produce hydrogen through

[1] CSIR-National Physical Laboratory, Dr K.S. Krishnan Marg, New Delhi-110012, India.
[2] Academy of Scientific and Innovative Research (AcSIR), Ghaziabad-201002, India.
* Corresponding author: singhvn@nplindia.org

the photocatalyst producing hydrogen and simultaneously removing a wide range of organic pollutants [11]. This book chapter tells the audience about the mechanism used in the process, and also briefly explains the composites and other materials that are being experimented for the same process through Table 14.1. Later it discusses about the further scope in this field.

2. Historical Overview

Before the formulation of band theory, a group of researchers around 1960s observed that a few semiconductors, such as ZnO and TiO_2 show some activity under UV light, for example, absorbing of H_2O or oxygen molecule on the surface, which also depends on the surface parameter and other things. This phenomenon is now explained by the band theory. In the journal Kogyo *Kagaku Zasshi*, two researchers Kato and Masuo published the oxidation of tetralin with titanium dioxide in UV light. Kato and Masuo were the first who showed a significant invention of photoinduced catalytic reaction of titanium dioxide in the UV region [10]. After that, many other researchers and scientist worked on the same field, but not much advancement was shown, except the same UV assisted photocatalytic reaction.

In early 70s, two Japanese scientists Honda and Fujishima found that electrolysis of water under UV light from TiO_2 as a photo anode with Pt electrode as cathode was at a lower voltage than the normal electrolysis [12]. At the end of the decade, two more scientists Schrauzer and Guth showed the same process in powdered form of titanium dioxide mixed with a little amount of Pt [13]. This led to the phenomenon of charge separation in photocatalysis, which is also called photo electrochemical process.

After these, efforts are being made to achieve the photocatalysis systems which have better conversion efficiency. New methods are being developed so that it comes under visible light region using doping, new system design making new composites, nanoparticles, and using different methods. Now worldwide, around 600 papers are published yearly, and it has become a new scope for scientific forum.

3. Types

Photocatalytic reaction occurs via two types—heterogeneously and homogeneously. Both the technologies are promising, but in the past few years, heterogeneous photocatalysis has got much interest among researchers. This is because it has the potential to solve energy demands and seems to be more sustainable for the future.

3.1 Homogeneous Photocatalysis

If the reactants and photocatalysts are present in the same phase, it is called homogeneous photocatalysis [14]. The main reason why researchers study this specific field is so that they can find out how toxic elements in water can degrade and help us to reduce the toxicity of water [15]. To remove the pollutants in water, they are first oxidized, which in turn gives a very powerful hydroxide ion (OH) or a radical [15, 16]. A few examples are photodecomposition of hydrogen peroxide, photolysis of ozone and transition metal oxides, etc. [17–19].

3.2 *Heterogeneous Photocatalysis*

In Heterogeneous photocatalysis, the reactants and the photocatalyst are in different phases. The semiconductor surface acts as a free path for the photocatalytic activity. This process is not only limited to the solid-liquid transitions, but also to liquid–gas transitions [20–22]. The phenomenon is used in solving a wide range of scientific problems, such as CO_2 reduction, water-splitting, and anti-dust surfaces [14, 23, 24].

4. Mechanism

The photocatalytic reaction is a multistep process happening in a sequence. The first step of this includes photons falling on the surface of the semiconductor. These photons have different energies coming from different wavelengths of the solar spectrum [25]. Now the semiconductor has a certain energy band gap. The electrons in the valance band are excited by the photon of equal or greater energy than the band gap, leaving a hole behind. This electron hole pair has a very small lifetime, in the order of 8–9 ns. They recombine in the bulk or on the surface before they help in photocatalytic reaction. Few of the e-h pair which reaches the surface oxidize or reduce the absorbent on the surface [1, 2, 25].

4.1 *For H Production*

Two conditions should be achieved in terms of band gap to achieve H splitting

i) energy level of conduction band should be lower than the hydrogen production, i.e., $E_{CB} < E_{H2/H2O}$.

ii) energy level of valance band should be higher than the oxidation [3], i.e., $E_{VB} < E_{O2/H2O}$.

After all the efforts, scientists are still not able to produce high output hydrogen splitting devices. The devices made till date have a very low efficiency. Mainly the following factors will work to increase the efficiency of these devices [16, 26].

i) Band gap—The band gap of TiO_2 is 3.3 eV, which lies near the UV region. To get the maximum output, we must be capable of using most of the solar spectrum, specially the visible region, for photocatalytic reaction. So we must choose such a material whose band gap lies in the visible region [27].

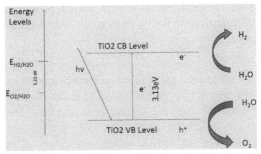

Figure 14.1. Mechanistic diagram of titanium dioxide water splitting.

ii) Absorbance—Not all the spectrum fallen on semiconductor is absorbed. Most of its part is reflected back or lost by some other means. To remove this hurdle, the surface area should be larger. For this, work is being done on nanoparticles, doped material, and nanotubes-based semiconductors [28].

iii) Lifetime of e-h pair—As already discussed, the e-h pair generated has a very rapid recombination. Achieving the larger life time is a herculean task for a good efficiency device. One method is to donate electrons repeatedly. Organic hole scavengers also work in this series, and result in increased quantum efficiency [29].

5. Methods for Synthesis

The melting point of TiO_2 is 1843°C. Due to this, the deposition of TiO_2 by conventional thermal deposition is a very tough task to achieve. The most commonly used deposition techniques are by spin coating, sputtering, electrophoretic deposition, dip coating, chemical vapor deposition, physical vapor deposition, etc.

i) Electrophoretic deposition

This method is the most widely used due to several reasons, because it gives a film of uniform thickness without any big setup. It consists of one cathode and an anode submerged in a charging solution mixed with depositing material, which is made of iodine most of the time. The charge particles flow towards the cathode and are deposited on the cathode when a dc voltage is applied. The thickness of the film depends on the concentration of charging solution, applied voltage, and area of film. The main disadvantage of this method is that we can deposit the film only on conducting surfaces, such as ITO plate, FTO plate. Deposition of the film on bare glass substrate is not possible.

ii) Hydro/solvothermal methods

Hydrothermal and solvothermal methods are similar kinds of methods. In hydrothermal method, a very high vapor pressure is applied to the material, and it is allowed to crystallize. In the method, different phases are obtained at different temperatures, and also the size of nanoparticles can be optimized by this method. The main disadvantage of this method is that deposition on the substrate can be very difficult, although in some cases, various nanorods are also formed on glass, etc. substrate by this method. On the other hand, the solvothermal method is applied on non-aqueous solvent, and its properties can be optimised conveniently as compared to the hydrothermal. Also, high temperature solvents can be used in it.

iii) Spin coating method

In this method, a transparent gel of titanium precursor is made. Then the substrate is rotated at the required rpm. A drop of liquid is cast over the substrate, and the same drop is dispersed over the surface as a fine film. This is a widely used method and is convenient for any surface. The properties of the film depend on the concentration of sol, rpm of substrate, and type of precursor used. In a typical experiment to make the gel, 0.75 ml of titanium isopropoxide precursor was added to 5 ml ethanol, and stirred for 10 minutes, after which we got a white solution. After all this, 0.05 ml of HCl was added to the same sol, as adding HCl will make the solution transparent.

iv) Dip coating technique

In dip coating technique, the substrate is allowed to dip in the sol and then it is taken out at an infinitesimally slow rate. The parameters on which the properties of film depend are force of inertia, viscous drag, gravitational force, and surface tension. The advantage of dip coating is that it is low in cost, the layer thickness can be adjusted, and it is easy to operate. The drawbacks of this technique are that the overall process is time consuming and it blocks the screen, which has a major impact in the final output.

6. Recent Modifications for Enhanced Result

In order to get better results, researchers are working on a number of aspects to find some enhanced result. To get some better results, various factors have contributed to various properties, such as light captivating properties, charge separation, and lifetime, high oxidizing and reduction reactions, and better quantum efficiency. These modifications and their effects are discussed below.

   i) Morphological effects—TiO_2 has three discrete phases—anatase, rutile, and brookite. Of all these, anatase has been found to be most suitable for photocatalytic behavior. Saravanan et al. reported the enhanced result when nanoparticles of smaller size are used as compared to rod-like structures [13]. It is believed that larger surface to volume gives better result. Not only in the case of water splitting, but water purification has also shown better results. Oxidation and reduction in photocatalytic behavior acts on the surface so the surface properties plays a big role in determining the output results [30–33].

  ii) Temperature effect—Thermal energy has a direct relation with the number of e-h recombination; the more the recombination, less the photocatalytic activity, and less will be the hydrogen production [34–36]. Malato and Rajeshwar have reported that beyond 80°C, the e-h recombination increased significantly, and subsequently, there is a decrease in photocatalytic efficiency [37, 38]. There is no need of temperature for photocatalytic activity for these systems. They work suitably at room temperature. The optimum range depends on the activation energy of the substance, and at zero degree Celsius, the activation energy was at its peak, reported Chatterjee and Dasgupta [39].

 iii) pH defines different properties, and researchers had shown that the oxidation and reduction of the pollutants is dependent on the pH. In photocatalytic reactions, sometimes acidic end products are formed. It was observed that in pH greater than 10, –OH ions counter these acidic products. In another scenario, when the pH was less than 5, the degradation was low for acidic medium r.

  iv) Preparation method: The preparations of semiconductor materials have an impact on various properties. Indris et al. reported that TiO_2 nanoparticles milled in high energy milling machines with milling time up to 40 minutes have their photocatalytic activity as a function of milling time [40]. This change in property was explained by two factors- one by increased surface area, and the other by change in electronic structure. In another experiment, Kang et al. reported that TiO_2 grinding with ethanol in the presence of air has shown a shift in absorption peak towards lower energy region [41].

7. Doping Methods and Outcomes

Several methods have been adopted to get better results, high absorbance, and high hydrogen yield, but no great success has been achieved [23, 42, 51, 52, 43–50]. To go one more step ahead, various doping techniques and other method have been performed.

Table 14.1. The table showing different materials used for the mechanism their method and brief outcome.

Sr. n.	Doping substance	Method used	Brief detail	Outcome	Reference
1	CNT	Hydrothermal-sonothermal	CNT-TiO_2 Nanocomposites were made after dispersing MWCNT in ethanol/H_2O, then sonication at 80°C. After this, it was added to titanium butoxide going through hydrothermal process after sonication. After this, various characterizations were done, such as TGA analysis, XRD, TEM, etc.	*For CO_2 reduction—* For CO_2 photocatalytic reduction a certain kind of experiment was done in closed gas recirculation quartz reactor and some improved result for CO_2 reduction was found. *For water splitting—* Hydrogen yield for various pH values was observed, and it was found that lower pH value favors the hydrogen yield. The highest yield was for pH 2, which was 69.41 μ mol $g^{-1} h^{-1}$	[53]
2	rGO	Aqueous processing	Reduced graphene oxide was made by modified hummer's method and a composite of titania and rGO was made in the ratio of 70:30 and then coated on various kinds of substrates, i.e., rubber, polycarbonate, and glass. After sample preparation various testing and characterizations were performed	Experiment result showed there is an increased surface area due to the presence of graphene. The band gap was also reduced to 2.7 eV. *Strain sensing properties:-* the composite has also shown a gauge factor of 15, which indicates that it can also be used for sensing of medium range loads	[54]

Table 14.1 Contd. ...

...Table 14.1 Contd.

Sr. n.	Doping substance	Method used	Brief detail	Outcome	Reference
3	Graphene	Hydrothermal method, spin coating	First TiO_2 nanoflowers were made on FTO substrate by hydrothermal method. A similar method was used to prepare graphene quantum dots, after which spin coating method was adopted for the sensitization of TiO_2 on GQD in different parameters	Photocurrent measurement showed the GQDs-x/TiO_2 nanoflowers shows higher photocurrent voltage at higher value of x when the bias voltage was ranging from 0.4 to 0.9 V	[55]
4	Au-CNT composites	sol-gel method	Gold–CNT Nanocomposites were made by the reduction of gold with the help of sodium citrate. The reduced gold was added in CNT and sonicated further for the CNT–gold reaction. Different samples were taken in different amounts of gold. Preparation of catalyst–gold–CNT composite and titanium butoxide was dissolved in ethanol and then stirred. After this, it was further dried and various samples of TiO_2–Au–CNT were achieved	Raman spectra showed that D–band and G–band were at 1343 and 1576 cm^{-1}. The intensity ratios I_D/I_G were found to be lowest in 5% gold-CNT composite, which was due to the decrease in oxygen functionalities. A decrease in intensity was shown at adsorption band while studying the FTIR spectra. After further investigation it was found that gold and CNT behave like electron receptors and improve the photocatalytic activity	[56]
5	Pt-graphene	Alcohol-thermal method Photo-deposition method	Graphene oxide was dispersed in DI water followed by ultra sonication for 30 minutes. The product was then further added with TiO_2. The resulting product is graphene-TiO_2 The Pt-TiO_2 and Pt-GN/TiO_2 was developed by the photo deposition method, using ethylene glycol as a reducing agent. $H_2PtCl\cdot6H_2O$ was used as Pt precursor	A specifically designed apparatus was used to determine the hydrogen gas. The reactor consists of a quartz annular reactor attached with 8 W mercury lamp. The amount of hydrogen production efficiency was enhanced from 1.91 to 4.71 mmol h^{-1} g^{-1} with the increase in the Pt value	[57]

Table 14.1 Contd. ...

...Table 14.1 Contd.

Sr. n.	Doping substance	Method used	Brief detail	Outcome	Reference
6	CNT/Ce-TiO$_2$	Modified sol–gel method	For sample preparation, titanium isopropoxide was used as a titanium precursor, 30 ml of which was added to 10 ml ethanol. In the same solution, Ce in the range of 3.0 wt% to 7.0 wt% was added. Another solution, which consists of 20 ml ethanol, 5 ml DI water, 2 ml HCl, and 1–6 wt% was added drop-wise in the titanium mixture. After that it was stirred for 12 h till a sol came into the picture. The sol was dried and calcinated to get a powder form. Three samples with different amounts of Ce were obtained from the same method	The X-ray diffraction, UV-Vis diffuse reflectance spectra, and photocatalytic output were analyzed. It was observed that CNT and Ce modified TiO$_2$ has a greater photocatalytic hydrogen production. This may be attributed to the fact that Ce could have captured the photo induced electrons and holes, and would have produced a hindrance in electron hole recombination	[45]

8. Setup and Characterizations

After more than four decades of research on water splitting, there is no solid agreement on the comparison and standards among research community. Large disparities between different methods to calculate the efficiency and hydrogen output are seen. The setup which is made for the production are self-designed and are different, even having some wrong practices to determine the efficiency definition. For example the 'photonic efficiency' is defined based on incident photons, but when the experiment of monochromatic light is performed, the term 'photonic yield' should be used. To make a setup, it will need three key things [58]. These are a gas chromatography, a photo reactor, and a light source [58, 59].

- **For GC:** One has to be careful in selecting the carrier gas. Helium can not be used because He and H$_2$ have almost similar retention times due to their similar thermal conductivities. Instead, nitrogen or Ar can be used. Detectors are TCD and PLOT column.

- **Photoreactor:** Usually a quartz reactor with septum for purging is used. It is always customized for photocatalytic H$_2$ production, and hence there is no fixed design. The only thing is that it must be leak-proof and be able to absorb light. If it's an online GC setup, then reactor design varies, or else if it's offline, then the septum provided can be used for extracting the products via gas-tight syringe.

- **Light source:** It depends on catalyst to catalyst and experimental conditions. One can use only visible light spectrum or AM1.5. The solar simulator can serve both the purposes where you can use a cut-off filter to make the light source limited to the visible spectrum, and if not, then it simulates the entire solar light. The pictorial representation of a customized setup is also given below in Figure 14.2 [60].

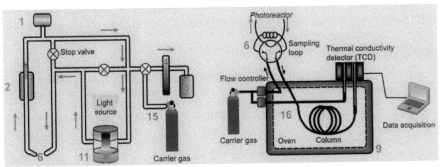

Figure 14.2. The figure explains a setup of measurement. The setup may be different at different places [60]. The basic components of the design are (1) capacitor manometer, (2) recirculating pump, (3) light source, (4) cold trap for the liquid nitrogen, (5) mag. stirrer, (6) Sampling loop, (7) chiller pump, (8) vacuum pump, (9) GC, (10) Light guide, (11) Photocatalyst, (12) lab jack, (13) Water window, (14) water cooler, (15) carrier gas inlet, (16) GC column.

9. Further Scope

A lot of work still needs to be done so that TiO_2 photocatalytic water splitting can have industrial applications. There are a few areas in which lots of science can be done with the help of optimization as well as new advancements. These are as following:

i) **Electron transport channel to TiO_2**—TiO_2 doesn't show good photocatalytic activity in visible light region. It can be increased by providing electron transport channel to it. Graphene can be considered as a potential candidate for this. Some more work in optimizing has to be done to improve visible light activity [61].

ii) **Reduction of wide band gap**—The band gap of TiO_2 is wide. Some metal and non-metal doping will lead to the decrease in band gap. Few scientists have already published a red shift of 436.4 nm in the absorption band by doping N, La [43]. Other similar approaches can lead to some new findings.

iii) **Finding new sacrificial agents**—Charge recombination is one of the basic drawbacks which can be certainly reduced to a certain amount with the help of sacrificial agents. There is a need to find out new sacrificial agents, such as silver nitrate, which acts exactly in a reverse manner to hole scavenger. This can further reduce the change in free energy (ΔG) [62].

10. Summary

The present review throws light on the present state of photocatalysis and the efforts carried out in the last few decades with their classification. It has also pointed out the disparities in calculation and measurement. The recent modification and hurdle has been presented through the table. The photocatalytic water splitting research has not got the required amount of attention in scientific community. A big part of the science behind it is yet to be explored. To bring this technology to an industrial level, some high abruption material, composites are to be developed further.

References

[1] Tang, J., J.R. Durrant and D.R. Klug. 2008. Mechanism of photocatalytic water splitting in TiO₂. Reaction of water with photoholes, importance of charge carrier dynamics, and evidence for four-hole chemistry. J. Am. Chem. Soc., 130(42): 13885–13891. doi: 10.1021/ja8034637.

[2] Yang, X. and M.B. Hall. 2010. Mechanism of water splitting and oxygen-oxygen bond formation by a mononuclear ruthenium complex. J. Am. Chem. Soc., 132(1): 120–130. doi: 10.1021/ja9041065.

[3] Walter, M.G., E.L. Warren, J.R. McKone, S.W. Boettcher, Q. Mi, E.A. Santori and N.S. Lewis. 2010. Solar water splitting cells. Chem. Rev., 110(11): 6446–6473. doi: 10.1021/cr1002326.

[4] Gross, M.A., A. Reynal, J.R. Durrant and E. Reisner. 2014. Versatile photocatalytic systems for H₂ generation in water based on an efficient DuBois-type nickel catalyst. J. Am. Chem. Soc., 136(1): 356–366. doi: 10.1021/ja410592d.

[5] Silija, P., Z. Yaakob, V. Suraja, N.N. Binitha and Z.S. Akmal. 2012. An enthusiastic glance in to the visible responsive photocatalysts for energy production and pollutant removal, with special emphasis on titania. Int. J. Photoenergy, doi: 10.1155/2012/503839.

[6] Chandraboss, J.K.V.L. and S. Senthilvelan. 2016. Synthesis and characterization of InWO₃-TiO₂ nanocomposite material and multi application. World Sci., 58: 97–121.

[7] Etacheri, V., C. Di Valentin, J. Schneider, D. Bahnemann and S.C. Pillai. 2015. Visible-light activation of TiO₂ photocatalysts: Advances in theory and experiments. J. Photochem. Photobiol. C Photochem. Rev., 25: 1–29. doi: 10.1016/j.jphotochemrev.2015.08.003.

[8] Parvathy, S., S. Saranya and S. Sivakumar. 2019. Semiconductor as efficient photocatalyst: structural designing, characteraization and investigation of photocatalytic efficiency. The Open Access J. Sci. Tech., vol. 7, Article ID 101265, 13 pages, doi: 10.11131/2019/101259.

[9] Ranjith Kumar, D., K.S. Ranjith, Y. Haldorai, A. Kandasami and R.T. Rajendra Kumar. 2019. Nitrogen-implanted ZnO nanorod arrays for visible light photocatalytic degradation of a pharmaceutical drug acetaminophen. ACS Omega, 4(7): 11973–11979. doi: 10.1021/acsomega.9b00557.

[10] Schneider, J., M. Matsuoka, M. Takeuchi, J. Zhang, Y. Horiuchi, M. Anpo and D.W. Bahnemann. 2014. Understanding TiO₂ photocatalysis: Mechanisms and materials. Chem. Rev., 114(19): 9919–9986. doi: 10.1021/cr5001892.

[11] Drahansky, M., M. Paridah, A. Moradbak, A. Mohamed, F. abdulwahab taiwo Owolabi, M. Asniza and S.H. Abdul Khalid. 2016. Photocatalytic degradation of organic pollutants in water. Intech, vol. i, no. tourism, p. 13. doi: http://dx.doi.org/10.5772/57353.

[12] Fujishima, A. and K. Honda. 1972. Electrochemical photolysis of water at a semconductor electrode. Nature, 238(5358): 38–40. doi: 10.1038/238038a0.

[13] Schrauzer, G.N. and T.D. Guth. 1977. Photolysis of water and photoreduction of nitrogen on titanium dioxide. J. Am. Chem. Soc., 99(22): 7189–7193,. doi: 10.1021/ja00464a015.

[14] Bonin, J., M. Robert and M. Routier. 2014. Selective and efficient photocatalytic CO₂ reduction to CO using visible light and an iron-based homogeneous catalyst. J. Am. Chem. Soc., 136(48): 16768–16771. doi: 10.1021/ja510290t.

[15] Prihod'ko, R.V. and N.M. Soboleva. 2013. Photocatalysis: Oxidative processes in water treatment. J. Chem., doi: 10.1155/2013/168701.

[16] Cermenati, L., P. Pichat, C. Guillard and A. Albini. 1997. Probing the TiO_2 photocatalytic mechanisms in water purification by use of quinoline, photo-fenton generated OH radicals and superoxide dismutase. J. Phys. Chem. B, 101(14): 2650–2658. doi: 10.1021/jp962700p.

[17] Yi, J., C. Bahrini, C. Schoemaecker, C. Fittschen and W. Choi. 2012. Photocatalytic decomposition of H_2O_2 on different TiO_2 surfaces along with the concurrent generation of HO_2 radicals monitored using cavity ring down spectroscopy. J. Phys. Chem. C, 116(18): 10090–10097. doi: 10.1021/jp301405e.

[18] Figueredo, M.A., E.M. Rodríguez, M. Checa and F.J. Beltran. 2019. Ozone-based advanced oxidation processes for primidone removal in water using simulated solar radiation and TiO_2 or WO_3 as photocatalyst. Molecules, 24(9). doi: 10.3390/molecules24091728.

[19] Sharma, S., A. Gulabani and S. Gautam. 2019. Photocatalytic degradation in silver doped TiO_2. Sādhanā, 44(9): 2–7. doi: 10.1007/s12046-019-1191-0.

[20] Ibhadon, A.O. and P. Fitzpatrick. 2013. Heterogeneous photocatalysis: Recent advances and applications. Catalysts, 3(1): 189–218. doi: 10.3390/catal3010189.

[21] Friedmann, D., A. Hakki, H. Kim, W. Choi and D. Bahnemann. 2016. Heterogeneous photocatalytic organic synthesis: State-of-the-art and future perspectives. Green Chem., 18(20): 5391–5411. doi: 10.1039/c6gc01582d.

[22] Colmenares, J.C. and R. Luque. 2014. Heterogeneous photocatalytic nanomaterials: Prospects and challenges in selective transformations of biomass-derived compounds. Chem. Soc. Rev., 43(3): 765–778. doi: 10.1039/c3cs60262a.

[23] Rusinque, B. 2018. Hydrogen production by photocatalytic water splitting under near-UV and visible light using doped Pt and Pd TiO_2. Catalysts 2019, 9: 33.

[24] Isaifan, R.J. et al. 2017. Improved self-cleaning properties of an efficient and easy to scale up TiO_2 thin films prepared by adsorptive self-assembly. Sci. Rep., 7(1): 1–9. doi: 10.1038/s41598-017-07826-0.

[25] McEvoy, J.P., J.A. Gascon, V.S. Batista and G.W. Brudvig. 2005. The mechanism of photosynthetic water splitting. Photochem. Photobiol. Sci., 4(12): 940–949. doi: 10.1039/b506755c.

[26] John Moma and Jeffrey Baloyi. 2016. Modified titanium dioxide for photocatalytic modified titanium dioxide for photocatalytic applications. Intech, vol. i, no. tourism, p. 13. doi: http://dx.doi.org/10.5772/57353.

[27] Dette, C., M.A. Pérez-Osorio, C.S. Kley, P. Punke, C.E. Patrick, P. Jacobson, F. Giustino, S.J. Jung and K. Kern. 2014. TiO_2 anatase with a bandgap in the visible region. Nano Lett., 14(11): 6533–6538. doi: 10.1021/nl503131s.

[28] Liu, S.H. and H.R. Syu. 2012. One-step fabrication of N-doped mesoporous TiO_2 nanoparticles by self-assembly for photocatalytic water splitting under visible light. Appl. Energy, 100: 148–154. doi: 10.1016/j.apenergy.2012.03.063.

[29] Chowdhury, P., G. Malekshoar and A.K. Ray. 2017. Dye-sensitized photocatalytic water splitting and sacrificial hydrogen generation: Current status and future prospects. Inorganics, 5(2). doi: 10.3390/inorganics5020034.

[30] Rashad, S., A.H. Zaki and A.A. Farghali. 2019. Morphological effect of titanate nanostructures on the photocatalytic degradation of crystal violet. Nanomater. Nanotechnol., 9: 1–10. doi: 10.1177/1847980418821778.

[31] Wei, J., X. Wen and F. Zhu. 2018. Influence of surfactant on the morphology and photocatalytic activity of anatase TiO_2 by solvothermal synthesis. J. Nanomater., doi: 10.1155/2018/3086269.

[32] Yu, J., G. Dai and B. Cheng. 2010. Effect of crystallization methods on morphology and photocatalytic activity of anodized TiO_2 nanotube array films. J. Phys. Chem. C, 114(45): 19378–19385, doi: 10.1021/jp106324x.

[33] Suhaimy, S.H.M., C.W. Lai, H.A. Tajuddin, E.M. Samsudin and M.R. Johan. 2018. Impact of TiO_2 nanotubes' morphology on the photocatalytic degradation of simazine pollutant. Materials (Basel)., 11(11): 1–15. doi: 10.3390/ma11112066.

[34] Li, Y., Y. Peng, L. Hu, J. Zheng, D. Prabhakaran, S. Wu, T.J. Puchtler, M. Li, K. Wong and R.A. Taylor. 2019. Photocatalytic water splitting by N-TiO_2 on MgO (111) with exceptional quantum efficiencies at elevated temperatures. Nat. Commun., 10(1): 1–9. doi: 10.1038/s41467-019-12385-1.

[35] Clarizia, L., D. Russo and I. Di Somma. 2017. Hydrogen generation through solar photocatalytic processes: a review of the configuration and the properties of effective metal-based semiconductor nanomaterials. Energies, 10: 1624 doi: 10.3390/en10101624.

[36] Mishra, P.R. and O.N. Srivastava. 2008. On the synthesis, characterization and photocatalytic applications of nanostructured TiO_2. Bull. Mater. Sci., 31(3): 545–550. doi: 10.1007/s12034-008-0085-2.

[37] Malato, S., P. Fernández-Ibáñez, M.I. Maldonado, J. Blanco and W. Gernjak. 2009. Decontamination and disinfection of water by solar photocatalysis: Recent overview and trends. Catal. Today, 147(1): 1–59. doi: 10.1016/j.cattod.2009.06.018.

[38] Rajeshwar, K., M.E. Osugi, W. Chanmanee, C.R. Chenthamarakshan, M.V.B. Zanoni, P. Kajitvichyanukul and R. Krishnan-Ayer. 2008. Heterogeneous photocatalytic treatment of organic dyes in air and aqueous media. J. Photochem. Photobiol. C Photochem. Rev., 9(4): 171–192, doi: 10.1016/j.jphotochemrev.2008.09.001.

[39] Chattergee, D. and S. Dagupta. 2017. Visible light induced photocatalytic degradation of congo red. J. Photochem. Photobiol., 6(6): 1200–1209. doi: 10.13140/RG.2.1.3848.1364.

[40] Indris, S. et al. 2005. Preparation by high-energy milling, characterization, and catalytic properties of nanocrystalline TiO_2. J. Phys. Chem. B, 109(49): 23274–23278. doi: 10.1021/jp054586t.

[41] Kang, I.C., Q. Zhang, S. Yin, T. Sato and F. Saito. 2008. Preparation of a visible sensitive carbon doped TiO_2 photo-catalyst by grinding TiO_2 with ethanol and heating treatment. Appl. Catal. B Environ., 80(1–2): 81–87. doi: 10.1016/j.apcatb.2007.11.005.

[42] Umebayashi, T., T. Yamaki, H. Itoh and K. Asai. 2002. Band gap narrowing of titanium dioxide by sulfur doping. Appl. Phys. Lett., 81(3): 454–456. doi: 10.1063/1.1493647.

[43] Zhang, X., G. Zhou, H. Zhang, C. Wu and H. Song. 2011. Characterization and activity of visible light-driven TiO_2 photocatalysts co-doped with nitrogen and lanthanum. Transit. Met. Chem., 36(2): 217–222. doi: 10.1007/s11243-010-9457-8.

[44] Teka, T. 2015. Current state of doped-TiO_2 photocatalysts and synthesis methods to prepare TiO_2 films: A review. Int. J. Technol. Enhanc. Emerg. Eng. Res., 3(01): 14–18.

[45] Shaari, N., S.H. Tan and A.R. Mohamed. 2012. Synthesis and characterization of CNT/Ce-TiO_2 nanocomposite for phenol degradation. J. Rare Earths, 30(7): 651–658. doi: 10.1016/S1002-0721(12)60107-0.

[46] Li, S., P. Xue, C. Lai, J. Qiu, M. Ling and S. Zhang. 2015. Pseudocapacitance of amorphous TiO_2 nitrogen doped graphene composite for high rate lithium storage. Electrochim. Acta, 180: 112–119. doi: 10.1016/j.electacta.2015.08.099.

[47] Dholam, R., N. Patel, M. Adami and A. Miotello. 2009. Hydrogen production by photocatalytic water-splitting using Cr- or Fe-doped TiO_2 composite thin films photocatalyst. Int. J. Hydrogen Energy, 34(13): 5337–5346. doi: 10.1016/j.ijhydene.2009.05.011.

[48] Akhavan, O., R. Azimirad, S. Safa and M.M. Larijani. 2010. Visible light photo-induced antibacterial activity of CNT-doped TiO_2 thin films with various CNT contents. J. Mater. Chem., 20(35): 7386–7392. doi: 10.1039/c0jm00543f.

[49] Wang, G., H. Wang, Y. Ling, Y. Tang, X. Yang, R.C. Fitzmorris, C. Wang, J.Z. Zhang and Y. Li. 2011. Hydrogen-treated TiO_2 nanowire arrays for photoelectrochemical water splitting. Nano Lett., 11(7): 3026–3033. doi: 10.1021/nl201766h.

[50] Zhang, Y., Z.-R. Tang, X. Fu and Y.-J. Xu. 2010. TiO_2 graphene nanocomposites for gas-phase photocatalytic degradation of volatile aromatic pollutant: is. ACS Nano, 4(12): 7303–14. doi: 10.1021/nn1024219.

[51] Saravanan, R., N. Karthikeyan, S. Govindan, V. Narayanan and A. Stephen. 2012. Photocatalytic degradation of organic dyes using ZnO/CeO_2 nanocomposite material under visible light. Adv. Mater. Res., 584: 381–385. January, doi: 10.4028/www.scientific.net/AMR.584.381.

[52] Police, A.K.R., M. Chennaiahgari, R. Boddula, S.V.P. Vattikuti, K.K. Mandari and B. Chan. 2018. Single-step hydrothermal synthesis of wrinkled graphene wrapped TiO_2 nanotubes for photocatalytic hydrogen production and supercapacitor applications. Mater. Res. Bull., 98: 314–321. August 2017, doi: 10.1016/j.materresbull.2017.10.034.

[53] Olowoyo, J.O., M. Kumar, S.L. Jain, J.O. Babalola, A.V. Vorontsov and U. Kumar. 2019. Insights into reinforced photocatalytic activity of the CNT-TiO_2 nanocomposite for CO_2 reduction and water splitting. J. Phys. Chem. C, 123(1): 367–378. doi: 10.1021/acs.jpcc.8b07894.

[54] Syed, M.A., T.S. Muahammad, F. Muhammad, A. Sharjeel and M. Isna. 2018. Hybrid effect of TiO_2/reduced graphene oxide based composite for photo-catalytic water splitting & strain sensing. Key Eng. Mater., 778: 144–150. KEM, doi: 10.4028/www.scientific.net/KEM.778.144.

[55] Bellamkonda, S., N. Thangavel, H.Y. Hafeez, B. Neppolian and G. Ranga Rao. 2019. Highly active and stable multi-walled carbon nanotubes-graphene-TiO_2 nanohybrid: An efficient non-noble metal photocatalyst for water splitting. Catal. Today, no. September, pp. 120–127. doi: 10.1016/j.cattod.2017.10.023.

[56] Chinh, V.D., L.X. Hung, L. Di Palma, V.T.H. Hanh and G. Vilardi. 2019. Effect of carbon nanotubes and carbon nanotubes/gold nanoparticles composite on the photocatalytic activity of TiO_2 and TiO_2 -SiO_2. Chem. Eng. Technol., 42(2): 308–315. doi: 10.1002/ceat.201800265.

[57] Nguyen, N.-T., D.-D. Zheng, S.-S. Chen and C.-T. Chang. 2017. Preparation and photocatalytic hydrogen production of Pt-Graphene/TiO_2 composites from water splitting. J. Nanosci. Nanotechnol., 18(1): 48–55. doi: 10.1166/jnn.2018.14556.

[58] Qureshi, M. and K. Takanabe. 2017. Insights on measuring and reporting heterogeneous photocatalysis: Efficiency definitions and setup examples. Chem. Mater., 29(1): 158–167. doi: 10.1021/acs.chemmater.6b02907.

[59] Ni, M., M.K.H. Leung, D.Y.C. Leung and K. Sumathy. 2007. A review and recent developments in photocatalytic water-splitting using TiO_2 for hydrogen production. Renew. Sustain. Energy Rev., 11(3): 401–425. doi: 10.1016/j.rser.2005.01.009.

[60] Obata, K., L. Stegenburga and K. Takanabe. 2019. Maximizing hydrogen evolution performance on Pt in buffered solutions: mass transfer constrains of H_2 and buffer ions. J. Phys. Chem. C, 123(35): 21554–21563. doi: 10.1021/acs.jpcc.9b05245.

[61] Wan, D.Y., Y.L. Zhao, Y. Cai, T.C. Asmara, Z. Huang, J.Q. Chen, J. Hong, S.M. Yin, C.T. Nelson, M.R. Motapothula. 2017. Electron transport and visible light absorption in a plasmonic photocatalyst based on strontium niobate. Nat. Commun., 8: 1–9. doi: 10.1038/ncomms15070.

[62] Rajaambal, S., K. Sivaranjani and C.S. Gopinath. 2015. Recent developments in solar H_2 generation from water splitting. J. Chem. Sci., 127(1): 33–47. doi: 10.1007/s12039-014-0747-0.

Index